Synthetic Gyrolite

Kęstutis Baltakys · Tadas Dambrauskas

Synthetic Gyrolite

A Dual Solution for Wastewater Treatment and Portland Cement

Springer

Kęstutis Baltakys
Department of Silicate Technology
Kaunas University of Technology
Kaunas, Lithuania

Tadas Dambrauskas
Department of Silicate Technology
Kaunas University of Technology
Kaunas, Lithuania

ISBN 978-3-031-49261-7 ISBN 978-3-031-49259-4 (eBook)
https://doi.org/10.1007/978-3-031-49259-4

English language editor: Armandas Rumšas

This Springer imprint is published by the registered company Springer Nature Switzerland AG
The registered company address is: Gewerbestrasse 11, 6330 Cham, Switzerland

Paper in this product is recyclable.

Preface

The challenges of the twenty-first century arise from a growing population and increasing societal needs, leading to excessive material consumption and environmental pollution. Achieving carbon neutrality is crucial, but waste generation remains a significant contributor to pollution. Fortunately, European scientific and technical companies possess the necessary capabilities for sustainable recycling and material development. However, this transformation requires political will, financial support, and collaboration among public authorities, companies, investors, and researchers. Scientists play a vital role in exploring and applying new and conventional materials for green technologies.

This monograph aims to provide a comprehensive understanding of calcium silicate hydrates. With over forty crystalline varieties, these compounds are classified based on their Ca/SiO_2 molar ratios, ranging from 0.44 to 3.00. Among them, low-basicity (Ca/SiO_2 = 0.6–1.5) and high-basicity (Ca/SiO_2 = 1.5–3.0) calcium silicate hydrates are particularly significant in cement chemistry. These hydrates have diverse applications, serving as adsorbents for organic and inorganic effluents, bioactive and heat insulation materials, fillers for rubber, plastics, paper, and paint, as well as immobilizers of heavy and radioactive metal ions. While all calcium silicate hydrates contribute to research and industry, gyrolite [$Ca_{16}Si_{24}O_{60}(OH)_8 \cdot 14H_2O$] stands out due to its exceptional adsorption capacity compared to others.

Gyrolite, a rare mineral classified as a phyllosilicate, can be synthesized under hydrothermal conditions using mixtures containing calcium oxide and silicon dioxide. The synthesis process relies on compound reactivity and specific temperature ranges, typically between 120 and 200 °C. The duration of synthesis varies depending on the reactivity of silicon-containing compounds. Gyrolite can form within a range of hours, with different types of silicon dioxide impacting the speed of formation. However, there is limited information regarding the synthesis of gyrolite using industrial waste, as well as its porosity, structure, and adsorption capacity for heavy metal ions. To address these gaps, this monograph takes a comprehensive approach to systematically explore the synthesis and application of gyrolite. It offers a summarized theoretical background and provides technological recommendations for future studies.

The monograph consists of five chapters, each delving into different aspects of synthetic gyrolite. It begins by offering a general description and classification of these compounds, including newly discovered ones. The synthesis conditions and properties of high purity hydroxyledgrewite are extensively discussed. The monograph further investigates the influence of synthesis duration on the properties of synthetic gyrolite, examining changes in the crystallite size, water content, surface area, and pore volume. Additionally, it explores the potential use of silica gel waste as a silica source for gyrolite synthesis, considering the impact of foreign ions. The study emphasizes the significance of adsorbing heavy metal ions and evaluates the adsorption capabilities of synthetic gyrolite, highlighting the need for cost-effective and efficient adsorbents. Furthermore, the application of synthetic gyrolite in ordinary Portland cement is explored, analyzing its effects on hydration and compressive strength. The monograph concludes by providing technological recommendations for gyrolite production and its application in cement.

The authors extend their heartfelt gratitude to the esteemed reviewers: Prof. Dr. Ing. Martin T. Palou (Slovak Academy of Sciences), Prof. Habil. Dr. Aivaras Kareiva (Vilnius University), and Asst. Prof. Dr. Ing. Pavel Šiler (Brno University of Technology), for their invaluable contributions. Their valuable remarks, professional criticism, kind comments, and dedicated time and effort have greatly enhanced the quality of this monograph.

Kaunas, Lithuania

Kęstutis Baltakys
Tadas Dambrauskas

Contents

1 General Understanding of Calcium Silicate Hydrates 1
1.1 Classification and the Main Properties of Calcium Silicate Hydrates 1
1.2 Newly Discovered or Synthesized Calcium Silicate Hydrates 12
1.3 Hydrothermal Synthesis and Properties of Hydroxyledgrewite 15
1.3.1 Hydrothermal Synthesis of Hydroxyledgrewite 15
1.3.2 Porosity of Synthetic and Calcined Hydroxyledgrewite Samples 22
1.4 Summary 30
References 31

2 Synthesis and Properties of Gyrolite 41
2.1 Natural Gyrolite: Historical View, Formation, and Main Properties 41
2.2 Crystal Structure of Gyrolite 45
2.3 Background Information About Hydrothermal Synthesis of Gyrolite 48
2.4 Effect of Hydrothermal Synthesis Duration at 200 °C Temperature on the Properties of Gyrolite 51
2.4.1 Effect of Hydrothermal Synthesis Duration on the Stability and Crystallinity of Gyrolite 51
2.4.2 Effect of Hydrothermal Synthesis Duration on the Porosity of Gyrolite 57
2.5 Summary 73
References 73

3 Solid Waste. Application of AlF_3 Production Waste to Gyrolite Synthesis 79
3.1 Influence of Additives on the Synthesis and Structure Properties of Synthetic Gyrolite 79
3.2 Formation of AlF_3 Production Waste, Its Application, and Properties 84

3.2.1 Solid Wastes and Their Utilization 85
3.2.2 Formation and Properties of By-Product of AlF_3 Production 87
3.3 Influence of AlF_3 Production Waste on the Formation of Gyrolite Family Compounds 99
3.4 Stability of Synthetic Calcium Silicate Hydrates with CaF_2 at 600–1000 °C Temperature 115
3.5 Influence of $Ca(NO_3)_2$ Additive on the Formation and Porosity of Gyrolite Family Compounds 120
3.5.1 Influence of $Ca(NO_3)_2$ on the Formation of Gyrolite Family Compounds 120
3.5.2 Porosity of Synthetic Gyrolite with Intercalated Nitrate Anions 127
3.6 Summary 131
References 131

4 Application of Synthetic Gyrolite for Adsorption 139
4.1 Water Pollution with Heavy Metal Ions 139
4.2 Types of Adsorbents 141
4.3 Synthetic Gyrolite Application for Single Metal Ions Adsorption Process 147
4.3.1 Literature Review on Synthetic Gyrolite Application for Adsorption Process 147
4.3.2 Synthetic Gyrolite Application for Zn^{2+} Ions Adsorption Process 149
4.3.3 Synthetic Gyrolite Application for Fe^{3+} Ions Adsorption Process 171
4.3.4 Synthetic Gyrolite Application for Ni^{2+}, Co^{2+}, and Mn^{2+} Ions Adsorption Process 178
4.4 Synthetic Gyrolite Application for Simultaneous Metal Ions Adsorption Process 185
4.4.1 Simultaneous Adsorption of Co^{2+}, Ni^{2+}, Cu^{2+}, Zn^{2+}, Mn^{2+}, and Fe^{3+} Ions by Synthetic Gyrolite 185
4.4.2 Simultaneous Adsorption of Cd^{2+}, Cu^{2+}, and Zn^{2+} Ions by Synthetic Gyrolite 190
4.4.3 Influence of EDTA on Adsorption of Cd^{2+}, Cu^{2+}, and Zn^{2+} Ions by Synthetic Gyrolite 196
4.5 Summary 204
References 205

5 Utilization of Synthetic Gyrolite in Ordinary Portland Cement 211
5.1 Environmental Impact of Ordinary Portland Cement 211
5.2 Supplementary Cementitious Materials and Their Influence on Ordinary Portland Cement Hydration 217
5.3 Effect of Synthetic Gyrolite Additive on Portland Cement Hydration 223

5.3.1 Influence of Synthetic Gyrolite on the Heat Flow of Cement Paste 223
5.3.2 Influence of Synthetic Gyrolite on Portlandite Quantity in Hydration Products 229
5.3.3 Kinetics and Modeling of the Hydration Process 232
5.3.4 Peculiarities of the Hydration Process Mechanism 237
5.3.5 Influence of Synthetic Gyrolite on the Compressive Strength and Mineralogical Composition of Hardened OPC 238
5.4 Utilization of Synthetic Gyrolite with Incorporated Cd^{2+} Ions in Ordinary Portland Cement 243
5.4.1 Synthetic Gyrolite Application for Cd^{2+} Ions Adsorption Process 243
5.4.2 Influence of Synthetic Gyrolite with Incorporated Cd^{2+} Ions on Ordinary Portland Cement Hydration 248
5.5 Technological Recommendations for the Production and Application of Synthetic Gyrolite 253
5.6 Summary 256
References 257

Author Index 263

Subject Index 265

About the Authors

Prof. Dr. Kęstutis Baltakys has Doctor of Science in the field of technological sciences, chemical engineering, Professor of Kaunas University of Technology (KTU), Vice-Rector for Research of Kaunas University of Technology. His main research interests include the development, structural properties, and application of calcium silicates and their derivatives with other non-organic oxides for innovative technologies. He has carried out and systematized extensive theoretical and applied research and has obtained new scientific results which have an impact and significance for the development of high technologies and the creation of new technologies that can be implemented in Lithuanian conditions as well as elsewhere. Prof. Dr. K. Baltakys has participated in and led three international and 25 national projects. He has been cooperating with international partners from the European Union and the USA for several years, with whom he has signed R&D cooperation agreements. Prof. Dr. K. Baltakys has published more than 270 scientific publications in scientific periodicals, including 138 publications with the citation index in the *Clarivate Analytics WoS* database, a scientific monograph, three textbooks, and two educational books. He has obtained two national patents. He has supervised served successfully defended doctoral dissertations, has been a member of seven doctoral dissertation defense committees, and is a member of the Doctoral Committee in Chemical Engineering and the Doctoral Committee in Civil Engineering at KTU. Prof. Dr. K. Baltakys has had internships at the *Institut National des Sciences Appliquées (INSA) de Rennes,*

France, at the *Technical University of Munich*, Germany, at *Riga Technical University*, Latvia, at the *University of Hradec Králové*, Czech Republic and at *Technical University of Cartagena (Universidad Politécnica de Cartagena)*, Spain. Prof. Dr. K. Baltakys has presented over 100 papers at international conferences in Latvia, France, Serbia, Belgium, Spain, and other countries.

Assoc. Prof. Dr. Tadas Dambrauskas With a Ph.D. in Chemical Engineering from Kaunas University of Technology received on 08 May 2019, Assoc. Prof. Dr. T. Dambrauskas boasts a solid academic background enriched by research experience in chemistry and chemical engineering. Assoc. Prof. Dr. T. Dambrauskas has been conducting research on the synthesis of functional silicate materials under hydrothermal conditions, their thermal stability, and potential applications (production of new cements, VOC oxidation, cosmetics, etc.).

With co-authors, Assoc. Prof. Dr. T. Dambrauskas has published 36 scientific publications in journals with a citation index included in the *Clarivate Analytics WoS* database and has obtained two Lithuanian patents. Assoc. Prof. Dr. T. Dambrauskas has managed more than 15 projects funded by the Lithuanian Academy of Sciences and private businesses. Also, he has been involved in the research of two international and four national, two business, and two institutional projects. In addition to research and teaching activities, Assoc. Prof. Dr. T. Dambrauskas has been involved in the activities of various study and science organizations.

Abbreviations

AAS	Atomic absorption spectrometry
AEEP	The European Phosphate Fertilizer Alliance
BET	Brunauer, Emmett, and Teller method
CNMNC	The Commission on New Minerals, Nomenclature and Classification
c_p	Specific heat capacity
cps	Counts per second
CSH	Calcium silicate hydrate
C_2ASH_8	Stratlingite
C_2S	Dicalcium silicate
C_3A	Tricalcium aluminate
C_3S	Tricalcium silicate
C_4AF	Tetracalcium aluminoferrite
C_4AH_{13}	Calcium aluminate hydrate
C/S	The molar ratio of CaO and SiO_2
DHEG	N,N-bis(2-hydroxyethyl)glycine
DSC	Differential scanning calorimetry
DSC_{max}	Peak of thermal effect determined by DSC
DTG	Derivative thermogravimetry analysis
DTG_{max}	Peak of thermal effect determined by DTG
DTPA	Diethylenetriaminepentaacetic acid
d-spacing	Distance between the parallel planes of atoms
EDDP	Ethylenediamine-N,N'-dipropionic acid
EDDS	Ethylenediamine-N,N'-disuccinic acid
EDTA	Ethylenediaminetetraacetic acid
EDX	Energy dispersive X-ray analysis
EGTA	Ethylene glycol bis (2-aminoethyl) tetraacetic acid
EU	European Union
FT-IR	Fourier-transform infrared spectroscopy
GDP	Gross domestic product
IFA	The International Fertilizer Association

IMA	International Mineralogical Association
IUPAC	International Union of Pure and Applied Chemistry
NMR	Nuclear magnetic resonance
NTA	Nitrilotriacetic acid
Me^{x+}	Metal cations with different charge
OPC	Ordinary Portland cement
PDF	Powder diffraction file
PDF-4+ 2021	Powder diffraction file database
Point B	Beginning of the linear part in N_2 adsorption at 77 K isotherm
PTFE	Polytetrafluoroethylene
p/p_0	Relative pressure
S_{BET}	Specific surface area determined by the BET method
SCMs	Supplementary cementitious materials
SEM	Scanning electron microscopy
SGW-F	Silica gel with incorporated F^- and Al^{3+} ions
STA	Simultaneous thermal analysis
S0	Sample with 0 mol% of Ca^{2+} ions replaced by calcium from $Ca(NO_3)_2$ solution
S5	Sample with 5 mol% of Ca^{2+} ions replaced by calcium from $Ca(NO_3)_2$ solution
S10	Sample with 10 mol% of Ca^{2+} ions replaced by calcium from $Ca(NO_3)_2$ solution
S20	Sample with 20 mol% of Ca^{2+} ions replaced by calcium from $Ca(NO_3)_2$ solution
S1 and S2	Tetrahedral silicate layers
TEM	Transmission electron microscopy
TGA	Thermogravimetry analysis
T_{onset}	Beginning of the thermal effect
T_{peak}	Peak of the thermal effect
T_{end}	End of the thermal effect
w/c	Water to cement ratio
XRD	Powder X-Ray diffraction analysis
XRF	X-Ray fluorescence analysis
α'_L-C_2S	α'_L-Ca_2SiO_4 or α'_L dicalcium silicate
β-C_2S	β-Ca_2SiO_4, larnite, or beta dicalcium silicate
γ-C_2S	γ-Ca_2SiO_4 or gamma dicalcium silicate
$\Sigma X_{e(exp)}$	Measured equilibrium adsorption capacity
$\Sigma X_{e(cal)}$	Calculated equilibrium adsorption capacity.

Chapter 1
General Understanding of Calcium Silicate Hydrates

The first chapter of the monograph reports the general description, and classification of calcium silicate hydrates. According to the classification, 45 calcium silicate hydrates can be categorized into seven groups. The classification was proposed by Roy and Taylor and was later expanded and fine-tuned by a number of other scientists. Notably, Richardson made the last update on the classification in 2008. However, new calcium silicate hydrates have been discovered or synthesized during the last decade, which are therefore not included in the classification yet. Thus, in this chapter, the description of the newly discovered or synthesized compounds (hydroxyledgrewite, trabzonite, chegemite, etc.) is presented. Also, the synthesis conditions of high purity hydroxyledgrewite and its main properties are extensively described.

1.1 Classification and the Main Properties of Calcium Silicate Hydrates

The formula of calcium silicate hydrates (CSH) denotes the correlation of the constituent components: xCaO·ySiO$_2$·pH$_2$O (indexes: x, y, p indicate the number of moles). They are synonymously called *silicic acid salt* [3]. Many of these compounds occur in the nature as hydrothermal alteration products of calcium carbonate rocks and as vesicle fillings in basalts. Other calcium silicate hydrates can be produced either by air curing (e.g. in cement or concrete), or hydrothermally by high-pressure steam curing in an autoclave [3–8]. The crystallization reactions of these calcium silicate hydrates depend on the hydrothermal synthesis conditions (duration and temperature), the composition of the initial mixture [CaO/SiO$_2$ (C/S) in terms of the molar ratio and the additives being used], the properties of the raw materials (the granulometric composition (grading), the strain), the mixing intensity, and a number

K. Baltakys and T. Dambrauskas, *Synthetic Gyrolite*,
https://doi.org/10.1007/978-3-031-49259-4_1

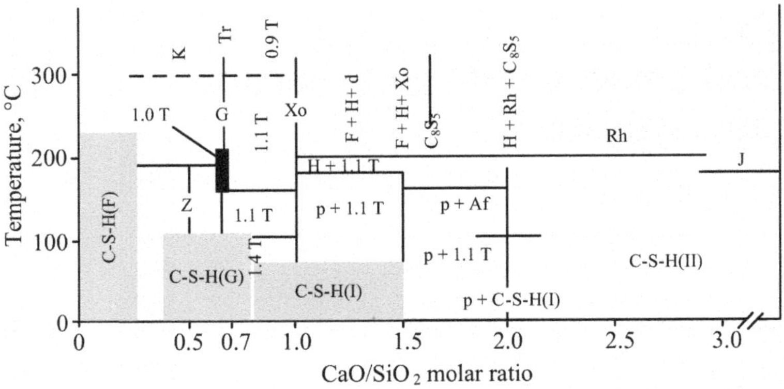

Fig. 1.1 Formation of C–S–H in ternary $CaO–SiO_2–H_2O$ system under hydrothermal conditions. Indexes: 0.9 T–0.9 nm tobermorite; 1.0 T–1.0 nm tobermorite; 1.1 T–1.1 nm tobermorite; 1.4 T–1.4 nm tobermorite; K–K-phase; Tr—truscottite; Xo—xonotlite; H—hillebrandite; F—foshagite; d—kilchoanite; Af—afwillite; J—Jaffeite; p—portlandite; G—gyrolite; Rh—calcium chondrodite

of other factors. For these reasons, the sequence of the formation of calcium silicate hydrates represented by scientists somewhat differs [4, 9–16]. The formation of calcium silicate hydrates in pure systems under hydrothermal conditions is presented in Fig. 1.1.

In terms of classification, there are more than forty crystalline calcium silicate hydrates covering the C/S ratios between 0.44 and 3.00 (Figs. 1.1 and 1.2). Low-basicity (C/S = 0.6–1.5) and high-basicity (C/S = 1.5–3.0) calcium silicate hydrates are the main compounds in the cement chemistry (Fig. 1.2) [5, 8, 11, 17]. The interest in calcium silicate hydrates has been constantly increasing as their application is denoted by a fairly broad range: as heat-insulating materials [18]; as adsorbents for CO_2 or the purification of water [19–21]; as fillers [22–24]; for the production of bioactive materials [25–27] and environmentally friendly cementitious materials [28–31]; and in other novel and traditional fields [8, 27, 32, 33]. Despite the fact that chemistry has known calcium silicate hydrates for more than a century, and even though their extensive research has already been conducted, their chemistry is still not fully understood [8, 17, 34]. Thus, synthetic calcium silicate hydrates are the subject of numerous scientific studies.

The mineralogical composition and structure of calcium silicate hydrates depend on the synthesis conditions and various cations and anions which can intercalate into the crystal lattice; thus, the investigation of the structure and properties of CSH is complicated. The complexity of the research on calcium silicate hydrates can be revealed by analyzing their structures, especially C–S–H(I) and C–S–H(II). The first structural model of the presently discussed compounds was presented by Bernal et al. [35, 36]. These authors investigated cement hydration and proposed the C–S–H(I) and C–S–H(II) structure model based on portlandite and monomeric silicate groups (layered C–S–H model). The suggested stoichiometry of C–S–H(I) and C–S–H(II) was $Ca[SiO_2(OH)_2][Ca(OH)_2]_x$. This model was supplemented by Shpynova et al.

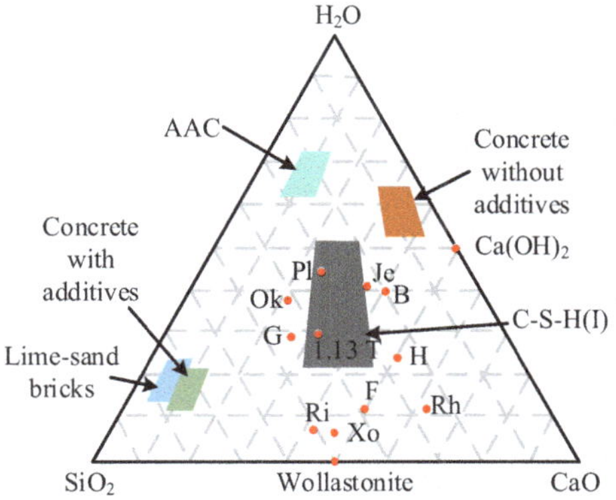

Fig. 1.2 Ternary $CaO–SiO_2–H_2O$ system with some C–S–H and the composition of common building materials. Indexes: 1.13 T–1.13 nm tobermorite; Xo—xonotlite; H—hillebrandite; F—foshagite; Je—jennite; G—gyrolite; Rh—calcium chondrodite; Ok—okenite; Pl—plombierite; B—bultfonteinite; Ri—Riversideite

[37] and Grudemo et al. [38]. The characterization of calcium silicate hydrates with ^{29}Si NMR allowed to reject these models because experimental results showed that the length of the silicate chain follows $3n - 1$, where n is an integer [39, 40]. It is worth noting that the early attempts at the determination of the structure of calcium silicate hydrates allowed determining the relationship between C–S–H gel and tobermorite [35, 41].

The main motifs of the contemporarily used C–S–H(I) and C–S–H(II) structure models were presented by Taylor and Howison [42]. These authors proposed a structure model based on tobermorite with finite silicate chains. The silicate chains feature the dreierketten arrangement (Fig. 1.3), i.e. the smallest repeating unit consists of one bridging and two pairings tetrahedra [43–45]. The silicate chain shares two oxygens with a layer of CaO (the *pairing tetrahedra*) and one with a Ca^{2+} and H_2O interlayer (the *bridging tetrahedra*). Based on the dreierketten arrangement and the experimental findings, Taylor and Howison postulated that the C/S ratio in tobermorite can be increased from 0.83 up to 1.5 by the omission of the bridging silica tetrahedra and the replacement by calcium ions in the interlayer [42].

Later on, Kurczyk and Schwiete proposed a new C–S–H gel model based on tobermorite while assuming that additional calcium hydroxide and water are presented in the interlayer of tobermorite [46, 47]. This model allowed to explain the high C/S ratio (> 1.80) of C–S–H gel. Other authors suggested that specific arrangements [48] or a solid solution of tobermorite and portlandite leads to the high C/S ratio of C–S–H gel [49]. However, these models were rejected because they considered infinite silicate chains.

In 1980, Stade and Wieker proposed *dimeric* and *dimeric with polymeric chains* models based on the tobermorite structure with finite silicate chains [50, 51]. These models would be deemed valid if C/S were lower than 1.25, and the scholars assumed that the silicate chains should be fully protonated. Glasser et al. [52] extended the dimeric model and reached a C/S ratio of 1.4.

Taylor further investigated the structure of C–S–H and subsequently suggested a model that would consist of the structural components of jennite and 1.4 nm tobermorite [53, 54]. According to this model, the omission of the bridging silica tetrahedra

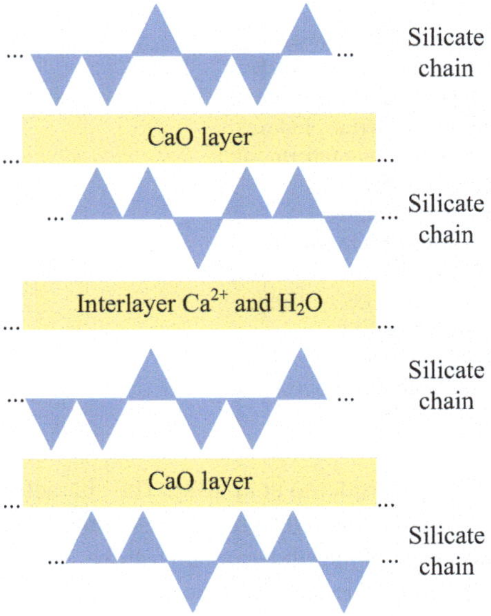

Fig. 1.3 Dreierketten structure of C–S–H gel

in tobermorite and the structure of jennite allows increasing the C/S ratio, i.e. that the C/S ratio is linearly proportional to the reciprocal of the silicate chain length.

Finally, Richardson and Groves [55–57] proposed a generalized model based on the stoichiometry of tobermorite, jennite, and calcium hydroxide. This model was similar to the previous models; however, the proposed model did not fix the level of protonation of the silicate chain. The generality of the model allowed for the specific inconsistencies of other models to be compensated [45]. According to this model, the general formula of C–S–H gel can be written as follows:

$$\left[\mathrm{Ca}_{2n}\mathrm{H}_{w}(\mathrm{Si}_{1-a}R_{a})_{3n-1}\mathrm{O}_{9n-2}\right]I^{c+}_{\frac{a}{c}(3n-1)}(\mathrm{OH})_{w+n(y-2)}\mathrm{Ca}_{\frac{ny}{2}}\cdot m\mathrm{H}_2\mathrm{O}$$

where the number of silanol groups is given by w, and the degree of protonation is defined as w/n. The term in the square brackets corresponds to the calcium and silicate layers, while the other terms are the extra ions and water. R refers to trivalent ions, such as aluminum, which substitute the bridging silicon ions, I represents the monovalent or divalent cations that ensure the electroneutrality after aliovalent substitutions, and c is the formal charge of these cations.

Due to the nature of the C–S–H formation and diversity, it is challenging to study their structure and properties. The critical point is that not all crystal structures of these compounds have been extensively explored [3, 41, 45, 58, 59]. Therefore, their classification has been constantly updated and improved. At present, a widely recognized classification of calcium silicate hydrates was presented by Taylor and Roy [60]. The current standard is based on the work by Richardson published in 2008 [3]. This classification consists of wollastonite, tobermorite, jennite, gyrolite, γ-C_2S, other C–S–H formed under hydrothermal synthesis conditions, and other high-temperature cement phases (Table 1.1). It should be foregrounded that the classification of calcium silicate hydrates can be continuously supplemented with newly discovered or synthesized compounds, the way it has been done with such positions as hydroxyledgrewite, trabzonite, etc.

The *wollastonite group of minerals* includes nekoite, okenite, xonotlite, foshagite, hillebrandite, wollastonite, and pectolite (Table 1.1). From the wollastonite group of calcium silicate hydrates, xonotlite [$Ca_6(Si_6O_{17})(OH)_2$] is of great practical importance and interest because this compound has been industrially produced as the main constituent for heat-insulating materials (Fig. 1.4a), building materials and artificial wood due to its high stability at high temperatures and its fibrous crystal form [61–65]. Xonotlite can be synthesized in stoichiometric CaO and SiO_2 mixtures (C/S = 1.0) under hydrothermal conditions at 190–400 °C temperature. The characteristic interplanar distance *d* of xonotlite equals to 0.7 nm. It was proven that the crystallization reactions of this compound strongly depend on the reactivity of SiO_2 as well as on the temperature and duration of the synthesis [18, 61, 64–66]. In addition, this author observed that xonotlite may be formed directly, without the formation of an intermediate mineral—1.13 nm tobermorite [67]. Wollastonite is the second compound in this group which has a wide range of applications [24, 26, 27, 68, 69]. This compound can be used for the production of low carbon cementitious materials or porcelain; as a coating material; or in biomedicine (Fig. 1.4b). In practice, wollastonite is produced by using various raw materials (diatomite, SiO_2 dust, slag, etc.) and methods (a solid-state reaction, two-step synthesis, sol–gel synthesis). Hillebrandite is not widely used in the industry; however, it can evidently be applied for the synthesis of stable and highly reactive β-C_2S [70, 71].

The *tobermorite group of minerals* are 1.4 nm tobermorite, 1.1 nm tobermorite, 1.13 nm tobermorite, 0.93 nm tobermorite, clinotobermorite, 0.9 nm clinotobermorite, oyelite, C–S–H(I) (C/S $\leq$ 1.5). Tobermorite, as a natural mineral, was discovered by Heddle (1880) at Tobermory, Scotland [72]. This compound forms in an alkaline environment under hydrothermal conditions. The investigations of natural tobermorite and other calcium silicate hydrates showed that they are identical to 'well-crystallized C–S–H(I)' [41, 73]. The main difference between synthetic and natural tobermorite is in their crystallinity. Tobermorite exhibits excellent properties, such as flame-retardant, high compressive strength, lightweight, etc., which makes it prominently important in the industry [15, 18, 41, 73–79]. Thus, this compound is being widely used for the construction of buildings (specifically, sand-lime brick, and autoclaved aerated concrete are used), heat-insulating, and fire-resistant materials (Fig. 1.5). These products can be synthesized by curing CaO and SiO_2 bearing materials under hydrothermal synthesis conditions at 170–210 °C for 5–8 h [15, 18, 77, 78]. Also, other synthesis methods, such as microwave or mechanochemical synthesis, can be successfully applied [74–76].

The *jennite group of minerals* includes jennite, metajennite, and C–S–H(II) (Table 1.1). The structure of the first two minerals is similar to that of poorly crystallized synthetic calcium silicate hydrates [41, 43, 53, 55]. Jennite, $Ca_9(Si_6O_{18})(OH)_6{\cdot}8H_2O$, is a fairly uncommon natural mineral which, in most cases, is present in the nature in contact with 14 Å tobermorite [80, 81]. It can be synthesized

Table 1.1 Supplemented classification of calcium silicate hydrates

Name of group	Name of mineral	Molar ratio of C/S	Formula of mineral
Wollastonite	Nekoite	0.5	$Ca_3Si_6O_{15}\cdot7H_2O$
	Okenite	0.56	$[Ca_8(Si_6O_{16})(Si_6O_{15})_2(H_2O)_6]^{4-}[Ca_2(H_2O)_9\cdot3H_2O]^{4+}$
	Pectolite	1.0	$Ca_2NaHSi_3O_9$
	Wollastonite 1A	1.0	$Ca_3Si_3O_9$
	Xonotlite	1.0	$Ca_6Si_6O_{17}(OH)_2$
	Foshagite	1.33	$Ca_4(Si_3O_9)(OH)_2$
	Hillebrandite	2.0	$Ca_2(SiO_3)(OH)_2$
Tobermorite	1.1 nm tobermorite	0.67	$Ca_4Si_6O_{15}(OH)_2\cdot5H_2O$
	1.13 nm tobermorite	0.75	$Ca_{4.5}Si_6O_{16}(OH)\cdot5H_2O$
	0.93 nm tobermorite	0.83	$Ca_5Si_6O_{16}(OH)_2$
	1.4 nm tobermorite	0.83	$Ca_5Si_6O_{16}(OH)_2\cdot7H_2O$
	Clinotobermorite	0.83	$Ca_5Si_6O_{17}\cdot5H_2O$
	Clinotobermorite 9 Å	0.83	$Ca_5Si_6O_{16}(OH)_2$
	Oyelite	1.25	$Ca_{10}B_2Si_8O_{29}\cdot12.5H_2O$
	C–S–H(I)	≤ 1.5	
Jennite	Jennite	1.5	$Ca_9Si_6O_{18}(OH)_6\cdot8H_2O$
	Metajennite	1.5	$Ca_9Si_6O_{18}(OH)_6\cdot8H_2O$
	C–S–H(II)	≥ 1.5	
Gyrolite	K-phase	0.44	$Ca_7Si_{16}O_{38}(OH)_2$
	Z-phase	0.56	$Ca_9Si_{16}O_{40}(OH)_2\cdot14H_2O$
	Truscottite	0.58	$Ca_{14}Si_{24}O_{58}(OH)_8\cdot2H_2O$

(continued)

Table 1.1 (continued)

Name of group	Name of mineral	Molar ratio of C/S	Formula of mineral
	Gyrolite	0.66	$Ca_{16}Si_{24}O_{60}(OH)_8 \cdot 14H_2O$ $NaCa_{16}Si_{23}AlO_{60}(OH)_8 \cdot 14H_2O$
	Reyerite	0.67	$Na_2Ca_{14}Si_{22}Al_2O_{58}(OH)_8 \cdot 6H_2O$
	Reyerite	0.58	$Ca_{14}Si_{24}O_{58}(OH)_8 \cdot 6(H_2O)$
γ-C_2S	Calcium chondrodite	2.5	$Ca_5[SiO_4]_2(OH)_2$
	Kilchoanite	1.5	$Ca_6(SiO_4)(Si_3O_{10})$
	C_8S_5	1.6	$Ca_8(SiO_4)_2(Si_3O_{10})$
	γ-C_2S	2.0	$\gamma\text{-}Ca_2SiO_4$
Other calcium silicate phases	Afwillite	1.5	$Ca_3(SiO_3OH)_2 \cdot 2H_2O$
	α-C_2SH	2.0	$Ca_2(HSiO_4)(OH)$
	Cuspidine	2.0	$Ca_4(F_{1.5}(OH)_{0.5})Si_2O_7$
	Dellaite	2.0	$Ca_6(Si_2O_7)(SiO_4)(OH)_2$
	Jaffeite	3.0	$Ca_6[Si_2O_7](OH)_6$
	Killalaite	1.6	$Ca_{6.4}(H_{0.6}Si_2O_7)_2(OH)_2$
	Poldervaartite	2.0	$Ca(Ca_{0.67}Mn_{0.33})(HSiO_4)(OH)$
	Rosenhahnite	1.0	$Ca_3Si_3O_8(OH)_2$
	Suolunite	1.0	$CaSiO_{2.5}(OH) \cdot _{0.5}H_2\ O$
	Tilleyite	2.5	$Ca_5Si_2O_7(CO_3)_2$
	C_3S hydrate	3.0	$Ca_6(Si_2O_7)(OH)_6$
Other high-temperature cement phases	Bicchulite	0.67	$Ca_2(Al_2SiO_6)(OH)_2$
	Fukalite	2.0	$Ca_4(Si_2O_6)(CO_3)(OH)_2$
	Katoite Hydrogarnet	0.94	$Ca_{1.46}AlSi_{0.55}O_6H_{3.78}$

(continued)

Table 1.1 (continued)

Name of group	Name of mineral	Molar ratio of C/S	Formula of mineral
	Rustumite	2.0	$Ca_{10}(Si_2O_7)_2(SiO_4)Cl_2(OH)_2$
	Scawtite	1.17	$Ca_7(Si_6O_{18})(CO_3)\cdot 2H_2O$
	Strätlingite	0.62	$Ca_2Al(AlSi)_{1.11}O_2(OH)_{12}\cdot 2.25H_2O$
Newly discovered or synthesized calcium silicate hydrates	Hydroxyledgrewite	2.25	$Ca_9(SiO_4)_4(OH)_2$
	Trabzonite	1.33	$Ca_4[Si_3O_9(OH)]OH$
	Chegemite	2.33	$Ca_7(SiO_4)_3(OH)_2$
	Spurrite	2.5	$Ca_5(SiO_4)_2(CO_3)$
	Galuskinite	2.33	$Ca_7(SiO_4)_3(CO_3)$

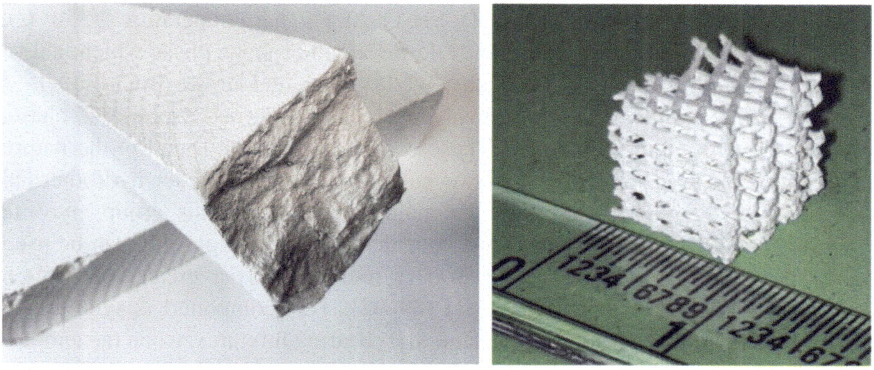

a b

Fig. 1.4 Heat insulation block based on xonotlite (**a**, photo by the authors) and scaffold with 20 wt% of wollastonite (**b**, photo by the authors)

a b

Fig. 1.5 Aerated autoclaved concrete blocks (**a**, photo ID:1326053465, *Source* iStock by Getty Images) and roofing sheets (**b**) (with permission from UAB Eternit Baltic) based on CSH

in suspensions of $Ca(OH)_2$ and hydrated silica at 60–100 °C temperatures [12, 77, 78, 82, 83]. The structure of jennite is used as a prototype of C–S–H(II), which is one of the most important components of hydrated cement [41, 43, 53, 55]. For this reason, many authors have investigated and refined the crystal structure of jennite. The ideal molar C/S ratio of jennite is equal to 1.5, which is higher than that of 14 Å tobermorite (with its C/S ratio of 1). According to the scientific literature, jennite consists of three structural modules: tilleyite ribbons of an edge-sharing calcium octahedral; dreierketten silicate chains running along [010] direction; an additional calcium octahedral between the presently discussed chains or on inversion centers [41, 43, 53, 55]. This compound loses 4 water molecules at 70–90 °C, and, as a result, the atomic environment of the Ca interlayer changes [84]. Such a structure of jennite is not stable, and it shrinks in the *c* direction, which leads to the recrystallization to a stable compound metajennite, $Ca_9Si_6O_{18}(OH)_6 \cdot 8H_2O$. At a higher temperature (~350 °C), metajennite dehydrates and subsequently forms an amorphous phase which, above 800 °C, recrystallizes to calcium silicates, wollastonite, and larnite (β-Ca_2SiO_4).

The *gyrolite group of minerals* involves gyrolite, fedorite, K-phase, Z-phase, reyerite, and truscottite (Table 1.1). These compounds can be found in the nature (in volcanic and basaltic rocks, voids or gaps) or be formed under hydrothermal conditions [85–88]. The interest in the main compound of this group, gyrolite [$Ca_{16}Si_{24}O_{60}(OH)_8 \cdot 14H_2O$], has grown considerably because gyrolite can be used as an adsorbent for the removal of impure ions and has a greater adsorption capacity in comparison with other CSH [89–92]. This compound is synthesized under hydrothermal conditions in CaO and SiO_2 bearing mixtures, when the temperature of the synthesis is higher than 120 °C and lower than 200 °C [85, 93–97]. At higher temperatures (> 200 °C), gyrolite is not stable, and truscottite [$Ca_{14}Si_{24}O_{58}(OH)_8 \cdot 2H_2O$] is formed. According to some sources, the metastable gyrolite may be identified even at temperatures up to 270 °C [85, 94, 98, 99]. The formation of gyrolite strongly depends on the nature of SiO_2 sources: by using amorphous silicon dioxide, gyrolite forms after 32 h of hydrothermal treatment at 200 °C; and, while using quartz, it can be obtained even after 168 h [100]. It should be noted that the low solubility rate of quartz leads to the significant excess of Ca^{2+} ions in the solution and the formation of higher basicity calcium silicate hydrates, tobermorite and xonotlite.

The *γ-Ca_2SiO_4 group minerals* are calcium chondrodite, kilchoanite, γ-Ca_2SiO_4, and C_8S_5 (Table 1.1). The literature data on the properties and formation of these minerals is relatively limited. These compounds are found as natural minerals or formed at hydrothermal treatment conditions [10, 101, 102]. The most investigated compounds are kilchoanite and γ-Ca_2SiO_4. The structure of kilchoanite $Ca_6(SiO_4)(Si_3O_{10})$, (C/S = 1.5) has an equal number of isolated silicon tetrahedra (Q0) and finite triple chains [103–108]. The triple chains consist of one Q_2 tetrahedron and two Q_1 tetrahedra. Thus, in the structure of kilchoanite, there are ¼Q_0, ½Q_1 and ¼Q_3 tetrahedra. These silicon tetrahedra are connected to sixfold and eightfold coordinated calcium atoms. Garbev et al. [108] determined that the degree of polymerization in kilchoanite directly depends on the temperature of formation. Roy et al. [103, 107] proved that kilchoanite synthesized under hydrothermal conditions at 715 °C is anhydrous, but Speakman et al. [104] found that kilchoanite formed at lower temperatures contains structural water. Taylor [106] and Mitsuda [109] announced their finding that kilchoanite can be prepared from lime and quartz mixtures at 280–600 °C and at 500–1000 bars. Later on, Dambrauskas et al. [10] determined that, under hydrothermal conditions, kilchoanite (Fig. 1.6a) formed only at higher temperatures than 175 °C. The pure compound (> 96 wt%) was obtained by treating stoichiometric (C/S = 1.5) CaO and amorphous silica mixtures at 200 °C for 48 h (Fig. 1.6b). The increment of the synthesis duration and temperature led to the formation of C_8S_5 and xonotlite. Also, these authors proved that this compound is stable till 1100 °C, whereas, at higher temperatures, it recrystallizes to its polymorph rankinite ($Ca_3Si_2O_7$).

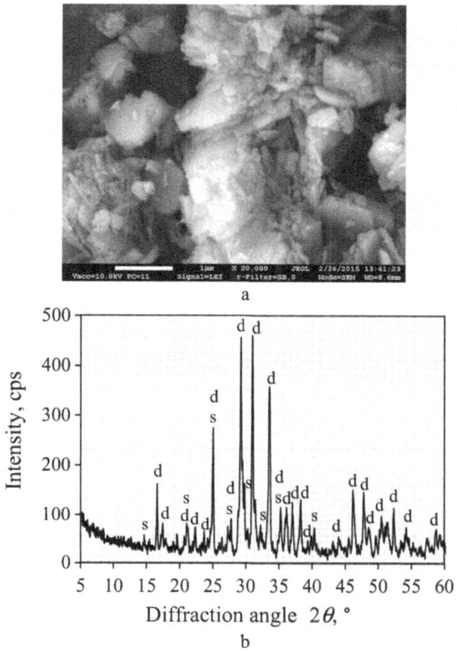

Fig. 1.6 SEM micrograph (**a**) and XRD pattern (**b**) of pure (> 96 wt%) synthetic kilchoanite

Other calcium silicate hydrate compounds are afwillite, cuspidine, α-C_2SH, dellaite, jaffeite, killalaite, poldervaartite, rosenhahnite, suolunite, tilleyite (Table 1.1). Most of these compounds are less significant for the chemistry of cement and binding materials [3]. For example, afwillite [$Ca_3(SiO_3OH)_2{\cdot}2H_2O$] is an incombustible and heat-insulating compound [110–112]. This compound forms by the hydrothermal reaction (at temperatures above 200 °C), the mechanochemical synthesis, and by the hydration of calcium silicates. Afwillite accelerates the hydration of tricalcium silicate and reduces the drying shrinkage of hardened samples. Another compound of this group, which is widely used in the steel industry as a mold flux, is cuspidine [$Ca_4(F_{1.5}(OH)_{0.5})Si_2O_7$] (Fig. 1.7a) [113–116]. This compound is a rare mineral, and it is synthesized by hydrothermal or solid-state reactions. Recently, the synthesis, properties, and structure of α-C_2SH [$2CaO{\cdot}SiO_2{\cdot}H_2O$] (Fig. 1.7b) (a relatively frequently found compound formed during the hardening of autoclaved calcium silicate products or the hydration of G-oil well cement pastes [117]) regained interest because, on its basis, an environmentally friendly cementitious material—*Celitement*—has been created [118–121]. The production of this material is based on the hydrothermal synthesis of α-C_2SH, and on its mechanochemical grinding together with quartz sand and/or thermal activation. Several papers suggest that α-C_2SH crystallizes quite easily in CaO and silicic acid or CaO and amorphous silica mixtures under hydrothermal conditions within the temperature range of 150–200 °C [122, 123]. On the other hand, crystallization proceeds more passively when quartz is being used as a source of SiO_2 [124–127]. Also, α-C_2SH can be obtained by

a b

Fig. 1.7 SEM micrograph of synthetic cuspidine (**a**) and synthetic α-C_2SH (**b**)

treating C_2S polymorphs under hydrothermal conditions [122, 128]. According to some sources, the formation of α-C_2SH can be accelerated by adding α-C_2SH crystals to the initial mixture as crystallization centers, or by reducing the molar ratio of the initial mixture to 1.75 [129, 130].

Other high-temperature cement phases involve bicchulite, fukalite, katoite hydrogarnet, rustumite, scawtite, and strätlingite (Table 1.1). The compounds of this group are less significant for the industry, and only strätlingite and katoite hydrogarnet a given a great interest in the cement chemistry. Strätlingite is one of the main hydration products of sulfoaluminate phase-rich cements [131, 132]. Meanwhile, katoite hydrogarnet forms during the cement hydration of calcium aluminate, calcium sulfoaluminate, or other substances [133, 134].

1.2 Newly Discovered or Synthesized Calcium Silicate Hydrates

The classification of calcium silicate hydrates can be supplemented with newly discovered or synthesized compounds, such as hydroxyledgrewite, trabzonite, chegemite, etc. Trabzonite {$Ca_4[Si_3O_9(OH)]OH$} [135, 136] and chegemite [$Ca_7(SiO_4)_3(OH)_2$] [137] are rare natural calcium silicate hydrates approved by the *International Mineralogical Association* (IMA) in 1983 and 2008, respectively. Sarp and Burri (1986) described trabzonite as a calcium silicate hydrate whose formula was expressed as $Ca_4[Si_3O_{10}]\cdot 2H_2O$ [136]. In 2012, Armbruster et al. investigated the structure of trabzonite and suggested a revised chemical formula, specifically, $Ca_4[Si_3O_9(OH)]OH$ [135]. These authors stated that trabzonite and foshagite are polymorphs (in their opinion, foshagite features wollastonite-like silicate chains, while trabzonite does not). It is worth noting that this compound was not reported as a product of cement hydration. According to the literature data, chegemite was discovered in 2007 in Northern Caucasus [137]. This calcium silicate hydrate is a calcium and hydroxyl analog of humite. Chegemite forms in high-temperature skarns by forming aggregates with grain sizes of < 5 mm. The structure and relationship with other minerals were described by Galuskin et al. [137].

The ternary system of $CaO–SiO_2–CO_2$ consists of three compounds: spurrite [$Ca_5(SiO_4)_2(CO_3)$], tilleyite [$Ca_5(Si_2O_7)(CO_3)_2$], and galuskinite [$Ca_7(SiO_4)_3(CO_3)$] [138, 139]. These silicates do not contain water in their structures; however, they are related to calcium silicate hydrates and their chemistry. For this reason, spurrite was earlier involved in the classification of calcium silicate hydrates. Spurrite was found in 1908 [140], yet it was only described in 1960 [138]. The structure of this compound was revised by Grice [139]. Spurrite is an orthosilicate which forms in high-temperature skarns. This compound can be obtained during clinker calcination in the presence of CO_2 within the temperature range of 750–900 °C [141–145]. Some elements (fluorine, chlorine, and potassium) promote the formation of spurrite. Spurrite can be synthesized by heating the mixture of wollastonite and calcite or hydrothermally in a mixture of calcium oxide, silica gel, and oxalic acid (a source of CO_2). In order to obtain pure spurrite, the temperature of hydrothermal synthesis should be higher than 500 °C, while the pressure of CO_2 should be equal to 1000 bar [140].

Tilleyite is a sorosilicate and the second mineral in the $CaO–SiO_2–CO_2$ system [139, 146–148]. This compound was discovered in 1933 and synthesized in 1959. The crystal structure of tilleyite was determined by Smith [149] and refined by Lottisnathan and Smith [146], and Grice [139]. The conditions for the formation of spurrite and tilleyite are similar; thus, tilleyite can be obtained by calcining a mixture of wollastonite and calcite or hydrothermally in a mixture of calcium oxide, silica gel, and oxalic acid (a source of CO_2) [142]. Tilleyite requires a higher temperature of hydrothermal synthesis (> 650 °C) than spurrite (> 500 °C). Perez et al. [148] found that tilleyite has a polymorph at a pressure above 8 GPa.

The third mineral of carbonated calcium silicate is galuskinite [150, 151]. This silicate mineral was approved by *The Commission on New Minerals, Nomenclature and Classification* (CNMNC) of the *International Mineralogical Association* as recently as in 2011. The mineral is named in honor of the mineralogists Galuskina and Galuskin. According to Lazic et al. [150], galuskinite is a retrograde product of skarn alteration. Galuskinite was not observed in any previous studies of cement chemistry or the $CaO–SiO_2–CO_2$ system.

The classification of calcium silicate hydrates can be concluded with hydroxyledgrewite [$Ca_9(SiO_4)_4(OH)_2$] which was discovered by Galuskin et al. in 2009 [152], approved by IMA in 2012, and synthesized for the first time by Dambrauskas and Baltakys [1]. Hydroxyledgrewite [$Ca_9(SiO_4)_4(OH)_2$] is the hydroxyl analogue of edgrewite [$Ca_9(SiO_4)_4F_2$] which was named in honor of Edward S. Grew [152]. These minerals were discovered in one of the seven large xenoliths of carbonate silicate skarn within the Upper Chegem caldera, Northern Caucasus, Kabardino-Balkaria, Russia [137, 152]. According to Galuskin et al. [153], xenoliths are perceived to be terrigenous-carbonate rocks metamorphosed to calc-silicate skarn under conditions of the sanidinite facies (temperature > 900 °C and pressure ~100–200 MPa). Edgrewite and hydroxyledgrewite were formed at somewhat lower temperatures during retrogression following crystallization of larnite skarns [152]. Bultfonteinite, hillebrandite, jennite, and chegemite-fluorchegemite are the most widespread minerals associated with edgrewite and hydroxyledgrewite.

Galuskin et al. [152] also described the physical and optical properties, the chemical composition and the crystal structure of natural edgrewite and hydroxyledgrewite. The outlined minerals show combinations of pinacoids and rhombic prisms. Edgrewite and hydroxyledgrewite form a solid-solution $Ca_9(SiO_4)_4(F,OH)_2$, in which the proportion of the edgrewite end-member $Ca_9(SiO_4)_4F_2$ ranges from 31 wt% at F = 1.52 wt% to 74 wt% at F content = 3.64 wt%.

Hydroxyledgrewite belongs to the nesosilicates or orthosilicates group [152]; these silicates are denoted by the lowest degree of polymerization possible, namely, C–S–H phases with isolated silicon tetrahedra (Q^0) [108]. The humite group [154] is a subgroup of nesosilicates with the general formula $An(SiO_4)m(F,OH)_2$ ('*A*' shows *n* moles of cation, '*m*' refers to the moles of SiO_4). Meanwhile, the given group formula for calcium members can be written as $n{\cdot}Ca_2SiO_4 + Ca_3(SiO_4)(F,OH)_2$, where Ca_2SiO_4 presents the 'calcio-olivine,' and $Ca_3(SiO_4)(F,OH)_2$ stands for the hypothetical 'calcio-norbergite.' When $n = 3$, the formula of edgrewite $Ca_9(SiO_4)_4F_2$ and hydroxyledgrewite $Ca_9(SiO_4)_4(OH)_2$ (Ca/Si = 2.25:1) is obtained [152, 155]. According to Galuskin et al. [152], the structure projection of edgrewite along {100} is similar to clinohumite. Meanwhile, the hydrogen bonds of hydroxyledgrewite are related/connected with O9 and/or F9. The space group of hydroxyledgrewite is P21/b11 (no. 14) with monoclinic lattice (Fig. 1.8) and the following parameters: a = 0.50672 nm, b = 1.13545 nm, c = 1.53941 nm, α = 100.587°, V = 870.63 [152].

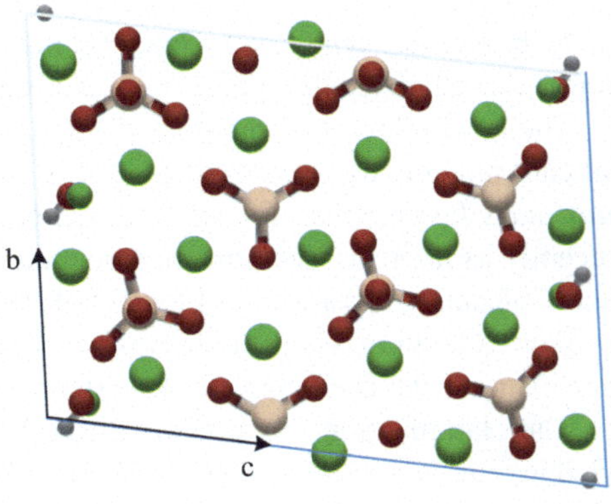

Fig. 1.8 Lattice structure and composition of the mineral hydroxyledgrewite (PDF No. 04-019-1437)

1.3 Hydrothermal Synthesis and Properties of Hydroxyledgrewite

The synthesis of the hydroxyledgrewite sample was based on the hydrothermal method. Finely ground $SiO_2 \cdot nH_2O$ (*Reaktiv*, Russia, loss of ignition—5.19 wt%) and calcium oxide produced by burning calcium hydroxide (*Reaktiv*, Russia) at 550 °C for 1 h (with the quantity of free CaO being equal to 98.7 wt%) were used as the starting materials. The molar ratio CaO/SiO_2 of the primary mixture was equal to 2.25, while the water/solid ratio was maintained at 10.0. Hydrothermal synthesis was carried out in unstirred suspensions in 25 ml volume PTFE cells which were placed in an autoclave produced by *Parr Instruments* (Germany). The duration of the synthesis at a temperature of 200 °C was equal to 8, 16, 48, and 72 h. The required temperature was reached within 2 h. After the synthesis, the suspensions were filtered off, the products were rinsed with ethanol to prevent carbonization of the obtained materials, dried in a vacuum chamber at 50 °C ± 5 temperatures for 24 h, and sieved through a sieve with an 80 μm mesh.

1.3.1 Hydrothermal Synthesis of Hydroxyledgrewite

It was determined that, in the mixture with CaO/SiO_2 equal to 2.25, after 8 h of isothermal curing at 200 °C, portlandite hardly reacted: intensive diffraction maximums characteristic of portlandite (PDF No. 00-044-1481; *d*-spacing—0.492; 0.262; 0.193 nm) were identified (Fig. 1.9, curve 1). Together with unreacted portlandite, a broad basal reflection was observed within the range of diffraction angles between 26 and 37° that is characteristic of semicrystalline type compounds. After 16 h of the synthesis, the interaction degree between the raw materials increased because intensive diffraction peaks characteristic of dibasic calcium silicate hydrate—hydroxyledgrewite (PDF No. 04-019-1437; *d*-spacing—0.303; 0.282; 0.191 nm)—formed in the products (Fig. 1.9, curve 2). In spite of that, relatively intensive diffraction peaks of unreacted portlandite were observed nevertheless.

According to the literature [3, 10, 12, 14, 15, 78, 100, 126], in mixtures with a molar ratio of CaO/SiO_2 higher than 2.0, afwillite, hillebrandite, α-C_2SH, and other higher basicity calcium silicate hydrates should be dominant in the products. However, previous researchers never managed to synthesize hydroxyledgrewite. Probably, the differences appear due to the nature and different reactivity of the raw materials.

It was determined that after 48–72 h of hydrothermal treatment, only traces of unreacted portlandite were present in the synthesis products (Fig. 1.9, curves 3 and 4). Meanwhile, the intensity of the diffraction peaks characteristic of hydroxyledgrewite slightly increased. It was determined that all the peaks characteristic of synthesized hydroxyledgrewite correspond (± 0.001 nm) to PDF-4+ 2021 entry PDF No. 04-019-1437 (Table 1.2). Since there is no significant difference between the products

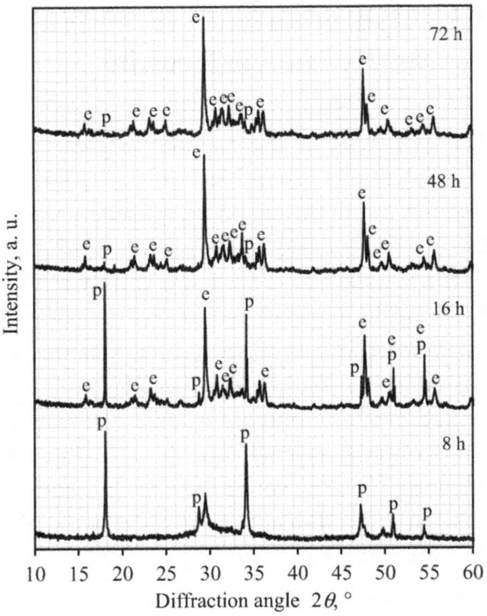

Fig. 1.9 XRD patterns of the synthesis products of hydrothermal treatment at 200 °C temperature. Indexes: p—portlandite; e—hydroxyledgrewite

obtained after 48 and 72 h of the synthesis, for the characterization of hydroxyledgrewite, a sample synthesized after 48 h was used. It was calculated that the mean crystallite size of hydroxyledgrewite is equal to 52 ± 3 nm (R_{exp}—6.43, R_{wp}—11.06, GOF—1.72).

The FT-IR spectrum of the stnthetic hydroxyledgrewite sample is presented in Fig. 1.10. According to the literature [152], the FT-IR spectrum of hydroxyledgrewite was only analyzed by Galuskin et al. However, the authors did not mark the individual peaks for the characteristic vibrations. In this work, the data obtained by Galuskin et al. [152] and the spectra of dibasic calcium silicate hydrates [156, 157] were used to analyze the FT-IR spectra of hydroxyledgrewite. Thus, the results of the FT-IR analysis may be not completely accurate. The intensive duplet at ~514 and ~558 cm^{-1} can be assigned to the bending vibrations of δ(Si–O–Si) and δ(O–Si–O) (Fig. 1.10). In the range characteristic of ν_s(Si–O–Si) vibrations, four absorption bands at ~616, ~654, ~708, and ~764 cm^{-1} were identified. According to the literature [156, 157], the absorption bands within the frequency interval of 800–1000 cm^{-1} can be assigned to symmetrical ν_s(O–SiO–) stretching vibrations, while the absorption bands within a higher frequency interval (1000–1300 cm^{-1}) can be assigned to asymmetrical ν_{as}(Si–O–Si) stretching vibrations. The intensive duplet in the frequency interval of 3400–3600 cm^{-1} can be assigned to OH^- stretching vibrations. This data is in good agreement with the statements of Galuskin et al. [152] who determined that, within the frequency interval of 3400–3600 cm^{-1}, hydroxyledgrewite is characterized by two absorption bands at 3486 and 3554 cm^{-1}. Meanwhile, edgrewite ($Ca_9(SiO_4)_4F_2$) is characterized by three overlapping bands at 3543, 3551, and 3558 cm^{-1}.

Table 1.2 Main peak characteristics of synthesized hydroxyledgrewite and entry

Estimated parameters		PDF No. 04-019-1437
2θ, deg	*d*-spacing, nm	*d*-spacing, nm
15.897	0.557	0.558
21.469	0.414	0.414
23.337	0.381	0.381
23.690	0.375	0.375
25.185	0.353	0.354
29.293	0.305	0.304
29.462	0.303	0.303
29.644	0.301	0.300
30.799	0.290	0.289
31.496	0.284	0.286
31.652	0.282	0.282
32.363	0.276	0.276
33.743	0.265	0.266
35.695	0.251	0.251
36.227	0.248	0.248
47.577	0.191	0.191
48.026	0.189	0.189

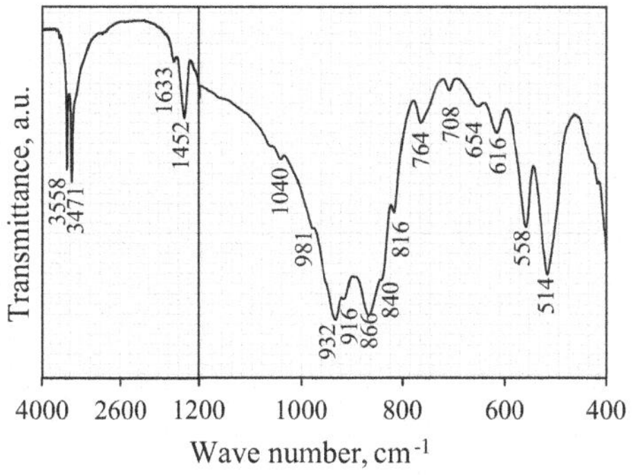

Fig. 1.10 FT-IR spectrum of synthetic hydroxyledgrewite sample

The absorption band at 1452 cm^{-1} is uncharacteristic of hydroxyledgrewite. This absorption band is usually characteristic of $\nu(CO_3^{-2})$ vibrations in calcium carbonate [156, 157]. Probably, a small amount of carbonate was formed due to carbonization when the products were dried in an air-conditioned chamber (50 °C, 24 h). It is worth noting that the diffraction peaks of the presently mentioned compound were not detected in the XRD patterns (Fig. 1.9).

With the objective to determine the thermal stability of hydroxyledgrewite, simultaneous thermal analysis (Table 1.3 and Fig. 1.11) and high-temperature in-situ XRD were performed (Fig. 1.12). An in-situ XRD experiment was performed in a high-temperature camera *MTC-hightemp* within the 25–1200 °C temperature range. In the DSC curve, the first endothermic effect (30–250 °C) can be assigned to the removal of adsorption water in the sample (Fig. 1.11) because the intensity of the diffraction peaks of hydroxyledgrewite remains unchanged till 675 °C (Fig. 1.12). According to the TGA data, the sample contains as little as 0.25 wt% of humidity (Table 1.3 and Fig. 1.11). The sharp endothermic effect at 451 °C temperature reflects the decomposition of unreacted portlandite (Fig. 1.11). It was calculated that, during the decomposition of the latter compound, the synthesis products lose ~0.36% of the mass, which is equivalent to 1.60 wt% of portlandite (Table 1.3).

It was also established that the synthetic product does not lose mass (except for the removal of humidity and the decomposition of portlandite) until the temperature of calcination is reached above 600 °C (Fig. 1.11). For this reason, the third endothermic effect (620–720 °C), in which ~2.2% of the mass was lost (Table 1.3), can be assigned to the decomposition of hydroxyledgrewite. The STA data is in good agreement with

Table 1.3 Thermal characteristics of sample of hydroxyledgrewite

Removal of adsorption/crystallization water			Decomposition of portlandite		
Peak, °C	Δm*, %	ΔQ*, J/g	Peak, °C	Δm, %	ΔQ, J/g
125	0.25	47.11	451	0.36	10.97
Decomposition of hydroxyledgrewite			Recrystallization of C_2S polymorphs		
Peak, °C	Δm, %	ΔQ, J/g	Peak, °C	Δm, %	ΔQ, J/g
705	2.2	20.50	861	0.15	11.49

* Δm—loss on ignition; ΔQ—heat of the thermal effect

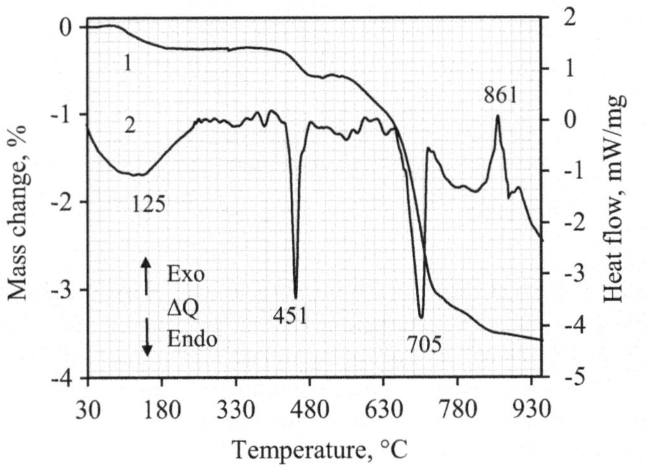

Fig. 1.11 STA curves (1—TGA; 2—DSC) of sample of synthetic hydroxyledgrewite

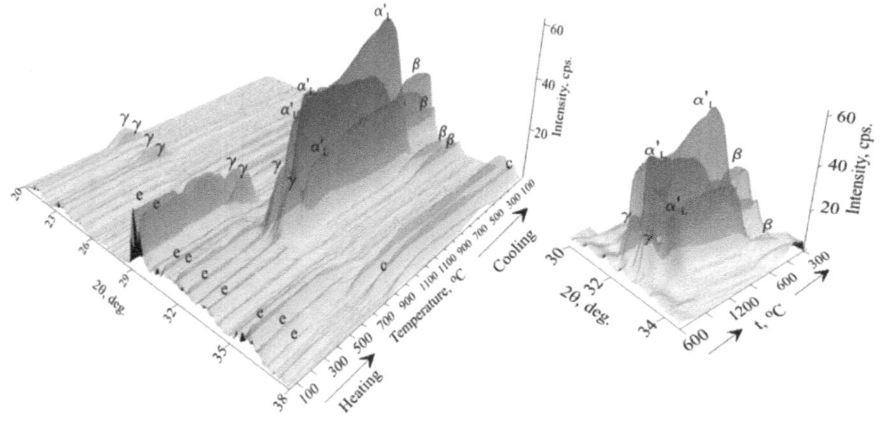

Fig. 1.12 In-situ XRD patterns of synthetic hydroxyledgrewite when the temperature of calcination is 25–1200 °C. Indexes: e—hydroxyledgrewite; γ—γ-C_2S; α'_L—α'_L-C_2S; β—β-C_2S; l—CaO

the results of high-temperature in-situ XRD: hydroxyledgrewite recrystallized to γ-C_2S and CaO at 675 °C temperature (Fig. 1.12). Since the molar ratio of CaO/SiO_2 in hydroxyledgrewite (2.25) is higher than in γ-C_2S (2.0), reflections of calcium oxide (PDF 0-043-1001; *d*-spacing—0.278; 0.241 nm) are visible in the XRD pattern. Finally, the exothermic effect at 861 °C can be assigned to the recrystallization of C_2S polymorphs (Fig. 1.12).

According to the data of high-temperature in-situ XRD, hydroxyledgrewite becomes metastable at 675 °C temperature and fully recrystallizes to γ-C_2S at 800 °C (Fig. 1.12). By increasing the temperature of calcination to 860 °C, γ-C_2S starts recrystallizing to α'_L-C_2S. At 1000 °C, diffraction peaks characteristic of γ-C_2S disappear; meanwhile, the intensity of the diffraction peaks characteristic of α'_L-C_2S increase. The presented results are in good agreement with the data obtained by Garbev et al. [59, 128, 158] and Toraya et al. [159]. It is worth specifying that α'_L-C_2S remained stable till 1200 °C and recrystallized into β-C_2S (*d*-spacing—0.278; 0.274; 0.261; 0.240 nm) after cooling (500 °C).

It is possible to assume that, during the experiments of hydrothermal synthesis and the calcination of synthetic hydroxyledgrewite samples, the phase formation and transformation occurred in the following sequence:

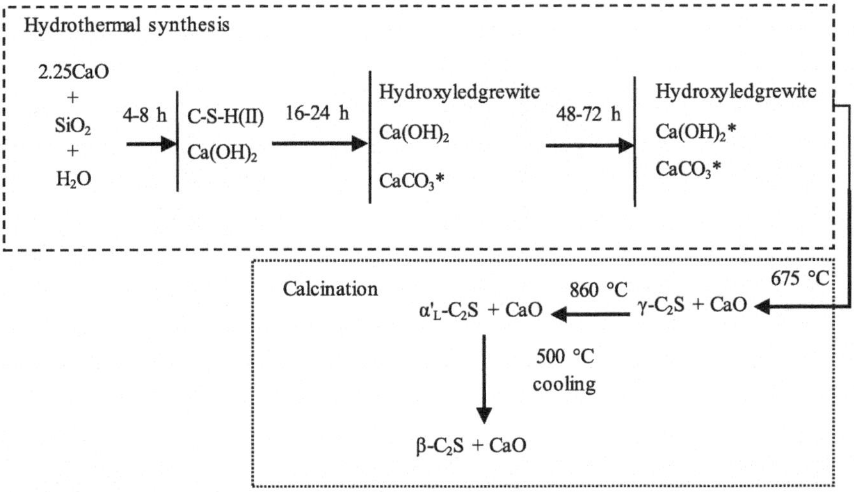

* – traces of the compound could be present in the products

The thermal stability of materials is one of the most important thermal parameters; however, other parameters, such as the specific heat capacity (c_p), thermal conductivity, thermal expansion, etc., can also determine the possibility to use the material. The specific heat capacity is the heat capacity per unit molecular weight of the substance, which shows the ability of materials to absorb heat from the surroundings [160–162]. The specific heat capacity of the hydroxyledgrewite sample at a constant pressure was determined as a function of temperature by the sapphire method (Fig. 1.13) [160]. According to the method, the c_p values of the sample were obtained in three stages (Fig. 1.13a):

(1) The first curve (baseline) is obtained by heating two empty crucibles.
(2) The second curve is recorded with the crucible filled with Al_2O_3.
(3) The third curve is recorded with the crucible filled with the hydroxyledgrewite sample.

The values of c_p were calculated by using the following equation:

$$c_p = \left(\frac{\Delta y_2 \cdot m'}{\Delta y_1 \cdot m}\right) \cdot c'_p, \qquad (1.1)$$

where Δy_1 is the distance from the baseline curve to that of the sample (Fig. 1.13a); Δy_2 is the distance from the baseline curve to the one of sapphire (Al_2O_3); m' is the mass of the sapphire; m is the mass of the sample; c'_p is the specific heat capacity of Al_2O_3.

The c_p values of the synthetic hydroxyledgrewite sample were determined within a temperature range of 25–420 °C. The presently discussed values cannot be determined in a higher temperature range because the decomposition of the traces of portlandite (422–452 °C) distorts the results. It was determined that, at 25 °C, the

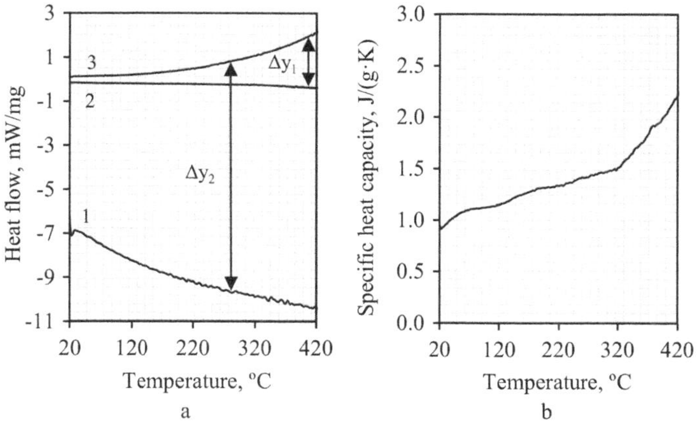

Fig. 1.13 Principal scheme of the application of the sapphire method **a** 1—the sapphire curve; 2—the sample curve; 3—the baseline curve; changes of c_p of hydroxyledgrewite **b** within the temperature range of 20–420 °C

c_p value of the hydroxyledgrewite sample is equal to 0.928 J/(g K), which increases monotonically in a continuous manner by increasing the temperature up to 320 °C (Table 1.4 and Fig. 1.13b). Meanwhile, in a higher temperature range up to 420 °C, c_p values grow rapidly. In general, particles of hydroxyledgrewite move more vigorously at higher temperatures than at lower temperatures. Hence, a particular quantity of hydroxyledgrewite at higher temperatures requires more energy to increase the temperature by the same number of grades than at lower temperatures; for this reason, the specific heat capacity increases with temperature.

Since hydroxyledgrewite is a hydrated calcium silicate, it does not exhibit hydraulic activity; i.e. it does not react with water by binding fillers (sand, gravel, etc.) in concrete. To achieve hydraulic activity, hydrogen bonds should be destroyed by depositing reactive fragments on a silicate surface. According to the literature, the hydrogen bonds in calcium silicate hydrates can be destroyed by milling and, subsequently, calcining. Usually, during the milling process, sand or other hard materials are used. For this reason, a hydroxyledgrewite sample was mixed with granite cutting waste (1:1 by mass), milled in a vibrating cup mill (*Pulverisette 9*) at 950 rpm for 5 min and thermally treated at a different temperature for 1 h. In order to test the effectiveness of the presently mentioned method, microcalorimetry analysis was performed, which is among the most accurate methods to monitor the global reaction process through the rate of heat production. The total amount of the released heat

Table 1.4 The c_p values of synthetic hydroxyledgrewite sample at different temperatures

T, °C	25	50	100	150	200	250	300	350	400	420
c_p, J/(g·K)	0.928	1.048	1.128	1.230	1.318	1.392	1.470	1.705	2.028	2.212

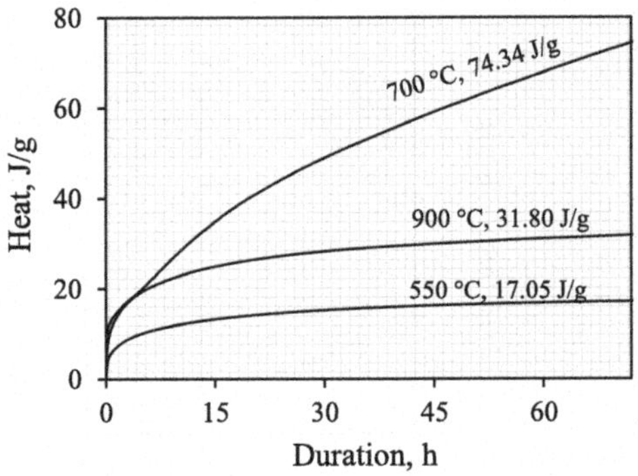

Fig. 1.14 Cumulative heat of hydroxyledgrewite samples after activation

was calculated on the grounds of a unit weight of hydroxyledgrewite; thus, the rates can be compared with each other or with the literature data.

Data on the heat release (J/g) of treated hydroxyledgrewite samples is presented in Fig. 1.14. It was determined that, after milling and calcination at 550 °C, hydroxyledgrewite did not show hydraulic activity because the amount of the total cumulative heat after 72 h of hydration was equal only to 17.05 J/g. The obtained data is in good agreement with STA (Fig. 1.11) and high-temperature in-situ XRD (Fig. 1.12) results, which shows that hydroxyledgrewite is stable up to ~620 °C. As expected, when the calcination temperature was increased to 700 °C, the amount of cumulative heat increased more than 4 times, i.e. to 74.34 J/g (Fig. 1.14). The increment in heat can be explained by the formation of dicalcium silicate (Figs. 1.11 and 1.12). However, the further increment of the calcination temperature (900 °C) yielded a negative effect on the amount of the released heat (Fig. 1.14). Probably, dicalcium silicate which is formed in a higher temperature is characterized by a lower activity. According to the literature [59, 69, 118–122, 126], α-C_2SH is the most commonly used dibasic calcium silicate hydrate in the production for environmentally friendly cementitious materials. The amount of the total cumulative heat after activation of α-C_2SH is equal to 40–230 J/g. It should be noted that the amount of heat depends on the conditions of synthesis and activation. Based on the obtained results and those of the literature, it can be stated that hydroxyledgrewite can be applied for the production of binder materials; however, the strength properties as well as the economic and environment impact should still be evaluated additionally.

1.3.2 Porosity of Synthetic and Calcined Hydroxyledgrewite Samples

Porosity is one of the most important characteristics of materials, while nitrogen adsorption–desorption is one of the main experimental techniques to determine the specific surface area, the shape of pores, as well as some other parameters [163–169]. The nitrogen adsorption/desorption isotherms of a synthetic hydroxyledgrewite sample are presented in Fig. 1.15. According to IUAPC [168], the adsorption isotherm of a hydroxyledgrewite sample can be classified as Type II (a characteristic feature is the absence of the Gurvich Plateau toward saturation) or Type IV (a characteristic feature is the existence of the hysteresis loop). This uncertainty occurs due to the presence of macropores in the sample which cause the unrestricted multilayer adsorption on the surface. Type IV adsorption isotherm is characteristic for materials with mesopores, i.e. whose pore diameter varies within the 2–50 nm range. A similar shape of adsorption isotherms is also detected in other calcium silicate hydrate samples [170–172], and it can be used for the calculations of the surface area (by using the BET equation) and the dominant shape of pores [168, 173].

It was observed that the adsorption isotherm increases sharply till p/p_0~0.05, which shows that small pores (diameter < 5 nm) may still be present in the sample [164–169]. According to the literature, a monolayer of the adsorbed nitrogen forms in the range of relative pressure at ~0.05–0.35. In the higher interval of relative pressures ($p/p_0 > 0.4$), the adsorption and desorption isotherms did not coincide; i.e. the hysteresis loop is observed. According to IUPAC [168], hysteresis loop can be classified as Type H3, which shows that slit-shape mesopores form/occur between disordered aggregates of plate-like particles. Finally, the rapidly increasing adsorption isotherm above $p/p_0 = 0.9$ can be explained by the existence of macropores or due to multilayer adsorption on a nonporous surface.

The TEM micrograph of the hydroxyledgrewite sample showed that needle-like and plate-like crystallites were dominant in the sample (Fig. 1.16a). It was measured

Fig. 1.15 Adsorption (1) desorption (2) isotherms of synthetic hydroxyledgrewite sample

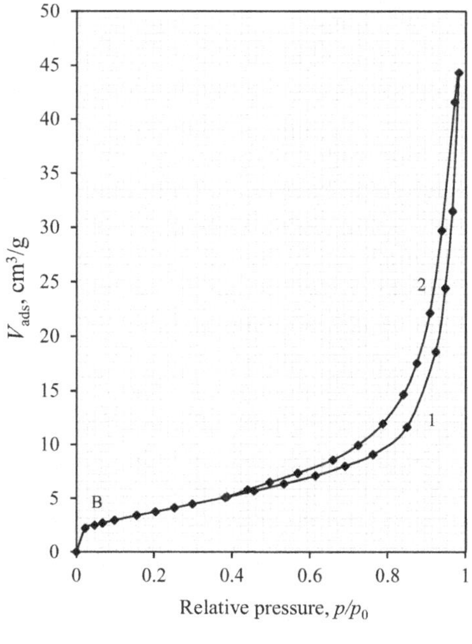

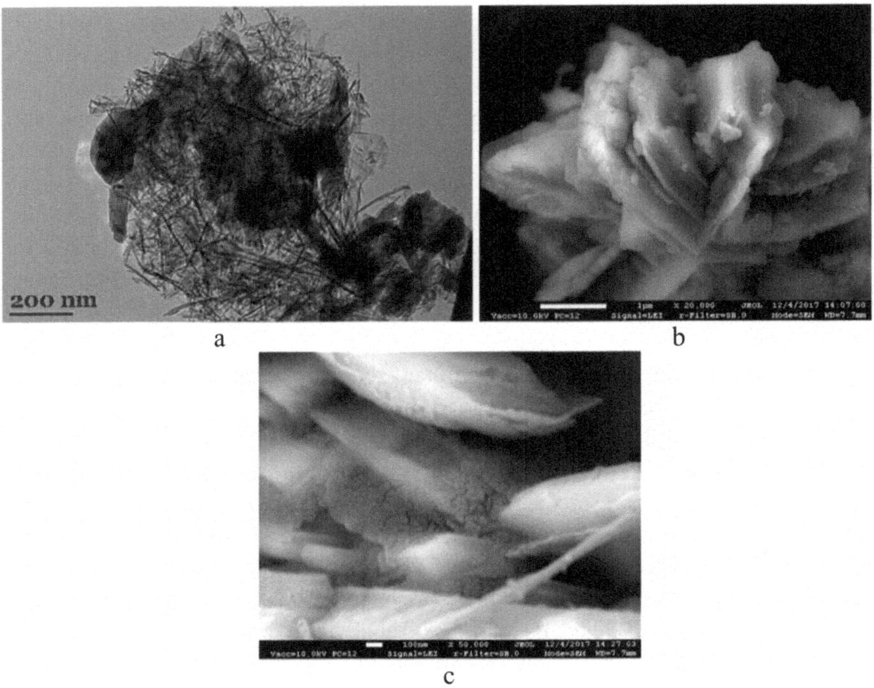

Fig. 1.16 TEM (**a**) and SEM (**b**, **c**) micrographs of synthetic hydroxyledgrewite sample

that the crystallites are characterized by a similar width (~2–5 nm) and length (~80–120 nm). The size of the crystallites is in good agreement with the XRD results, and their calculated mean size was equal to 52 $\pm$ 3 nm. It is clearly seen that, between the crystallites of the sample, mesopores were formed (2–10 nm).

It was determined that particles of synthetic hydroxyledgrewite are of the irregular form with the average size of 2–6 μm (Fig. 1.16b). In addition, SEM images showed that macropores (pore diameter > 50 nm) may form between particles of hydroxyledgrewite. By using 50,000–150,000X magnification, it is clearly seen that the particles of hydroxyledgrewite are mesopores (Fig. 1.16c). The results of TEM and SEM confirmed that the adsorption isotherm of the hydroxyledgrewite sample cannot be clearly assigned either to Type II or Type IV (Fig. 1.15).

The specific surface area (S_{BET}) of the hydroxyledgrewite sample was calculated with the Brunauer–Emmett-Teller (BET) method [166–169, 174–180]. The description of the method is presented in Chap. 2, Eq. 2.2.

The BET equation of the hydroxyledgrewite sample yields a linear plot in the range of the relative pressures between $0.05 \leq p/p_0 \leq 0.35$ (Fig. 1.17). It was calculated that the value of the correlation coefficient R^2 of the obtained straight line is equal to 0.9995. The parameters calculated by using the BET equation are presented in Table 1.5.

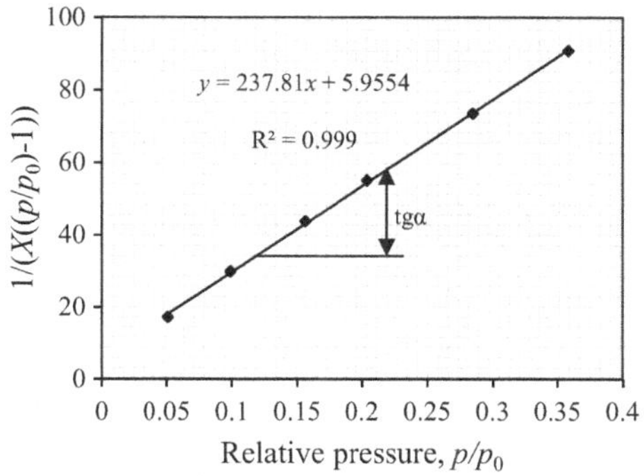

Fig. 1.17 Isotherm of N_2 adsorption at 77 K in BET plot of hydroxyledgrewite sample

Table 1.5 Calculated parameters of the hydroxyledgrewite sample by using BET equation

Mass, g	BET equation constants		S_{BET}, m^2/g	C_{BET} constant	Capacity of monolayer X_m, g	Correlation coefficient R^2
	Slope, $S = tg\alpha$	Intercept, I				
0.1051	237.81	5.955	14.30	40.93	0.0041	0.995

According to the BET theory [166–169, 174–180], constant C_{BET} is related to the monolayer adsorption energy and should be equal to 50–150 if no weak interaction or no chemisorption occurs. In general, if the value of the C_{BET} constant is > 50, the knee of the isotherm is sharp, and its Point B (indicating the completion of the monolayer) is fairly well defined. It was determined that the value of C_{BET} is slightly lower than 50 (Table 1.5). In this case, Point B cannot be identified as a single point on the isotherm (monolayer and multilayer adsorption overlap); however, if the value of this constant is more than 10, it can be considered as being acceptably high, and the errors obtained by using the BET equation are insignificant. The calculation revealed that the capacity of monolayer X_m and the values of S_{BET} of the hydroxyledgrewite sample are equal to 0.0041 g and 14.30 m^2/g, respectively (Table 1.5).

The visual inspection of the existence of pores is complicated; thus, in order to determine the absence or presence of different pores, the t-plot model (Eqs. 2.3–2.5), the corrected Kelvin equation combined with the scheme developed by Orr et al. (Eqs. 2.6–2.17), and some other models can be used [166, 167, 175, 176]. The t-plot model is one of the most suitable models for the determination of micropores [166, 167, 175, 176]. By using this model, the experimental adsorbed volume was plotted versus the thickness (Ä) (Eqs. 2.3–2.5) of the adsorbed layer (Fig. 1.18).

The straight line (correlation coefficient $R^2 > 0.99$) of the t-plot curve confirmed the formation of a N_2 monolayer on the surfaces of the synthetic hydroxyledgrewite sample (Fig. 1.18). The intercept and the slope of the obtained t-plot curve can be used for the calculation of the micropore volume (V_{micro}) and the total surface area S_t of all the pores, respectively [166, 167, 175, 176]. The presently discussed parameters were calculated by using the following equations:

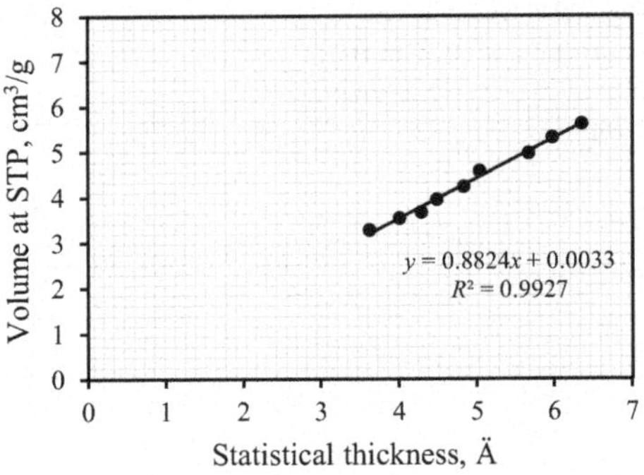

Fig. 1.18 t-plot of hydroxyledgrewite sample

$$V_{\text{micro}} = I \cdot 0.001547 \left(\text{cm}^3/\text{g}\right) \quad (1.2)$$

$$S_t = S \cdot 1.547 \left(\text{m}^2/\text{g}\right). \quad (1.3)$$

The surface area of micropores is equal to the difference between S_t and S_{BET}. If these values are equal, the sample does not contain any micropores. It was calculated that the hydroxyledgrewite sample represents a material with mesopores because the calculated micropore volume and area are lower than 5×10^{-6} cm³/g and 1.4 m²/g, respectively.

In order to determine the shape of mesopores as well as the dominant pores, the corrected Kelvin equation and the scheme as developed by Orr et al. were used (Eqs. 2. 6–2.17). The calculations were derived by using the slit-shape and the cylindrical pores models. The model is considered to be right as long as the difference between S_{BET} and $\sum A$ is not significant (under 10%). The values obtained by using different models are presented in Table 1.6.

The calculations showed that slit-shape pores between plate-like particles are dominant in the hydroxyledgrewite sample because, when using the slit-shape pores model, the obtained difference between S_{BET} and $\sum A$ is equal to as little as 2.15% (Table 1.6). For this reason, the total pore volume ($\sum V_{\text{p}}$) and the differential distributions of the volume pore sizes of the hydroxyledgrewite sample were calculated by using the slit-shape pores model. It was calculated that the value of $\sum V_{\text{p}}$ of the synthetic sample is equal to 0.069 cm³/g (Table 1.6).

Table 1.6 Data of ΣA calculations of hydroxyledgrewite sample

S_{BET}, m²/g	Results when using cylindrical pores model			Results when using slit-shape pores model			
	ΣA, m²/g	$\lvert S_{\text{BET}}-\Sigma A \rvert$, m²/g	$\lvert S_{\text{BET}}-\Sigma A \rvert$, %	ΣA, m²/g	$\lvert S_{\text{BET}}-\Sigma A \rvert$, m²/g	$\lvert S_{\text{BET}}-\Sigma A \rvert$, %	$\sum V_{\text{p}}$, cm³/g
14.30	22.20	7.90	55.25	14.00	0.3	2.15	0.069

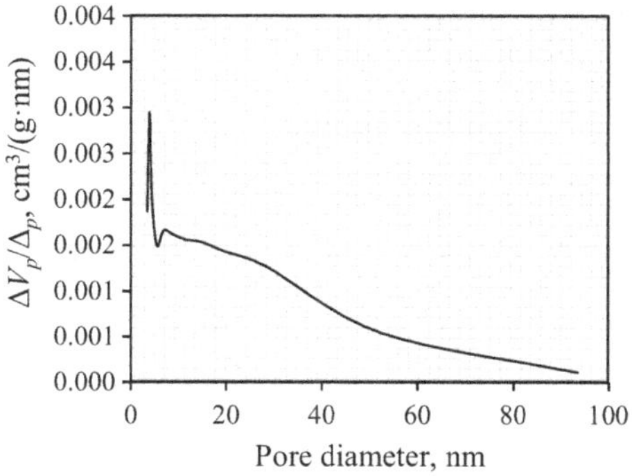

Fig. 1.19 Differential distributions of the volume sizes of pores of hydroxyledgrewite sample

It was obtained that mesopores are predominant in the structure of the hydroxyledgrewite sample (Fig. 1.19). The scattering of pores can be associated with the existence of pores in between the crystallites and the particles of these samples (Fig. 1.16). Also, macropores (> 50 nm) were observed in the sample, which explains the rapidly increasing adsorption isotherm above $p/p_0 = 0.9$ (Fig. 1.15).

The hydroxyledgrewite sample was calcined in a high-temperature furnace *Nabertherm LH 15/13* at 1000 °C temperature for 1 h. After calcination, hydroxyledgrewite recrystallized to β-C_2S and calcium oxide (Fig. 1.12). It was determined that calcination exerts significant influence on the porosity of the investigated sample (Figs. 1.15–1.23).

The adsorption isotherm of the calcined hydroxyledgrewite sample can be classified as Type II which is characteristic for nonporous materials (Fig. 1.20). According to the literature [168], the classical Type II isotherm does not contain a hysteresis loop. However, in the case of the calcined hydroxyledgrewite sample, the hysteresis loop, which can be classified as Type H3, was obtained. The presence of the hysteresis loop shows that, in the sample, there exist some slit-shape pores in between relatively nonporous plate-like particles [166–168]. It was established that the adsorption isotherm increases monotonically till $p/p_0 = 0.85$, which shows the absence of micropores in the sample. It should be remarked that the adsorption isotherm of the synthetic sample increased sharply above $p/p_0 = 0.9$ (Fig. 1.15), while, in the adsorption isotherm of the calcined sample, the increment is insignificant (Fig. 1.20). The smaller increment is associated with the decrease of macropores during calcination.

The previously discussed results are in good agreement with the data of TEM and SEM analysis. It was observed that, during calcination, the crystallites of the hydroxyledgrewite sample sintered by increasing their dimension (Figs. 1.16a, and 1.21a, b). It should be highlighted that, in the sample, the crystallites of a different size can be identified. Meanwhile, the SEM micrograph showed that, during calcination, the particles of hydroxyledgrewite decomposed by forming sintered particles of an uncertain form (Fig. 1.21c).

According to the TEM and SEM results, it can be stated that, after the calcination of the hydroxyledgrewite samples, micropores and/or mesopores exist mainly between

Fig. 1.20 Adsorption (1)–desorption (2) isotherms of calcined hydroxyledgrewite sample

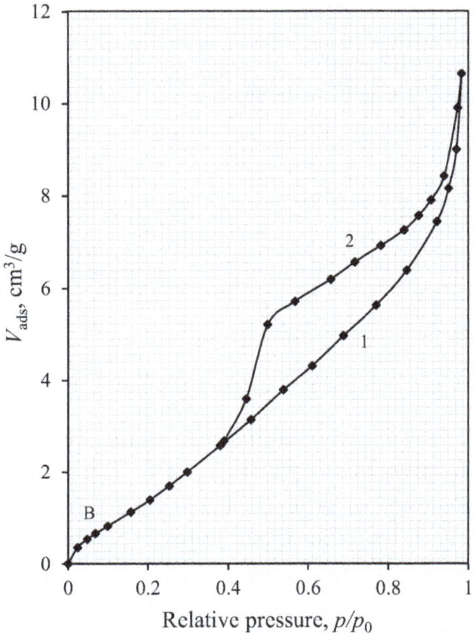

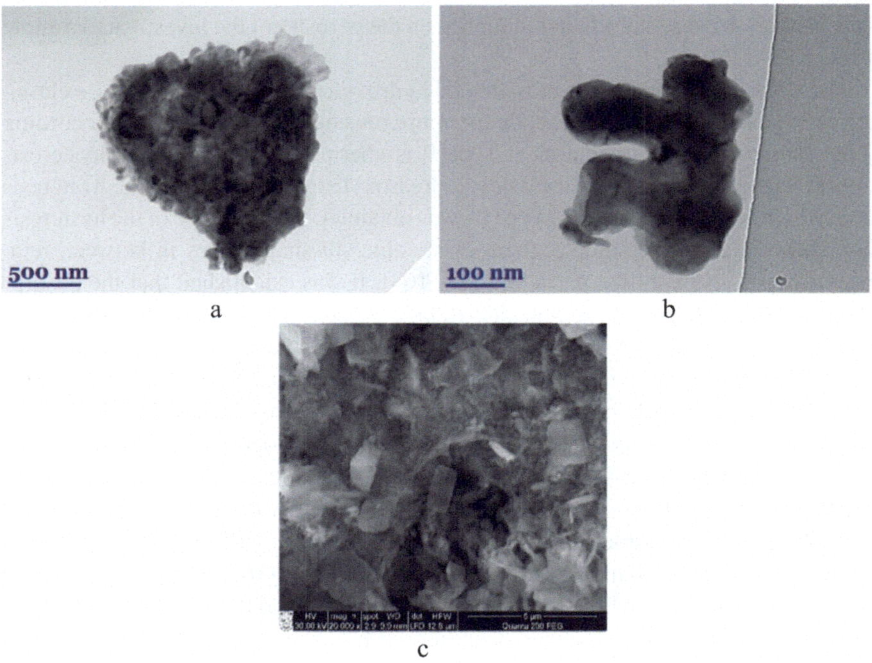

Fig. 1.21 TEM (**a**, **b**) and SEM (**c**) micrographs of calcined hydroxyledgrewite sample

nonporous particles. This data confirmed the classification of N_2 adsorption isotherms to Type II, and the existence of the hysteresis loop.

As mentioned above, the constant C_{BET}, which is related to the monolayer adsorption energy, should be equal to 50–150 if no weak interaction or no chemisorption occurs. In the case of the calcined hydroxyledgrewite sample, the obtained value of the C_{BET} constant was equal only to 19.54 (Table 1.6). The low value of this constant shows a greater interaction force in between the adsorbate molecules than the adsorbent–adsorbate values [166–168, 175, 176]. However, the value of the C_{BET} constant is more than 10; thus errors obtained by using the BET equation should be insignificant. The calculations revealed that the specific surface area of the calcined hydroxyledgrewite sample decreased by 1.5 times, i.e. from 14.30 to 9.53 m^2/g (Tables 1.5 and 1.7). The decrease of the latter value is related with the solid sintering of the particles.

In order to confirm the absence of micropores in the sample, the t-plot model was applied (Fig. 1.22, Eqs. 2.3–2.5). As in the previous case (Fig. 1.18), a linear plot was obtained ($R^2 = 0.999$) (Fig. 1.22), which confirms that the monolayer of N_2 on the surfaces of the calcined hydroxyledgrewite sample was formed. As expected, the t-plot model showed that the calcined sample does not contain any micropores (< 2 nm) because the values of the intercept of the t-plot are equal only to 0.0009, and the difference between S_t and S_{BET} is lower than 0.1 m^2/g (Fig. 1.22).

The corrected Kelvin equation and the scheme as developed by Orr et al. (Eqs. 2.6–2.17) were used to determine the shape of the dominant mesopores. As expected, the calculations by using the cylindrical pores model showed high disagreement between the values of ΣA and S_{BET} (~70%) (Table 1.8). Meanwhile, when using the slit-shape pores model, the obtained difference between the discussed values was

Table 1.7 Calculated parameters of calcined hydroxyledgrewite sample by using BET equation

Mass, g	BET equation constants		S_{BET}, m^2/g	C_{BET} constant	Capacity of monolayer X_m, g	Correlation coefficient R^2
	Slope, $S = tg\alpha$	Intercept, I				
0.1027	347.04	18.716	9.53	19.54	0.0027	0.999

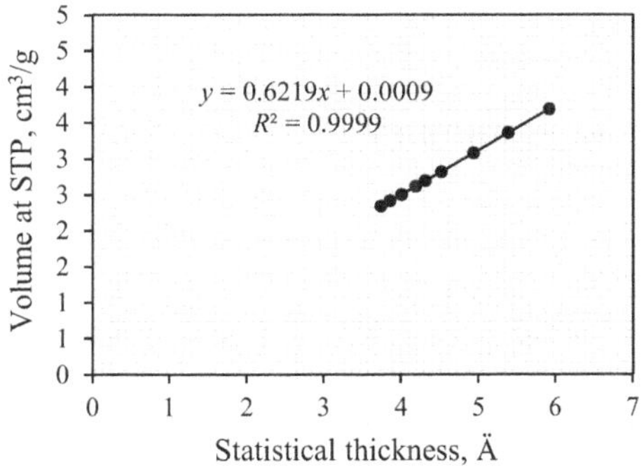

Fig. 1.22 t-plot of calcined hydroxyledgrewite sample

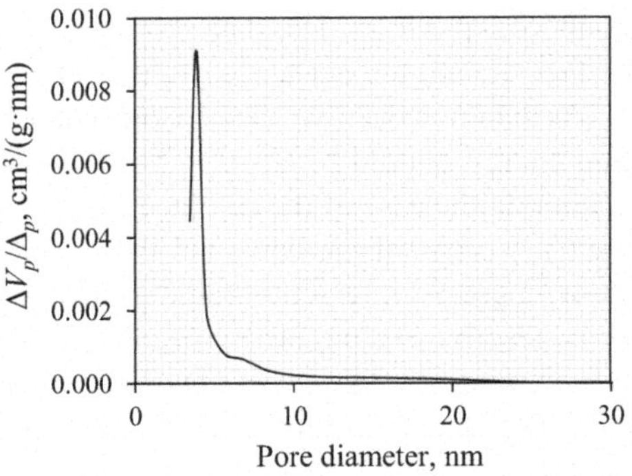

Fig. 1.23 Differential distributions of the volume sizes of pores of the calcined hydroxyledgrewite sample

Table 1.8 Data of ΣA calculations of calcined hydroxyledgrewite sample

S_{BET}, m²/g	Results when using cylindrical pores model			Results when using slit-shape pores model			
	ΣA, m²/g	$\lvert S_{BET}-\Sigma A\rvert$, m²/g	$\lvert S_{BET}-\Sigma A\rvert$, %	ΣA, m²/g	$\lvert S_{BET}-\Sigma A\rvert$, m²/g	$\lvert S_{BET}-\Sigma A\rvert$, %	$\sum V_p$, cm³/g
9.53	16.15	6.62	69.47	8.92	0.61	6.40	0.018

insignificant (6.4%). These results are in good agreement with the classification of the hysteresis loop as Type H3 (Fig. 1.20); the results thus confirm that slit-shape pores formed between nonporous plate-like particles. Further calculations showed that the value of $\sum V_p$ of the calcined sample is equal to 0.018 cm³/g (Table 1.8), which is more than 3.8 times less in comparison to the synthetic sample (Table 1.6).

The differential distributions of the volume pore sizes (Fig. 1.23) showed that, in the calcined hydroxyledgrewite sample, mesopores with a diameter of 3–5 nm are dominant. It should be highlighted that pores with a diameter higher than 20 nm were not observed in the sample. Meanwhile, in the case of a synthetic sample, macropores with a diameter higher than 50 nm were identified (Fig. 1.19). These results confirmed the data obtained in N_2 adsorption isotherms (the increment of the adsorption isotherm of the calcined sample above $p/p_0 = 0.9$ is significantly smaller in comparison to the data obtained after the synthesis).

1.4 Summary

When concluding the first chapter, it can be stated that the ternary system of calcium oxide, silicon dioxide, and water is a complex system which includes more than 50 compounds (Table 1.1). According to the main classification, calcium silicate hydrates are divided into seven groups: Wollastonite; Tobermorite; Jennite; Gyrolite; γ-C_2S; other calcium silicate phases; other high-temperature cement phases. However, this classification does not include the newly discovered or synthesized calcium silicate hydrates: hydroxyledgrewite, trabzonite, chegemite, spurrite, and galuskinite. Despite the fact that these compounds were introduced in this monograph, still, the classification of calcium silicate hydrates has not been completed yet. The classification can be supplemented by rarely found calcium silicate hydrates (riversideite [181, 182], stellerite [183, 184], bultfonteinite [185, 186], reinhardbraunsite [11, 187, 188], crestmoreite [181], aklimaite [189, 190], etc.); however, the literature concerning the data about these compounds is too scarce for generalizations.

Thus, calcium silicate hydrates have been known for more than 100 years, and extensive research on them has already been performed; however, their chemistry is still not fully understood. Due to the excellent properties of calcium silicate hydrates, most of them are used in the industry and are the topic of current academic research. The subsequent chapters of this monograph shall focus onto one calcium silicate hydrate—synthetic gyrolite—which is of major interest in science because of its outstanding absorption properties.

References

1. Dambrauskas T, Baltakys K (2018) Part 1: a study on the properties of synthetic hydroxyledgrewite. J Am Ceram Soc 101:3710–3718
2. Dambrauskas T, Baltakys K, Eisinas A, Kitrys S (2018) The specific surface area and porosity of synthetic and calcined α-C_2SH, kilchoanite and hydroxyledgrewite. Powder Technol 355:504–513
3. Richardson IG (2008) The calcium silicate hydrates. Cem Concr Res 38:137–158
4. Garbev K (2004) Struktur, Eigenschaften und quantitative Rietveldanalyse von hydrothermal kristallisierten Calciumsilikathydraten (CSH-Phasen). Disertation, Karlsruhe Research Center
5. Fan J, Cao D, Jing Z, Zhang Y, Pu L, Jing Y (2014) Synthesis and microstructure analysis of autoclaved aerated concrete with carbide slag addition. J Wuhan Univ Technol Mat Sci Edit 29:1005–1010
6. He Y, Zhao X, Lu L, Struble LJ, Hu S (2011) Effect of C/S ratio on morphology and structure of hydrothermally synthesized calcium silicate hydrate. J Wuhan Univ Technol Mat Sci Edit 26:770–773
7. Smalakys G, Siauciunas R (2018) The synthesis of 1.13 nm tobermorite from carbonated opoka. J Therm Anal Calorim 134:493–502
8. Tian H, Stephan D, Lothenbach B, Lehmann C (2021) Influence of foreign ions on calcium silicate hydrate under hydrothermal conditions: a review. Constr Build Mater 301:124071
9. Bernstein S (2011) Determination of reaction kinetics and mechanisms of 1.13 nm tobermorite by in-situ neutron diffraction. Dissertation, Ludwig Maximilian University of Munich
10. Dambrauskas T, Baltakys K, Eisinas A, Siauciunas R (2017) A study on the thermal stability of kilchoanite synthesized under hydrothermal conditions. J Therm Anal Calorim 127:229–238
11. Bezerra UT, Martinelli AE, Melo DM, de Araújo M, Antônio M, de Freitas OVG (2011) The strength retrogression of special class Portland oilwell cement. Cerâmica 57:150–154
12. Hong S, Glasser FP (2004) Phase relations in the CaO–SiO_2–H_2O system to 200 °C at saturated steam pressure 34:1529–1534
13. Yanagisawa K, Hu X, Onda A, Kajiyoshi K (2006) Hydration of β-dicalcium silicate at high temperatures under hydrothermal conditions. Cem Concr Res 36:810–816
14. Hu X, Yanagisawa K, Onda A, Kajiyoshi K (2006) Stability and phase relations of dicalcium silicate hydrates under hydrothermal conditions. J Ceram Soc Jpn 114:174–179

15. Meller N, Hall C, Phipps JS (2005) A new phase diagram for the $CaO–Al_2O_3–SiO_2–H_2O$ hydroceramic system at 200 °C. Mater Res Bull 40:715–723
16. Shaw S, Clark SM, Henderson CMB (2000) Hydrothermal formation of the calcium silicate hydrates, tobermorite ($Ca_5Si_6O_{16}(OH)_2·4H_2O$) and xonotlite ($Ca_6Si_6O_{17}(OH)_2$): an in situ synchrotron study. Chem Geol 167:129–140
17. John E, Matschei T, Stephan D (2018) Nucleation seeding with calcium silicate hydrate—a review. Cem Concr Res 113:74–85
18. Siauciunas R, Smalakys G, Eisinas A, Prichockiene E (2022) Synthesis of high crystallinity 1.13 nm tobermorite and xonotlite from natural rocks, their properties and application for heat-resistant products. Materials 15:3474
19. Liu SJ, Cui SP, Guo HX, Wang YL, Li N, Gan YL, Zheng Y (2020) Preparation of hydrated calcium silicate gel and its adsorption properties for Cu(II). Mater Sci Forum 993:1445–1449
20. Martín-Garrido M, Martínez-Ramírez S (2021) CO_2 adsorption on calcium silicate hydrate gel synthesized by double decomposition method. J Therm Anal Calorim 143:4331–4339
21. Bu J, Gonzalez Teresa R, Brown KG, Sanchez F (2019) Adsorption mechanisms of cesium at calcium-silicate-hydrate surfaces using molecular dynamics simulations. J Nucl Mater 515:35–51
22. Qiu Y, Cao S, Chen F, You S, Zhang Y (2020) Synthesis of calcium silicate as paper filler with desirable particle size from desilication solution of silicon-containing waste residues. Powder Technol 368:137–148
23. Konuklu Y, Ersoy O (2017) Fabrication and characterization of form-stable phase change material/xonotlite microcomposites. Sol Energy Mater Sol Cells 168:130–135
24. Bendary IM, Garcia IM, Collares FM, Takimi A, Samuel SMW, Leitune VCB (2020) Wollastonite as filler of an experimental dental adhesive. J Dent 102:103472
25. Myat-Htun M, Mohd Noor A, Kawashita M, Baba Ismail YM (2022) Tailoring mechanical and in vitro biological properties of calcium-silicate based bioceramic through iron doping in developing future material. J Mech Behav Biomed Mater 128:105122
26. Ismail H, Zakwan Zakri MN, Mohamad H (2022) A comparative study of physicomechanical and in vitro bioactivity properties of β-wollastonite/cordierite scaffolds obtained via gel casting method. Ceram Int 48:25495–25505
27. Ogur E, Botti R, Bortolotti M, Colombo P, Vakifahmetoglu C (2021) Synthesis and additive manufacturing of calcium silicate hydrate scaffolds. J Mater Res Technol 11:1142–1151
28. Meyer V, de Cristofaro N, Bryant J, Sahu S (2018) Solidia cement an example of carbon capture and utilization. Key Eng Mater 761:197–203
29. Siauciunas R, Hilbig H, Prichockiene E, Smigelskyte A, Takulinskas Z (2020) Accelerated carbonation of C_2SH based dense concrete. Ceram Int 46:29436–29442
30. Siauciunas R, Takulinskas Z, Prichockiene E, Selskiene A (2023) Hydrothermal synthesis of α-C_2SH and kilchoanite mixture and its application in CO_2 hardening mortar. Ceram Int 49:14864–14874
31. Wang X, Guo M, Ling T (2022) Review on CO_2 curing of non-hydraulic calcium silicates cements: mechanism, carbonation and performance. Cem Concr Compos 133:104641
32. Taweekarn T, Wongniramaikul W, Choodum A (2022) Removal and recovery of phosphate using a novel calcium silicate hydrate composite starch cryogel. J Environ Manage 301:113923
33. Liu X, Feng P, Cai Y, Yu X, Yu C, Ran Q (2022) Carbonation behavior of calcium silicate hydrate (C-S-H): its potential for CO_2 capture. Chem Eng J 431:134243
34. Qi F, Cao J, Zhu G, Sun J, Li S, Wu W, Zheng J, Li H, Zhang Y (2022) Crystallization behavior of calcium silicate hydrate in highly alkaline system: structure and kinetics. J Cryst Growth 584:126578
35. Bernal JD, Jeffery JW, Taylor HFW (1952) Crystallographic research on the hydration of Portland cement. A first report on investigations in progress. Mag Concr Res 4:49–54
36. Bernal JD (1952) The structures of cement hydration compounds. In: Proceedings of 3rd international symposium on chemistry of cement. Cement and Concrete Association, London, pp 216–236

37. Shpynova LG, Nabitovi ID, Belov NV (1967) Microstructure of alite cement stone (hydrated tricalcium silicate). Sov Phys Crystallogr 11:747
38. Grudemo Ä (1984) The crystal structures of cement hydration. A review and a new gel structure model. Nord Concr Res 49–62
39. Cong X, Kirkpatrick RJ (1996) ^{17}O MAS NMR investigation of the structure of calcium silicate hydrate gel. J Am Ceram Soc 79:1585–1592
40. Brough AR, Dobson CM, Richardson IG, Groves GW (1994) In situ solid-state NMR studies of Ca_3SiO_5: hydration at room temperature and at elevated temperatures using ^{29}Si enrichment. J Mater Sci 29:3926–3940
41. Duque-Redondo E, Bonnaud PA, Manzano H (2022) A comprehensive review of C-S-H empirical and computational models, their applications, and practical aspects. Cem Concr Res 156:106784
42. Taylor H, Howison JW (1956) Relationships between calcium silicates and clay minerals. Clay Miner Bull 3:98–111
43. Bonaccorsi E, Merlino S, Taylor HFW (2004) The crystal structure of jennite, $Ca_9Si_6O_{18}(OH)_6{\cdot}8H_2O$. Cem Concr Res 34:1481–1488
44. Chen JJ, Thomas JJ, Taylor HFW, Jennings HM (2004) Solubility and structure of calcium silicate hydrate. Cem Concr Res 34:1499–1519
45. Nonat A (2004) The structure and stoichiometry of C-S-H. Cem Concr Res 34:1521–1528
46. Kurczyk H (1961) Elektronenmikroskopische und thermochemische Untersuchungen über die Hydratation der Calciumsilikate $3CaO{\cdot}SiO_2$ und Beta-$2CaO{\cdot}SiO_2$ und den Einfluß von Calciumchlorid und Gips auf den Hydratationsvorgang. Dissertation, Technische Hochschule
47. Kurczyk HG, Schwiete HE (1962) Paper III-S13. Concerning the hydration products of $3CaO{\cdot}SiO_2$ and β-$2CaO{\cdot}SiO_2$. In: Chemistry of cement: proceedings of the fourth international symposium 1960, Washington, vol II. National Bureau of Standards Monograph 43, pp 1:349
48. Kantro DL, Brunauer S, Weise CH (1962) Development of surface in the hydration of calcium silicates. II. Extension of investigations to earlier and later stages of hydration. J Phys Chem 66:1804–1809
49. Fujii K, Kondo W (1983) Communications of the American Ceramic Society estimation of thermochemical data for calcium silicate hydrate (C-S-H). J Am Ceram Soc 66:C-220, C-221
50. Stade H, Wieker W (1980) Zum Aufbau schlecht geordneter Calciumhydrogensilicate. I. Bildung und Eigenschaften einer schlecht geordneten Calciumhydrogendisilicatphase. Z anorg allg Chem 466:55–70
51. Stade H (1980) Zum Aufbau Schlecht geordneter Calciumhydrogensilicate. II. Über eine aus Poly- und Disilicat bestehende Phase. Z anorg allg Chem 470:69–83
52. Glasser FP, Lachowski EE, Macphee DE (1987) Compositional model for calcium silicate hydrate (C-S-H) gels, Their solubilities, and free energies of formation. J Am Ceram Soc 70:481–485
53. Taylor HFW (1986) Proposed structure for calcium silicate hydrate gel. J Am Ceram Soc 69:464–467
54. Taylor H (1993) Nanostructure of C S H: current status. Adv Cem Based Mater 1:38–46
55. Richardson IG (2004) Tobermorite/jennite- and tobermorite/calcium hydroxide-based models for the structure of C-S-H: applicability to hardened pastes of tricalcium silicate, β-dicalcium silicate, Portland cement, and blends of Portland cement with blast-furnace slag, metakaolin, or silica fume. Cem Concr Res 34:1733–1777
56. Richardson IG, Groves GW (1993) The incorporation of minor and trace elements into calcium silicate hydrate (C S H) gel in hardened cement pastes. Cem Concr Res 23:131–138
57. Richardson IG, Groves GW (1992) Models for the composition and structure of calcium silicate hydrate (C S H) gel in hardened tricalcium silicate pastes. Cem Concr Res 22:1001–1010
58. Baltušnikas A, Šiaučiūnas R, Lukošiūtė I, Baltakys K, Eisinas A, Kriūkienė R (2015) Crystal structure refinement of synthetic pure gyrolite. Medziagotyra 21:111–116

59. Garbev K, Gasharova B, Stemmermann P (2014) A modular concept of crystal structure applied to the thermal transformation of α-C_2SH. J Am Ceram Soc 97:2286–2297
60. Taylor H, Roy DM (1980) Structure and composition of hydrates. In: Proceedings 7th international symposium on the chemistry of cement, Paris, p 1:112
61. Spudulis E, Šavareika V, Špokauskas A (2013) Influence of hydrothermal synthesis condition on xonotlite crystal morphology. Medziagotyra 19:190–196
62. Xu L, Sun P, Chen X, Li J, Zhai P, Zhang H, Zhang Z, Zhu W (2018) High aspect ratio $Ca_6Si_6O_{17}(OH)_2$ nanowires: green hydrothermal synthesis, formation mechanism, optical and photoluminescence properties. Powder Technol 335:360–370
63. Rizaal M, Nakajima K, Saito T, Osaka M, Okamoto K (2022) High-temperature gaseous reaction of cesium with siliceous thermal insulation: the potential implication to the provenance of enigmatic Fukushima cesium-bearing material. ACS Omega 7:29326–29336
64. Yanagisawa K, Feng Q, Yamasaki N (1997) Hydrothermal synthesis of xonotlite whiskers by ion diffusion. J Mater Sci Lett 16:889–891
65. Smalakys G, Siauciunas R (2022) Peculiarities of xonotlite synthesis from the raw materials with different SiO_2 activities. J Therm Anal Calorim 142:1671–1679
66. Nocuń-Wczelik W (1997) Effect of some inorganic admixtures on the formation and properties of calcium silicate hydrates produced in hydrothermal conditions. Cem Concr Res 27:83–92
67. Baltakys K, Prichockiene E (2010) Influence of CaO reactivity on the formation of low-base calcium silicate hydrates. Mater Sci Pol 28:295–304
68. Gineika A, Baltakys K, Dambrauskas T (2019) The application of silica gel waste for the two-step synthesis of wollastonite in temperature range of 200–950 °C. J Therm Anal Calorim 138:2263–2273
69. Sun H, He S, Wu P, Gao C, Feng P, Xiao T, Deng Y, Shuai C (2016) A novel MgO-CaO-SiO_2 system for fabricating bone scaffolds with improved overall performance. Materials 9:287
70. Georgescu M, Tipan J, Badanoiu A, Crisan D, Dragan I (2000) Highly reactive dicalcium silicate synthesised by hydrothermal processing. Cem Concr Compos 22:315–319
71. Tantawy MA, Shatat MR, El-Roudi AM, Taher MA, Abd-El-Hamed M (2014) Low temperature synthesis of belite cement based on silica fume and lime. Int Sch Res Not 2014
72. Heddle MF (1880) Preliminary notice of substances which may prove to be new minerals. Mineral Mag J Mineral Soc 4:119–123
73. Bonaccorsi E, Merlino S, Kampf AR (2005) The crystal structure of tobermorite 14 Å (plombierite), a C–S–H phase. J Am Ceram Soc 88:505–512
74. Tränkle S, Jahn D, Neumann T, Nicoleau L, Hüsing N, Volkmer D (2013) Conventional and microwave assisted hydrothermal syntheses of 11 Å tobermorite. J Mater Chem 1:10318–10326
75. Lamidi YD, Owoeye SS, Abegunde SM (2022) Preparation and characterization of synthetic tobermorite (CaO–Al_2O_3–SiO_2–H_2O) using bio and municipal solid wastes as precursors by solid state reaction. Bol Soc Esp Ceram V 61:76–81
76. Mi G, Saito F, Hanada M (1997) Mechanochemical synthesis of tobermorite by wet grinding in a planetary ball mill. Powder Technol 93:77–81
77. Meller N, Hall C, Kyritsis K, Giriat G (2007) Synthesis of cement based CaO–Al_2O_3–SiO_2–H_2O (CASH) hydroceramics at 200 and 250 °C: ex-situ and in-situ diffraction. Cem Concr Res 37:823–833
78. Meller N, Kyritsis K, Hall C (2009) The mineralogy of the CaO–Al_2O_3–SiO_2–H_2O (CASH) hydroceramic system from 200 to 350 °C. Cem Concr Res 39:45–53
79. Qu X, Zhao X (2017) Previous and present investigations on the components, microstructure and main properties of autoclaved aerated concrete—a review. Constr Build Mater 135:505–516
80. Müller D, Hochleitner R, Fehr KT (2016) Raman spectroscopic investigations of natural jennite from Maroldsweisach, Bavaria, Germany. J Raman Spectrosc 47:602–606
81. Carpenter AB, Chalmers RA, Gard JA, Speakman K, Taylor H (1966) Jennite, a new mineral. Am Min 51:56–74
82. Hara N, Inoue N (1980) Formation of jennite from fumed silica. Cem Concr Res 10:677–682

83. Komarneni S, Roy DM, Fyfe CA, Kennedy GJ (1987) Naturally occurring 1.4 nm tobermorite and synthetic jennite: characterization by ^{27}Al and ^{29}Si MASNMR spectroscopy and cation exchange properties. Cem Concr Res 17:891–895
84. Yu P, Kirkpatrick RJ (1999) Thermal dehydration of tobermorite and jennite. Concr Sci Eng 1:185–191
85. Meyer JW, Jaunarajs KL (1961) Synthesis and crystal chemistry of gyrolite and reyerite. Am Min 46:913–933
86. Kaneva EV, Shendrik RY, Radomskaya TA, Suvorova LF (2020) Fedorite from Murun alkaline complex (Russia): spectroscopy and crystal chemical features. Minerals 10:702
87. Gard JA, Luke K, Taylor HFW (1981) $Ca_7Si_{16}O_{40}H_2$, a new calcium silicate hydrate phase of the truscottite group. Cem Concr Res 11:659–664
88. Luke K, Taylor HFW, Kalousek GL (1981) Some factors affecting formation of truscottite and xonotlite at 300–350 °C. Cem Concr Res 11:197–203
89. Miyake M, Iwaya M, Suzuki T, Kakehi H, Mitsuda T (1990) Aluminum-substituted gyrolite as cation exchanger. J Am Ceram Soc 73:3524–3527
90. Noshita K, Nishi T, Yoshida T, Fujihara H, Saito N, Tanaka S (2000) Categorization of cement hydrates by radionuclide sorption mechanism. MRS Online Proceedings Library (OPL), p 663
91. Bankauskaite A, Baltakys K (2009) The sorption of copper ions by gyrolite in alkaline solution. Mater Sci Pol 27:899
92. Eisinas A, Ruginyte K, Baltakys K, Demcak S, Dambrauskas T, Balintova M, Stevulova N (2020) Cu^{2+} ion adsorption by synthetic mayenite and its thermal stability. Ceram Int 46:29429–29435
93. Eisinas A, Dambrauskas T (2023) Cu^{2+} ions adsorption by dense calcium silicate hydrate samples and their catalytic activity. Ceram Int 49:14886–14894
94. Różycka A, Kotwica Ł (2020) Effect of alkali on the synthesis of single phase gyrolite in the system CaO-quartz-H_2O. Constr Build Mater 239:117799
95. Okada Y, Masuda T, Ishida H (1995) Behavior of silicate anions on the formation processes of gyrolite followed by Si-29 NMR. JCS Jpn 103:124–127
96. Eisinas A, Dambrauskas T, Eisinaite V, Sipailiene A (2022) Synthesis of antibacterially active wollastonite by using dense calcium silicate hydrate samples. J Therm Anal Calorim 147:14163–14174
97. Jander W, Franke B (1941) Die Bildung von Calciumhydrosilikaten aus Calciumoxyd und Kieselsäuregel bei 300° und 350° und hohen Drucken. III. Mitteilung über hydrothermale Reaktionen. Z Anorg Allg Chem 247:161–179
98. Mackay AL, Taylor HFW (1953) Gyrolite. Mineral Mag J Mineral Soc 30:80–92
99. Speakman K (1968) The stability of tobermorite in the system CaO-SiO_2-H_2O at elevated temperatures and pressures. Mineral Mag J Mineral Soc 36:1090–1103
100. Siauciunas R, Baltakys K (2004) Formation of gyrolite during hydrothermal synthesis in the mixtures of CaO and amorphous SiO_2 or quartz. Cem Concr Res 34:2029–2036
101. Saidani S, Smith A, El Hafiane Y, Ben Tahar L (2018) Re-examination of the $\beta\rightarrow\gamma$ transformation of Ca_2SiO_4. J Eur Ceram 38:4756–4767
102. Buckle ER, Taylor H (1958) Calcium analogue of chondrodite. Am Min 43:818–823
103. Roy DM, Gard JA, Nicol AW, Taylor H (1960) New data for the calcium silicate, 'Phase Z.' Nature 188:1187–1188
104. Speakman K, Taylor H, Bennett JM, Gard JA (1967) Hydrothermal reactions of γ-dicalcium silicate. J Chem Soc A Inorg Phys Theor 1052–1060
105. Méducin F, Bresson B, Lequeux N, De Noirfontaine M, Zanni H (2007) Calcium silicate hydrates investigated by solid-state high resolution 1H and 29Si nuclear magnetic resonance. Cem Concr Res 37:631–638
106. Taylor H (1971) The crystal structure of kilchoanite, $Ca_6(SiO_4)(Si_3O_{10})$, with some comments on related phases. Mineral Mag 38:26–31
107. Roy DM (1958) Studies in the system CaO-Al_2O_3-SiO_2-H_2O IV; phase equilibria in the high-lime portion of the system CaO-SiO_2-H_2O. Am Mineral J Earth Planet Mater 43:1009–1028

108. Garbev K, Black L, Beuchle G, Stemmermann P (2002) Inorganic polymers in cement based materials. Wasser Geotechnologie 1:19–30
109. Mitsuda T (1972) Low-temperature synthesis of kilchoanite from quartz-lime mixtures by a new method. Mineral J 7:108–114
110. Shaw R (1953) The thermal expansion of afwillite. Acta Crystallogr 6:428–429
111. Davis RW, Young JF (1975) Hydration and strength development in tricalcium silicate pastes seeded with afwillite. J Am Ceram Soc 58:67–70
112. Bellmann F, Sowoidnich T, Horgnies M, Gartner E (2020) Basic mechanisms of afwillite seeding for acceleration of tricalcium silicate hydration. Cem Concr Res 132:106030
113. Merlino S, Perchiazzi N (1988) Modular mineralogy in the cuspidine group of minerals. Can Mineral 26:933–943
114. Gineika A, Baltakys K, Dambrauskas T, Siauciunas R (2023) Hydration properties and CO_2-curing of synthetic cuspidine samples. J Therm Anal Calorim 148:1511–1521
115. Gineika A, Siauciunas R, Baltakys K (2019) Synthesis of wollastonite from AlF3-rich silica gel and its hardening in the CO_2 atmosphere. Sci Rep 9:1–10
116. Watanabe T, Fukuyama H, Nagata K (2002) Stability of cuspidine ($3CaO{\cdot}2SiO_2{\cdot}CaF_2$) and phase relations in the $CaO–SiO_2–CaF_2$ system. ISIJ Int 42:489–497
117. Palou M, Živica V, Ifka T, Boháč M, Zmrzlý M (2014) Effect of hydrothermal curing on early hydration of G-oil well cement. J Therm Anal Calorim 116:597–603
118. Siauciunas R, Mikaliunaite J, Urbonas L, Baltakys K (2014) Tribochemical and thermal activation of α-C_2S hydrate as precursor for cementitious binders. J Therm Anal Calorim 118:817–823
119. Stemmermann P, Schweike U, Garbev K, Beuchle G (2011) "Celitement." A new sustainable hydraulic binder based on calcium hydrosilicates. In: 13th international congress on the chemistry of cement (ICCC), Madrid
120. Stemmermann P, Schweike U, Garbev K, Beuchle G, Möller H (2010) Celitement—a sustainable prospect for the cement industry. Cem Int 8:52–66
121. Garbev K, Beuchle G, Schweike U, Stemmermann P (2011) Hydration behavior of Celitement®: kinetics, phase composition, microstructure and mechanical properties. In: 13th international congress on the chemistry of cement (ICCC), Madrid
122. Ishida H, Yamazaki S, Sasaki K, Okada Y, Mitsuda T (1993) α-Dicalcium silicate hydrate: preparation, decomposed phase, and its hydration. J Am Ceram Soc 76:1707–1712
123. Yano T, Urabe K, Ikawa H, Teraushi T, Ishizawa N, Udagawa S (1993) Structure of α-dicalcium silicate hydrate. Acta Crystallogr C Struct Chem 49:1555–1559
124. Midgley HG, Chopra SK (1960) Hydrothermal reactions in the lime-rich part of the system CaO-SiO_2-H_2O. Mag Concr Res 12:19–26
125. Imlach BV, Taylor H (1972) Prolonged hydrothermal treatment of cement mixes. 1. Curing in water under saturated steam pressures at 140–170 degrees C. Trans J Br Ceram Soc 71:71
126. Siauciunas R, Baltakys K, Gendvilas R, Prichockiene E (2016) Synthesis of low-energy cement based on α-C_2SH. Adv Cem Res 28:241–250
127. Mitsuda T, Kobayakawa S, Toraya H (1986) Characterization of hydrothermally formed CSH. In: The 8th international congress on the chemistry of cement, Rio de Janeiro, vol 3, p 176
128. Garbev K, Gasharova B, Beuchle G, Kreisz S, Stemmermann P (2008) First observation of α-$Ca_2[SiO_3(OH)](OH)$–$Ca_6[Si_2O_7][SiO_4](OH)_2$ phase transformation upon thermal treatment in air. J Am Ceram Soc 91:263–271
129. Taylor H, Bessey GE (1950) Review of hydrothermal reactions in the system CaO-SiO_2-H_2O. Mag Concr Res 2:15–26
130. Kalousek GL, Logiudice JS, Dodson VH (1954) Studies on the lime-rich crystalline solid phases in the system lime—silica—water. J Am Ceram Soc 37:7–13
131. Shah V, Parashar A, Scott A (2022) Understanding the importance of carbonates on the performance of Portland metakaolin cement. Constr Build Mater 319:126155
132. Jeong Y, Hargis CW, Chun S, Moon J (2018) The effect of water and gypsum content on strätlingite formation in calcium sulfoaluminate-belite cement pastes. Constr Build Mater 166:712–722

133. Lee J, Yang HJ, Jang I, Kim SK, Jung D (2022) Comparison of calcium aluminate cements on hydration and strength development at different initial curing regimes. Case Stud Constr Mater 17:e01596
134. Goergens J, Manninger T, Goetz-Neunhoeffer F (2020) In-situ XRD study of the temperature-dependent early hydration of calcium aluminate cement in a mix with calcite. Cem Concr Res 136:106160
135. Armbruster T, Lazic B, Galuskina IO, Galuskin EV, Gnos E, Marzec KM, Gazeev VM (2012) Trabzonite, $Ca_4[Si_3O_9(OH)]OH$: crystal structure, revised formula, new occurrence and relation to killalaite. Mineral Mag 76:455–472
136. Sarp H, Burri G (1986) Trabzonite $Ca_4Si_3O_{10}$•$2H_2O$ a new hydrated calcium silicate. Schweizerische Mineralogische und Petrographische Mitteilungen 66:453
137. Galuskin EV, Gazeev VM, Lazic B, Armbruster T, Galuskina IO, Zadov AE, Pertsev NN, Wrzalik R, Dzierżanowski P, Gurbanov AG (2009) Chegemite $Ca_7(SiO_4)_3(OH)_2$—a new humite-group calcium mineral from the Northern Caucasus, Kabardino-Balkaria, Russia. Eur J Mineral 21:1045–1059
138. Smith JV, Karle IL, Hauptman H, Karle J (1960) The crystal structure of spurrite, $Ca_5(SiO_4)_2CO_3$. II. Description of structures. Acta Crystallogr 13:454–458
139. Grice JD (2005) The structure of spurrite, tilleyite and scawtite, and relationships to other silicate–carbonate minerals. Can Mineral 43:1489–1500
140. Wright FE (1908) On three contact minerals from Velardena, Durango, Mexico (gehlenite, spurrite, and hillebrandite). Am J Sci s4–26:545
141. Bolio-Arceo H, Glasser FP (1990) Formation of spurrite, $Ca_5(SiO_4)_2CO_3$. Cem Concr Res 20:301–307
142. Henmi C, Henmi K (1978) Synthesis of spurrite and tilleyite at low CO_2 partial pressure 9:106–110
143. Glasser FP (1973) The formation and thermal stability of spurrite, $Ca_5(SiO_4)_2CO_3$. Cem Concr Res 3:23–28
144. Tuttle OF, Harker RI (1957) Synthesis of spurrite and the reaction wollastonite + calcite $\leftrightarrow$ spurrite + carbon dioxide. Am J Sci 255:226–234
145. Zhang Y, Radha AV, Navrotsky A (2013) Thermochemistry of two calcium silicate carbonate minerals: scawtite, $Ca_7(Si_6O_{18})(CO_3)\cdot 2H_2O$, and spurrite, $Ca_5(SiO_4)_2(CO_3)$. Geochim Cosmochim Acta 115:92–99
146. Lottisnatham SJ, Smith JV (1970) Crystal structure of tilleyite: refinement and coordination. Zeitschrift für Kristallographie Cryst Mater 132:288–306
147. Larsen ES, Dunham KC (1933) Tilleyite, a new mineral from the contact zone at Crestmore, California. Am Mineral J Earth Planet Mater 18:469–473
148. Santamaria-Perez D, Ruiz-Fuertes J, Peña-Alvarez M, Chulia-Jordan R, Marqueño T, Zimmer D, Gutiérrez-Cano V, MacLeod S, Gregoryanz E, Popescu C (2019) Post-tilleyite, a dense calcium silicate-carbonate phase. Sci Rep 9:1–11
149. Smith JV (1953) The crystal structure of tilleyite. Acta Crystallogr 6:9–18
150. Lazic B, Armbruster T, Savelyeva VB, Zadov AE, Pertsev NN, Dzierzanowski P (2011) Galuskinite, $Ca_7(SiO_4)_3(CO_3)$, a new skarn mineral from the Birkhin gabbro massif, Eastern Siberia, Russia. Mineral Mag 75:2631–2648
151. Gao J, Wu X, Qin S (2015) The crystal chemistry and the compressibility of silicate-carbonate minerals: spurrite, galuskinite and tilleyite. Geosci Front 6:771–777
152. Galuskin EV, Lazic B, Armbruster T, Galuskina IO, Pertsev NN, Gazeev VM, Włodyka R, Dulski M, Dzierżanowski P, Zadov AE (2012) Edgrewite $Ca_9(SiO_4)_4F_2$-hydroxyledgrewite $Ca_9(SiO_4)_4(OH)_2$, a new series of calcium humite-group minerals from altered xenoliths in the ignimbrite of Upper Chegem caldera, Northern Caucasus, Kabardino-Balkaria, Russia. Am Mineral 97:1998–2006
153. Galuskin EV, Gfeller F, Savelyeva VB, Armbruster T, Lazic B, Galuskina IO, Többens DM, Zadov AE, Dzierżanowski P, Pertsev NN (2012) Pavlovskyite $Ca_8(SiO_4)_2(Si_3O_{10})$: a new mineral of altered silicate-carbonate xenoliths from the two Russian type localities, Birkhin massif, Baikal Lake area and Upper Chegem caldera, North Caucasus. Am Mineral 97:503–512

154. Dana JD, Dana ES, Gaines RV, Skinner HCW, Foord EE, Mason B, Rosenzweig A (1997) Dana's new mineralogy: the system of mineralogy of James Dwight Dana and Edward Salisbury Dana. Wiley-Interscience
155. Thompson JB (1978) Biopyriboles and polysomatic series. Am Mineral 63:239–249
156. Šiaučiūnas R, Baltakys K, Baltušnikas A (2007) Instrumental analysis of the silicate materials. Vitae Litera, Kaunas
157. Chukanov NV (2013) Infrared spectra of mineral species: extended library. Springer Science & Business Media
158. Garbev K, Stemmermann P, Black L, Breen C, Yarwood J, Gasharova B (2007) Structural features of C-S–H (I) and its carbonation in air—a Raman spectroscopic study. Part I: fresh phases. J Am Ceram Soc 90:900–907
159. Toraya H, Yamazaki S (2002) Simulated annealing structure solution of a new phase of dicalcium silicate Ca_2SiO_4 and the mechanism of structural changes from α-dicalcium silicate hydrate to α_L'-dicalcium silicate via the new phase. Acta Crystallogr B 58:613–621
160. Zadaviciute S, Bankauskaite A, Baltakys K, Eisinas A (2018) The study of C_P determination of hydrotalcite intercalated with heavy metal ions. J Therm Anal Calorim 131:521–527
161. Redkin A, Korzun I, Yaroslavtseva T, Reznitskikh O, Zaikov Y (2017) Isobaric heat capacity of molten halide eutectics. J Therm Anal Calorim 128:621–626
162. Naumov VN, Musikhin AE (2017) Fundamental characteristics of solids from low-temperature heat capacity. Comput Mater Sci 130:257–267
163. Choma J, Jagiello J, Jaroniec M (2021) Assessing the contribution of micropores and mesopores from nitrogen adsorption on nanoporous carbons: application to pore size analysis. Carbon 183:150–157
164. Rouquerol J, Avnir D, Fairbridge CW, Everett DH, Haynes JM, Pernicone N, Ramsay J, Sing K, Unger KK (1994) Recommendations for the characterization of porous solids (Technical Report). Pure Appl Chem 66:1739–1758
165. Arvaniti EC, Juenger MC, Bernal SA, Duchesne J, Courard L, Leroy S, Provis JL, Klemm A, De Belie N (2015) Determination of particle size, surface area, and shape of supplementary cementitious materials by different techniques. Mater Struct 48:3687–3701
166. Rouquerol J, Rouquerol F, Llewellyn P, Maurin G, Sing KS (2013) Adsorption by powders and porous solids: principles, methodology and applications. Academic Press
167. Naderi M (2015) Surface area: Brunauer–Emmett–Teller (BET). In: Progress in filtration and separation. Academic Press pp 585–608
168. Thommes M, Kaneko K, Neimark AV, Olivier JP, Rodriguez-Reinoso F, Rouquerol J, Sing KS (2015) Physisorption of gases, with special reference to the evaluation of surface area and pore size distribution (IUPAC Technical Report). Pure Appl Chem 87:1051–1069
169. Rudzinski W, Everett DH (2012) Adsorption of gases on heterogeneous surfaces. Academic Press
170. Zhou H, Lv W, Jiao J, Lu D, Li D (2021) Effect of Li^+ on the composition and morphology of C-A-S-H. Constr Build Mater 301:123982
171. Zhu X, He B, Xie M, Zhang Y, Jiang Z (2021) A comprehensive nitrogen adsorption measurement on the pore structure of calcium-silicate-hydrate subjected to cryogenic attack. Measurement 184:109941
172. Baltakys K, Eisinas A, Dizhbite T, Jasina L, Siauciunas R, Kitrys S (2011) The influence of hydrothermal synthesis conditions on gyrolite texture and specific surface area. Mater Struct 44:1687–1701
173. Sun Z, Huang D, Duan X, Hong W, Liang J (2020) Functionalized nanoflower-like hydroxyl magnesium silicate for effective adsorption of aflatoxin B1. J Hazard Mater 387:121792
174. Keller JU, Staudt R (2005) Gas adsorption equilibria: experimental methods and adsorptive isotherms. Springer Science & Business Media
175. Webb PA, Orr C (1997) Analytical methods in fine particle technology. Micromeritics Instrument Corporation
176. Lowell S, Shields JE, Thomas MA, Thommes M (2006) Characterization of porous solids and powders: surface area, pore size and density. Springer Science & Business Media

177. Gregg SJ, Sing KSW (1982) Adsorption, surface area and porosity, 2nd edn. Academic Press, London
178. García-Tejeda YV, Barrera-Figueroa V (2019) Least squares fitting-polynomials for determining inflection points in adsorption isotherms of spray-dried açaí juice (Euterpe oleracea Mart.) and soy sauce powders. Powder Technol 342:829–839
179. Zhou L, Bai S, Zhou Y, Yang B (2002) Adsorption of nitrogen on silica gel over a large range of temperatures. Adsorption 8:79–87
180. Storck S, Bretinger H, Maier WF (1998) Characterization of micro- and mesoporous solids by physisorption methods and pore-size analysis. Appl Catal A Gen 174:137–146
181. Taylor H (1953) Crestmoreite and riversideite. Mineral Mag J Mineral Soc 30:155–165
182. McConnell JD (1954) The hydrated calcium silicates riversideite, tobermorite, and plombierite. Mineral Mag J Mineral Soc 30:293–305
183. Fridriksson T, Neuhoff PS, Arnórsson S, Bird DK (2001) Geological constraints on the thermodynamic properties of the stilbite—stellerite solid solution in low-grade metabasalts. Geochim Cosmochim Acta 65:3993–4008
184. Galli E, Alberti A (1975) The crystal structure of stellerite. Bull Minéral 98:11–18
185. McIver EJ (1963) The structure of bultfonteinite, $Ca_4Si_2O_{10}F_2H_6$. Acta Crystallogr 16:551–558
186. Cairncross B (2017) The where of mineral names: bultfonteinite, bultfontein diamond mine, Kimberley, South Africa. Rocks Miner 92:578–581
187. Hamm HM, Hentschel G (1983) Reinhardbraunsite, $Ca_5(SiO_4)_2(OH,F)_2$, a new mineral—the natural equivalent of synthetic "calcio-chondrodite". N Jb Min Mh 119–129
188. Kirfel A, Hamm H, Will G (1983) The crystal structure of reinhardbraunsite, $Ca_5(SiO_4)_2$(OH, F)2, a new mineral of the calcio-chondrodite type. TMPM Tschermaks Petr Mitt 31:137–150
189. Zubkova NV, Pekov IV, Pushcharovsky DY, Zadov AE, Chukanov NV (2012) The crystal structure of aklimaite, $Ca_4[Si_2O_5(OH)_2](OH)_4{\cdot}5H_2O$. Zeitschrift für Kristallographie 227:452–455
190. Zadov AE, Pekov IV, Zubkova NV, Gazeev VM, Chukanov NV, Yapaskurt VO, Kartasheov PM, Galuskin EV, Galuskina IO, Pertsev NN (2013) Aklimaite, $Ca_4[Si_2O_5(OH)_2](OH)_4{\cdot}5H_2O$, a new natural hydrosilicate from Mount Lakargi, the Northern Caucasus, Russia. Geol Ore Deposits 55:541–548

Chapter 2
Synthesis and Properties of Gyrolite

The second chapter of the monograph starts with the comprehensive literature review on the natural gyrolite (properties, formation, and crystal structure) and the background information about hydrothermal synthesis of gyrolite. In this chapter, the dependence of the main properties (specific surface area, predominant pore size and their radial distribution, crystal size, and shape) of synthetic gyrolite on the duration of hydrothermal synthesis is described. It was found that gyrolite is stable till 168 h during hydrothermal treatment at 200 °C temperature in mixtures of calcium oxide and amorphous silicon dioxide. Furthermore, it was shown that, by prolonging the duration of hydrothermal synthesis from 32 to 168 h, the crystallite size of synthetic gyrolite increases from 44 to 95 nm, while the amount of the adsorbed and interlayer water in the structure of synthetic gyrolite as well as the specific surface area and total pore volume decrease.

Methodology of this chapter has been reproduced with permission from Ref. [1] ***Copyright (2019) Elsevier***.

2.1 Natural Gyrolite: Historical View, Formation, and Main Properties

Gyrolite (Fig. 2.1) is a rare mineral which was discovered on the Isle of Skye by Anderson in 1849, while the first investigation on it was published in journal of The London, Edinburgh, and Dublin Philosophical Magazine and Journal of Science in 1851 [2, 3]. The name of gyrolite was given from its spherical concentrations consisting of thin plates (in Greek γύρος, in Latin orbiculatus). According to Anderson [2] gyrolite is associated with apophyllite ($AB_4[Si_8O_{22}]_X \cdot 8H_2O$, where A = K^+, Na^+, NH_4^+, Cs^+; B = Ca^{2+}, Sr^{2+}; X = F^-, OH^-), stilbite [$NaCa_4(Si_{27}Al_9)O_{72} \cdot 28H_2O$], laumontite [$CaAl_2Si_4O_{12} \cdot 4H_2O$], and other zeolites as well as with amygdaloidal basalt. This author determined the basic crystallographic

K. Baltakys and T. Dambrauskas, *Synthetic Gyrolite*,
https://doi.org/10.1007/978-3-031-49259-4_2

a b

Fig. 2.1 White gyrolite globules on chalcedony matrix from India (**a**, photo ID: 178575777, *Source* iStock by Getty Images) and white gyrolite globules on prehnite from India (**b**, photo ID: 462207101, *Source* iStock by Getty Images)

properties of gyrolite: color—white, luster—vitreous or pearly, transparency—transparent; cleaves—perfect in one direction (parallel to the plates); hardness—between 3 and 4 by Mohs scale. The chemical formula of gyrolite proposed by Anderson was $2(CaO,SiO^3) + 3HO$.

In 1861, gyrolite was found in Annapolis County (Trap of the Bay of Fundy) Canada by How [4]. How found the chemical composition of gyrolite to be very similar to that suggested by Anderson. In the specimen investigated by How, a lower amount of calcium oxide and a higher amount of silica were obtained. Also, this sample contained impurities such as potassium, magnesium, and aluminum. How suggested that gyrolite is formed from apophyllite which is exposed to water saturated with calcium carbonate. Carbonates react with potassium silicates which are present in apophyllite, while water dissolves fluorine.

Later investigations on natural gyrolite were made by Heddle (1889 and 1891) and Clarke (1889). Heddle determined some crystallographic and optical parameters of a single plate of gyrolite specimen obtained in Treshnish Isles (Scotland) [5, 6]. Also, Heddle reported that mineral gyrolite can be found in between Mumbai to Pune (India) [7]. Clarke investigated the gyrolite specimen obtained in New Almaden Quicksilver Mine in California (USA) and compared it to the results obtained by How [8]. The obtained results allowed us to suggest the formula of gyrolite to be $Ca_2Si_3O_8{\cdot}3H_2O$, in which the molar ratio of CaO and SiO_2 is equal to 0.66. This ratio corresponds to the real molar ratio of gyrolite. Also, Clarke suggested that gyrolite is older than apophyllite (in contrast to How findings [4]), i.e. apophyllite formed from gyrolite. According to Clarke, gyrolite partially dissolves in water containing alkaline fluorides, and afterward it becomes sediment, thereby forming apophyllite [8].

Currie summarized previous findings on the localities of gyrolite in 1905 [9]. The author states that gyrolite is a characteristic mineral of the denser igneous rock of the Hebrides. Currie made some notes about the formation of gyrolite. He observed that gyrolite forms in basaltic rock with a higher number and larger steam cavities.

Also, the author agreed with the hypothesis of Clarke that apophyllite formed from gyrolite. Few years later (1910), Fleischamann found that gyrolite is widely spread in County Antrim (Ireland) [10]. The author confirmed previous findings that gyrolite is associated with zeolites and is widely distributed throughout the harder lavas. Also, Fleischamann determined that gyrolite specimens contain 13.06–13.44 wt% of water.

Christie (1925) made a more detailed investigation on gyrolite collected by Ribeiro in Mumbai (India) in 1922 [11]. Christie determined that gyrolite has a perfect platy cleavage while its crystals are flexible but not elastic. The hardness determined by Christie is about 3, which less compared to the data obtained by Anderson [2, 11]. The gravity of gyrolite is 2.388–2.390 g/cm^3. According to Christie [11], the luster of gyrolite is yellowish-white, the cleavage faces are pearly and shining, while the mammillary surface is dull. The cleavage flakes of gyrolite are transparent to translucent. Monocrystals of gyrolite are uniaxial, although cleavage plates sometimes show two axes. Gyrolite is an optically negative mineral. The chemical composition determined by Christie is similar to that presented earlier [2, 4, 7, 11]. The chemical formula of gyrolite discussed by Christie should be $2CaO{\cdot}3SiO_2 + 3H_2O$ or $Ca_4(Si_2O_7)_3H_{10}$. The double refraction of gyrolite is ~1.5645, which is a little higher compared to the first determination by Himmelbauer (1.540–1548) [11]. Also, Christie suggested that the crystallization sequence of gyrolite and the associated minerals is as follows: gyrolite, okenite, and apophyllite, calcite.

Further investigations on gyrolite were made only in 1938 by Flint et al. [12] and in 1953 by Mackay and Taylor [3]. Since Flint et al. investigated synthetic gyrolite, it will be discussed in the next subchapter. Mackay and Taylor [3] investigated a gyrolite sample collected in Mumbai (India). These authors discussed not only the chemical and optical properties of gyrolite, but also presented X-ray investigation of single and powder gyrolite as well as thermal stability till 1250 °C. It is worth mentioning that these authors made the first attempt to determine the crystal structure of gyrolite.

Natural gyrolite was also the subject of research by other scientists, which allowed us to understand the formation, properties, and structure of gyrolite [3, 9, 13–21]. Based on the findings of natural gyrolite, it can be summarized that natural gyrolite usually has incorporated sodium and aluminum ions; thus, its chemical formula is $NaCa_{16}Si_{23}AlO_{60}(OH)_8{\cdot}64(H_2O)$ [2] (Fig. 2.1). However, some scientists discovered that natural gyrolite does not contain Al^{3+} ions [14, 15]. According to Mackay and Taylor [3], the intercalated ions do not change significantly the structure of gyrolite; thus, the main structural parameters are the same.

The formation of gyrolite in nature is a complicated process. The first considerations about the formation of gyrolite were made by How [4] and Christie [11]. How suggested that gyrolite forms apophyllite during a material exchange process, while Christie proposed a different formation sequence (gyrolite $\rightarrow$ okenite and apophyllite $\rightarrow$ calcite). In later years, Seager [13] suggested that, primarily, calcite was formed, and later, in the voids of calcite, the formation of gyrolite proceeded. As a result, aluminum reacted earlier than the crystallization of gyrolite actually started. Detailed discussion on gyrolite and other CSH phases formation in the nature was made by

Koritnig [22]. According to Koritnig, gyrolite formed during various development stages of quartzite but not siliceous limestone. Based on the results of hydrothermal synthesis of calcium silicate hydrates obtained by other scientists and the facts of rock formation processes, normal pyrometamorphosis of quartzite proceeded. Later, then the basalt rock cooled down to 250 °C or even lower temperatures, and the transformation process of the quartzite inclusions into calcium silicate hydrates proceeded. During the presently discussed process, quartz reacted with a hot alkaline aqueous solution containing various ions (usually Ca^{2+}, K^{+}, F^{-} and occasionally Na^{+}, Al^{3+}, and Mg^{2+}) by forming apophyllite and CSH gel. During this process, silicic acid dissolves and is removed from quartz. This observation shows that the quartz of quartzite was first replaced by apophyllite and then or at the same time by CSH minerals. The formation of CSH minerals appears to involve an initial formation of an amorphous CSH gel. When CSH gel is formed, it starts to recrystallize to gyrolite, okenite [$Ca_{10}Si_{18}O_{46}\cdot 18H_2O$], or tacharanite [$Ca_{12}Al_2Si_{18}O_{33}(OH)_{36}$] with some residue of amorphous CSH gel. When quartzite is transformed into gyrolite, ~44 wt% of SiO_2 from the quartzite should be removed (by dissolving in aqueous mediums) and ~27 wt% of CaO and ~17 wt% of H_2O should be added (from alkaline aqueous solution containing various ions). Such replacement requires a significant mass transfer. Koritnig concludes that CSH formation from quartzite under hydrothermal conditions (temperature up to 250 °C and increased pressure) [22]. However, the formation mechanism is questionable because it is not explained how a large amount of water with dissolved calcium ions can reach the quartzite inclusions and remove the excess of silicic acid in such a dense rock as basalt without leaving any traces.

By summarizing the literature data about the formation of natural gyrolite, it can be stated that gyrolite is usually formed during a material exchange process in various rocks (basalt, druses, almond, country, etc.). Also, gyrolite can be formed in rhyolites, sediments, and some ore deposits under hydrothermal conditions. At the laboratory scale, gyrolite, whose chemical formula is given as $Ca_{16}Si_{24}O_{60}(OH)_8\cdot(14+x)H_2O$, can be synthesized by treating CaO and SiO_2 mixtures under hydrothermal conditions [16, 23–27]. Also, gyrolite can be synthesized by treating glass hydrothermally at 200 °C for 9–72 h [28–30].

Most commonly, gyrolite is associated with apophyllite, calcite, and calcium silicate hydrates (okenite, tobermorite, and xonotlite). Gyrolite is a rare mineral, which can be detected in around 100 localities. The main localities are in Scotland (Isle of Skye, Isle of Mull), Ireland (Antrim), Italy (Elba, Trentino-Alto Adige), Faroe Islands (Eysturoy, Streymoy), India (around Mumbai), etc.

The physical properties of natural gyrolite are [2, 3, 5, 6, 11, 16, 31, 32]:

- Luster—vitreous, pearly.
- Transparency—transparent, translucent, opaque.
- Color—white or colorless. If contains impurities, the color can be yellowish, brownish, grayish, green, etc.
- Tenacity—brittle while flakes flexible.
- Cleavage—perfect in one direction (parallel to the plates).

- Fracture—irregular, uneven.
- Hardness by Mohs scale—between 2.5 and 4.
- Density—measured: 2.39–2.51 g/cm^3, calculated 2.40–2.74 g/cm^3.
- Morphology—spherical, radial aggregates of platy/tabular/flaky crystals. Aggregates up to 5 cm.

The optical properties of natural gyrolite are [2, 3, 5, 6, 11, 16, 31, 32]:

- Type—uniaxial (–), or else it may be slightly biaxial (–) due to the bindings of flakes.
- Refraction indexes—$n_\alpha = 1.535$; $n_\beta = 1.548$; $n_\gamma = 1.549$.
- Birefringence—$\delta = 0.014$ (maximum).

2.2 Crystal Structure of Gyrolite

Gyrolite is a crystalline material, which, based on Dana 8 classification of minerals, belongs to the silicates group—Phyllosilicates Phyllosilicate Condensed Tetrahedral Sheets with Double and Single Layers (73.02.02c) [33, 34]. Gyrolite belongs to the triclinic crystal system and $\overline{1}$-Pinacoidal class (space group $P\overline{1}$) with lattice parameters $a = 9.74$ Å; $b = 9.74$ Å; $c = 22.40$ Å; $\alpha = 95.71°$; $\beta = 91.51°$; $\gamma = 120.01°$; and $Z = 1$ [16, 35]. The ratio of edge a:b:c = 1:1:2.3. The calculated volume of an unit cell is 1823.86 Å^3. The structural scheme of gyrolite is $OS_1OS_2XS_2OS_1O$, where O is the octahedral layer, S is the silicate layer, X is the interlayer.

The crystal structure of gyrolite was firstly investigated by Mackay and Taylor [3]. These authors used X-ray diffraction of single crystals and powder samples to determine the main parameters of the structure. Later investigations were performed by Strunz and Micheelsen [36], Meyer and Jaunarajs [23], Cann [19], and Eberhard and Hamid [37]. Although these scientists determined the main features of the gyrolite structure, their research had a number of errors. The most detailed and the most accurate investigation was made by Merlino [16]. The structure of gyrolite was also a topic in later years [35, 38–41]; however, new investigations just insignificantly supplemented the results obtained by Merlino.

Gyrolite belongs to the phyllosilicate group. Phyllosilicates consist of alternating stacking silicate layers (denoted as S) and calcium layers (denoted as O) [41–43]. Ca-octahedral layers are relatively regular whose composition is $[Ca_7O_{10}(OH)_4]^{4-}$. In this layer, the Ca-octahedra are connected to each other by the edges and form an endless repeating network of seven octahedra. These layers are connected on both sides to the two types of silicate tetrahedra through shared oxygen atoms. Layered silicates consist of two types of tetrahedral silicate layers (denoted S_1 and S_2) which are formed by connecting three ordinary oxygen atoms of neighboring tetrahedra and forming six-membered rings of tetrahedra. The composition of these silicate layers is the same $[Si_8O_{20}]^{8-}$; however, they have differently oriented silicon tetrahedra in rings. In the case of the S_1 sheet, silicate tetrahedra are forming two types of 6-membered rings (Fig. 2.2a). In the first type of rings, silicate tetrahedra are pointing

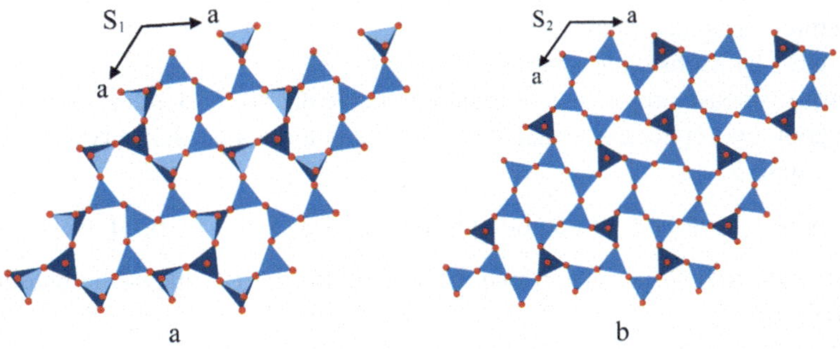

Fig. 2.2 Structure of S_1 **a** and S_2 **b** tetrahedral silicate layers

in alternating directions, i.e. up or down by forming the up-down-up-down-up-down rings. Each tetrahedra in the ring is additionally connected to the same type of tetrahedra by forming an infinite layer. The second type of rings is of the oval form, which has three consecutive tetrahedra facing up and three facing down. Meanwhile, the 6-membered hexagonal rings in S_2 sheets consist of the same directing (down) tetrahedra, which are additionally connected to oppositely directed tetrahedra (up) (Fig. 2.2b). This connection gives additional oval-shaped rings, in which two tetrahedra are facing up and the remaining four are facing down. According to Garbev et al. [41], nekoite can be classified as phyllosilicate. This mineral has a different structure of the silicate sheet compared to K-phase, Z-phase, truscottite, or gyrolite. The silicate sheets consist of wollastonite type condensed single chains. This silicate sheet is denoted as S_3. Generally, a silicate sheets network is called the T (tetrahedron) layer. The layered silicate is formed by connecting tetrahedral layers with octahedral layers. It is known that calcium layers with silicate layers can form TO (1:1) and TOT (2:1) structures [42]. In addition to the aforementioned T and S layers, an X layer can also form in the structure of phyllosilicates. When S_1–O–S_2 layers are arranged in the densest packing, they can be connected with each other through silicon tetrahedra or X sheet. X sheet is partially filled with Ca^{2+} and Na^+ cations.

According to Merlino [16], the crystal lattice of gyrolite consists of two types silicate layers (S_1 and S_2), Ca-octahedral layers (O), and the interlayer X. These layers are connected in the $S_1\overline{OS_2}XS_2OS_1$ sequence (Fig. 2.3). Such a structure of layers gives approximately 2.2 nm length. In the case of pure gyrolite, the composition of silicate and Ca-octahedral layers are $[Si_8O_{20}]^{8-}$ and $[Ca_7O_{10}(OH)_4]^{-4}$, respectively. X interlayer consists of H_2O and O^{2-} coordinated Ca-octahedra (coordinate bond). According to the structure model of gyrolite proposed by Merlin, the X interlayer is formed from octahedral atomic positions with a degree of occupancy of 1/3. Natural gyrolite usually contains impurities such as aluminum or sodium ions. Aluminum ions partially replace silicon in S_2 interlayers and thus its composition becomes $[Si_7AlO_{19}]^{7-}$. Meanwhile, sodium ions intercalate in the X interlayer. In this case, the interlayer consists of two symmetrically related Ca-octahedra and one Na-octahedron.

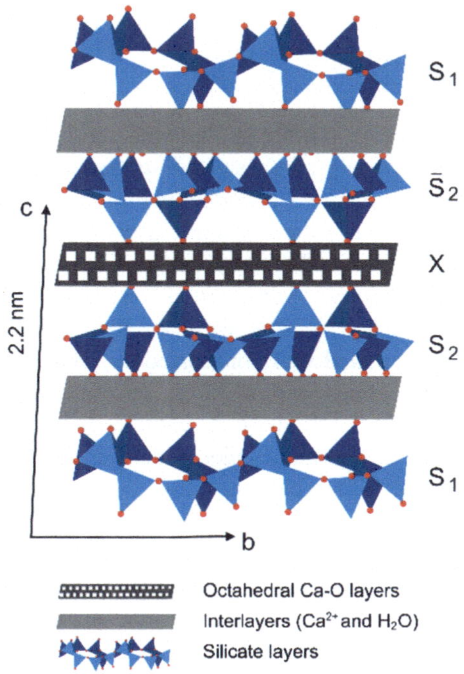

Fig. 2.3 Crystal lattice structure of gyrolite

As shown in Fig. 2.3, the silicate layers are connected through Ca-octahedral layers by forming a 'complex layer' perpendicular to the *c*-axis. These complex layers connect with the X interlayer. The distances between silicate layers are ~0.3 nm; i.e. they are sufficient to intercalate new ions while maintaining the neutral charge. For this reason, the adsorption properties of gyrolite mainly depend on the features of the silicate (S_1, S_2, and $\overline{S}_2$) layers and the X interlayer. However, some properties (such as the sorption and reaction rate, stability, activity, etc.) depend not only on the material's crystal structure but also on the morphology (particle size, specific surface area, the dominant pore shape and size) of the particles. Unfortunately, there still is lack of scientific data on the influence of the synthesis conditions on the gyrolite morphology.

2.3 Background Information About Hydrothermal Synthesis of Gyrolite

The hydrothermal synthesis of gyrolite is a complex and time-consuming process [23–27]. This compound is usually synthesized in suspensions consisting of 0.66 mol CaO and 1 mol of various forms of SiO_2 under hydrothermal conditions at temperatures of about 200 °C [44]. For the first time, gyrolite was synthesized by Flint et al. in 1938 [12]. They determined that, by treating CaO and SiO_2 mixtures with a molar ratio from 0.5 to 0.66 under hydrothermal conditions within the 150–300 °C temperature range for 6–42 days, gyrolite was the only calcium silicate hydrate that got formed. Later, this compound was synthesized by Jander and Franke [45] after 30 days of hydrothermal treatment at 300–350 °C temperature, and also by Mackay and Taylor [3] after 76 days of hydrothermal synthesis at 150 °C.

It was proven experimentally that, under hydrothermal conditions, gyrolite is stable within the 120–200 °C temperature range, while, at higher temperatures, it becomes metastable and starts to recrystallize to a related compound truscottite [$Ca_{14}Si_{24}O_{58}(OH)_8 \cdot 2H_2O$] [16, 23, 24]. According to some sources, metastable gyrolite can also be obtained at 270 °C. The recrystallization process of gyrolite into truscottite has been studied by many scientists [3, 23, 46]. These authors determined that gyrolite rearrangement to truscottite occurs during the formation of the intermediate phase whose *d*-spacing varies, and parameter *c* of the crystal lattice is lower than that of gyrolite ($c = 2.24$ nm) but higher than that of truscottite ($c = 1.883$ nm). This can be explained by the fact that gyrolite and truscottite are made up of the same, only differently located, structural layers. This was proven by further research when the structures of both compounds were fully described [16, 47].

Synthetic gyrolite forms only at temperatures above 120 °C; otherwise, a semicrystalline C–S–H gel is obtained [31]. However, if the synthesis is carried out for a long time, the C–S–H gel becomes metastable, and traces of gyrolite are formed.

The results of calcium silicate hydrate crystallization within the 150–180 °C temperature range obtained by Mitsuda et al. [48] showed that a sufficient isothermal curing time is a key factor which determines the dominant crystalline CSH phase in mixtures of varying composition: Assarsson's Z-phase or gyrolite dominates when C/S = 0.3–0.6, tobermorite dominates when C/S = 0.8, xonotlite dominates when C/S = 1, foshagite dominates when C/S = 1.3, and α-C_2SH dominates when C/S = 2. These results essentially fully overlap with the thermodynamic calculations reported by Babushkin et al. [49, 50] and Blanc et al. [51] (Fig. 2.4).

Roy and Harker [52], Harker [53], and Taylor (1964) [54] investigated the synthesis conditions of gyrolite, its stability and structural relationships with similar compounds. These authors proposed the formation diagrams of calcium silicate hydrates and found that gyrolite remains stable up to 210 °C. Meanwhile, Kalousek and Nelson [55] and Števula and Petrović [56] determined that gyrolite can be formed after hydrothermal treatment of a mixture consisting of dicalcium silicate ($2CaO \cdot SiO_2$) and SiO_2. Furthermore, Kalousek and Nelson established that gyrolite

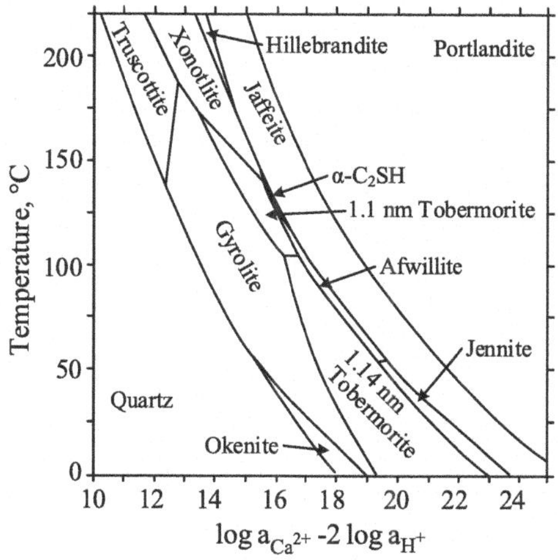

Fig. 2.4 Phase diagram in the $CaO–SiO_2–H_2O$ system according to Blanc et al. [51]. Reused with permission from Ref. [51] Copyright 2010 Elsevier

is stable till the temperature of 200 °C while, at 300 °C, the formation of truscottite is already observed [55].

Chan and Mitsuda [57] investigated the sequence of the formation of calcium silicate hydrates in a calcium oxide, quartz, and water system. They reported that CSH formation occurs in the following order: tobermorite gel, tobermorite, and, finally, xonotlite. The results showed that tobermorite gel (which formed at the beginning of the synthesis) directly recrystallized to tobermorite, and the intermediate phases Z-phase and/or gyrolite did not form. It can be explained by the low solubility rate of quartz, which leads to the abundance of Ca ions in the solution. In a later research, Hara and Inoue [58] supported the results obtained by Chan and Mitsuda.

A different sequence of the crystallization of calcium silicate hydrates under hydrothermal conditions in calcium oxide and amorphous silica mixtures with molar ratio $C/S < 1$ was obtained by Moertel [59]. At the beginning of the synthesis, the formation of the higher basicity calcium silicate hydrate α-C_2SH was observed. Afterward, this compound recrystallized to semicrystalline calcium silicate hydrates tobermorite or xonotlite, and the final product of the synthesis was gyrolite.

Other different results were obtained by Jauberthie, Esteoule, Temimi, and Laquerbe [60–63]. These authors established that pure gyrolite does not form even after 800 h of hydrothermal treatment at 210 °C. The final products of the synthesis always consisted of tobermorite and gyrolite. After further experiments, they outlined the sequence of the formation of calcium silicate hydrates in the following chains:

$$\text{Tobermorite gel} \rightarrow \text{Gyrolite gel} \rightarrow \text{Z-phase} \rightarrow \text{Gyrolite}$$

or

$$\text{Tobermorite gel} \rightarrow \text{Gyrolite gel} \rightarrow \text{Gyrolite} \rightarrow \text{Tobermorite}$$

It was thus found out that gyrolite gel recrystallization to Z-phase or gyrolite depends on the amount of lime in the initial mixture: If the amount of CaO is less than 37 wt%, gyrolite gel recrystallizes to Z-phase and gyrolite; whereas, if the amount of CaO varies from 37 to 42 wt%, gyrolite gel recrystallizes to tobermorite via an intermediate phase–gyrolite. The formed tobermorite is stable only in the presence of gyrolite, while Z-phase is metastable [60].

Shaw et al. [31, 64, 65] used high-energy synchrotron X-ray radiation to investigate mechanical, kinetic, and energetic processes taking place during the formation of CSH. They discovered that the formation of gyrolite within the 190–240 °C temperature range is a complex process involving three stages:

$$\text{amorphous gel} \rightarrow \text{C}-\text{S}-\text{H gel} \rightarrow \text{Z-phase} \rightarrow \text{gyrolite}$$

The first step involves the formation of the octahedral Ca–(O,OH) layer which acts as an additional silicate layer in the subsequent crystallization steps. C–S–H gel is always formed before the formation of crystalline CSH. In the second step, the dominant silicate chains in the disordered structural layers of C–S–H gel are rearranged and positioned in the layers that are characteristic of Z-phase. It should also be considered that the amount of calcium ions and water can vary in the C–S–H gel and in the structural interlayers of Z-phase.

In the final step, Z-phase recrystallizes to gyrolite. As established by Shaw et al. [31], the transition of C–S–H gel to gyrolite takes place as one continuous process, rather than two separate stages. Gard et al. [66] described the mechanism of this process, and this explanation has not been challenged yet.

It is known that the solubility of quartz in water is much lower than that of amorphous SiO_2; thus, the basicity and reactivity of the primary mixtures used by different scientists vary. As a result, Shaw et al. [31] presented different sequences of calcium silicate hydrate crystallization in comparison with other scientists working in this field [57–63]. In 1973, Gundlach observed that CSH formation proceeded slowly when quartz was being used as the starting compound [67].

The influence of SiO_2 modification on gyrolite's crystallization process was examined in detail by Siauciunas and Baltakys [68]. These authors established that, in mixtures of CaO and amorphous SiO_2 ($C/S = 0.66$), gyrolite does not form even after 168 h of hydrothermal treatment at 150 °C temperature. The increment of temperature had a positive effect on the formation of the latter compound: Pure gyrolite is obtained at 175 °C after 72 h, and after 32 h at 200 °C. On these grounds, the authors proposed the following sequence of gyrolite formation under hydrothermal conditions:

$$\begin{aligned} &\text{Ca(OH)}_2 + \text{amorphous SiO}_2 + \text{H}_2\text{O} \rightarrow \text{C}-\text{S}-\text{H gel} \\ &\rightarrow \text{C}-\text{S}-\text{H(I)} \rightarrow \text{Z-phase} \rightarrow \text{gyrolite} \end{aligned}$$

Meanwhile, in mixtures with quartz, Z-phase and gyrolite do not get formed even after 72 h of hydrothermal treatment within the 150–200 °C temperature range. Due to a low solubility rate of quartz, the molar ratio of CaO and SiO_2 increased; thus, the formation of dibasic CSH (α-C_2SH and C–S–H(II)) took place at the beginning of the synthesis. Finally, these compounds recrystallized into 1.13 nm tobermorite and xonotlite.

Gyrolite was synthesized by many scientists; however, this compound was obtained as a trace or in a mixture with other calcium silicate hydrates [17, 69–76]. These authors did not investigate the formation peculiarities of gyrolite.

2.4 Effect of Hydrothermal Synthesis Duration at 200 °C Temperature on the Properties of Gyrolite

2.4.1 Effect of Hydrothermal Synthesis Duration on the Stability and Crystallinity of Gyrolite

In order to determine the influence of the duration of hydrothermal treatment on the crystal stability of gyrolite and on the morphology of the particles, hydrothermal synthesis experiments in the stoichiometric composition (molar ratio of CaO/SiO_2 = 0.66) of the initial CaO and $SiO_2{\cdot}nH_2O$ mixture at 200 °C temperature were performed. The synthesis of gyrolite was carried out in unstirred suspensions (fine-grained $SiO_2{\cdot}nH_2O$; CaO) in stainless steel vessels for 32, 48, 72, 120, 168, 336, and 504 h. The products of the synthesis were filtered, rinsed with ethyl alcohol (so that the drying process can be ensured), dried at a temperature of 50 ± 5 °C, and then sieved through a sieve with a mesh width of 50 μm.

Detailed experimental and thermodynamic results of gyrolite formation during hydrothermal synthesis at 200 °C temperature for 8–72 h are presented in Chap. 3 3.3. *Influence of AlF_3 production waste on the formation of gyrolite family compounds.* It is worth noting that smoothed XRD patterns were used to calculate the characteristics of the diffraction peaks (the crystallite size, area, and intensity) which were being used in our experiments (Fig. 2.5). The smoothing was performed by using the X-ray line profile fitting program *XFIT* and the Savitzky–Golay polynomial filter.

It was determined that, after 32 h of isothermal curing, gyrolite is the only crystalline product formed in the CaO and $SiO_2{\cdot}nH_2O$ mixture (Fig. 2.5). In the X-ray diffraction pattern, all the characteristics of the diffraction peak of the latter compound (PDF No. 00-042-1452; *d*-spacing: 2.273, 1.126, 0.844, 0.420, 0.365, 0.280, and 0.224 nm) are identified. It was proven that, by extending the isothermal curing duration from 32 to 168 h, the intensity of the main diffraction peak characteristics of gyrolite (*d*-spacing: 2.273 nm) increased by more than 5.5 times (see Fig. 2.5 and Table 2.1). It is known that the intensity of diffraction peaks strongly depends on the orientation of crystallites. In order to avoid errors due to this phenomenon, the analysis was performed by using special backloading sample holders which were rotated during analysis.

After extending the duration of isothermal curing to 336 h, gyrolite became metastable because the diffraction peaks which are characteristic to truscottite (which is a member of the gyrolite group; PDF No. 00-017-0761; *d*-spacing: 2.047, 1.880,

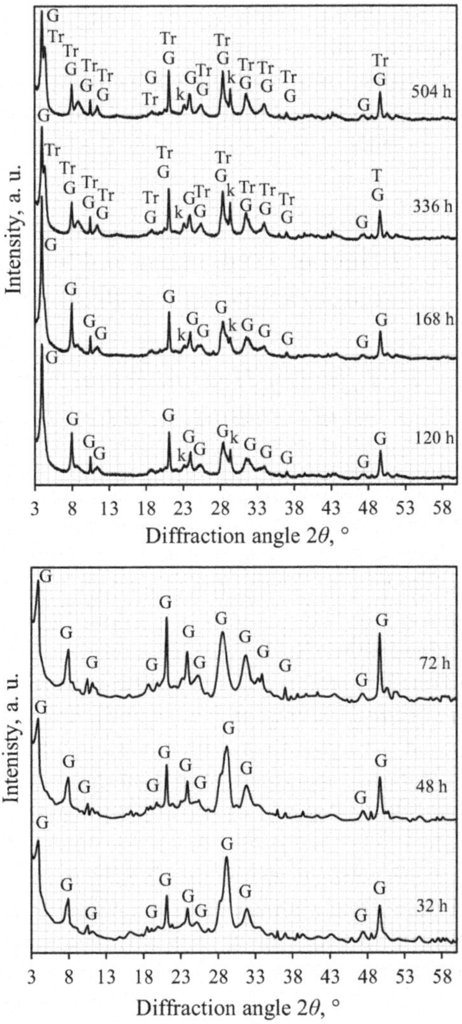

Fig. 2.5 XRD patterns of the synthesis products when the molar ratio of CaO and amorphous silica dioxide mixture and temperature were equal to CaO/SiO_2 = 0.66 and 200 °C, respectively. Indexes: G—gyrolite; Tr—truscottite; k—calcite

0.942, 0.769, 0.471, 0.314, 0.183, and 0.176 nm) were identified (Fig. 2.5). It should be noted that the main diffraction peak characteristic of the presently discussed compounds feature similar interplanar distances (gyrolite ~2.273 nm; truscottite ~2.047 nm), and thus a doublet peak was formed in the range of 3–5° diffraction angles (Figs. 2.5 and 2.6). It was established that the intensity of the main diffraction peak characteristic of synthetic gyrolite decreased from 130.6 cps (after 168 h of synthesis) to 94.6 cps (after 336 h of synthesis) (Table 2.1). This data conforms quite precisely to the thermodynamic calculations of gyrolite stability during hydrothermal synthesis (Fig. 3.24). Our calculations showed that gyrolite recrystallization reaction to truscottite (Eq. 3.48) is most probable because $\Delta_r G_T^0$ values were the lowest.

Table 2.1 Dependence of intensities of gyrolite and truscottite diffraction peaks on duration of hydrothermal synthesis

Duration of synthesis, h	Peak (*d*-spacing 2.273 nm) intensity of gyrolite, cps	Peak (*d*-spacing 2.040 nm) intensity of truscottite, cps
32	23.2	–
48	35.2	–
72	45.9	–
120	112.6	–
168	130.6	–
336	94.6	62.5
504	91.5	62.9

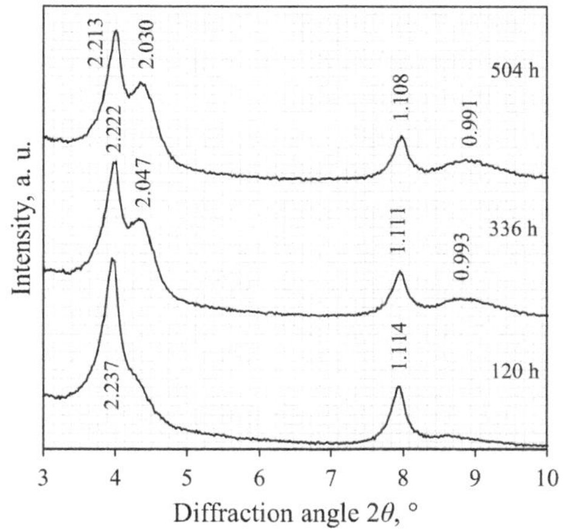

Fig. 2.6 XRD Patterns of the synthesis products after different durations of hydrothermal synthesis

It was thus determined that, when the duration of the synthesis was extended by 1.5 times (i.e. to 504 h), the intensities of the diffraction peaks characteristic to gyrolite and truscottite changed negligibly (Figs. 2.5, 2.6 and Table 2.1).

The main diffraction peaks of synthetic gyrolite and synthetic truscottite were used to calculate the crystallite size. The calculations were performed by using the Scherrer equation:

$$D_{hkl} = \frac{k_{hkl} \cdot \lambda}{\beta_{hkl} \cdot \theta}, \tag{2.1}$$

where λ is the wavelength of CuK_α radiation, θ is Bragg's diffraction angle, β_{hkl} is the full width at half maximum intensity, and k is the shape factor (the value used in this study was 0.94).

The calculations revealed that the crystallite size of gyrolite depends on the duration of hydrothermal curing (Fig. 2.7). It was proven that, after 32 h of synthesis, the crystallite size of gyrolite is equal to 43.8 nm. By extending the duration of

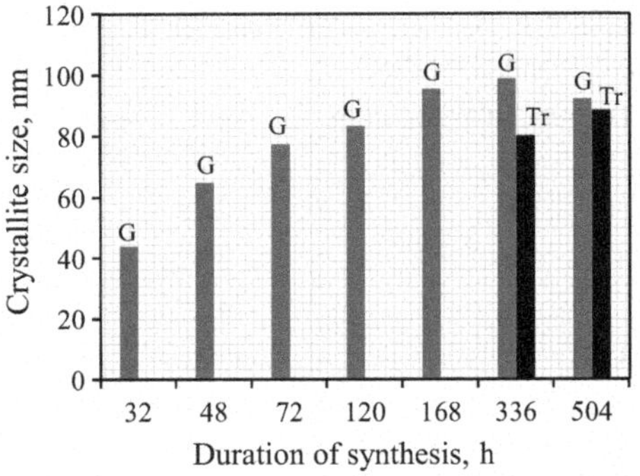

Fig. 2.7 Dependence of gyrolite and truscottite crystallite size on the duration of hydrothermal synthesis. Indexes: G—gyrolite; Tr—truscottite

hydrothermal curing to 336 h, the crystallite size increased by more than two times, i.e. to 98.7 nm (Fig. 2.7). A further extension of the synthesis duration (504 h) leads to a decrease of the crystallite size of gyrolite and an augmentation of that of truscottite from 80.1 to 88.4 nm.

The XRD results are only in partial agreement with the literature data. According to Flint et al. [12], gyrolite is the one and only synthesis product in the mixtures with the CaO/SiO_2 ratio equal to 0.5–0.66, when the duration of hydrothermal treatment within the temperature range of 150–350 °C was 6–42 days. Mackay and Taylor [3] obtained similar results. These authors state that gyrolite can be obtained after 76 days of isothermal treatment at 150 °C temperature. Later on, Garbev [32] synthesized gyrolite from C–S–H gel with C/S = 0.66 after 10 days at 200 °C and after 33 days at 220 °C. The results presented in the literature and obtained in this work could differ due to the reactivity of the raw materials, their finest particles, and some other parameters.

The XRD results were successfully verified with the data of STA analysis (Fig. 2.8 and Tables 2.2, 2.3). In the DSC curve, the endothermic effect within the temperature range of 80–200 °C is related to the loss of the adsorbed and interlayer water from the crystal structure of the synthesis products (Fig. 2.8 and Table 2.2). It was determined that, by prolonging the synthesis duration from 32 to 120 h, the heat of the presently discussed effect increased from 48.7 to 71.1 J/g, while the loss on ignition decreased from 5.4 to 3.9% (Table 2.2). A further prolongation of the synthesis duration (504 h) leads to the decrement of the heat from 71.1 to 54.2 J/g (Fig. 2.8 and Table 2.2). At the same time, the mass loss during this thermal conversion decreases from 3.9 to 3.1 wt% (Table 2.2).

The second endothermic effect at ~690 °C can be attributed to the decomposition of calcium carbonate (Fig. 2.8). It should be taken into consideration that calcium carbonate is not detected in the synthesis products synthesized for 32–72 h (Fig. 2.5).

The exothermic effect at ~870 °C is attributed to the recrystallization of gyrolite into wollastonite. It was estimated that, after 32 h of synthesis, the heat and mass loss of this effect were equal to 25.2 J/g and 0.3%, respectively (Table 2.3). By extending the hydrothermal treatment duration until 120 h, the heat of the exothermic effect nearly doubled; i.e. it went up to 49.9 J/g, while the mass loss remained

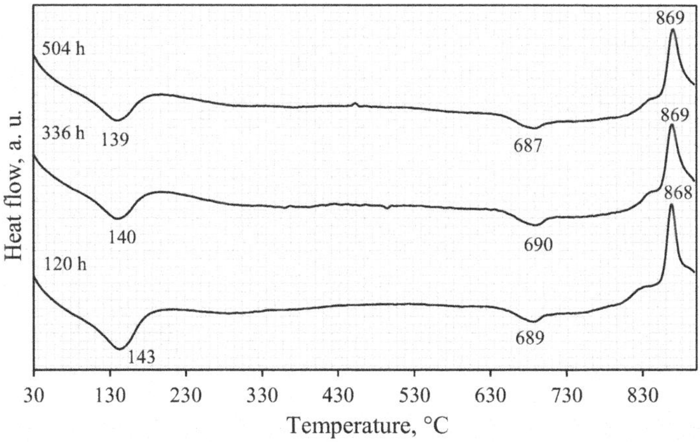

Fig. 2.8 DSC curves of synthesis products after different durations of hydrothermal synthesis

Table 2.2 Characteristics of the endothermic effect at 80–200 °C of synthesis products

Duration of hydrothermal treatment, h	T_{onset}, °C	T_{max}, °C	T_{end}, °C	Mass loss, %	Heat of reaction, J/g
32	108	138	176	5.4	48.7
48	108	144	176	5.0	55.8
72	110	143	175	4.7	70.9
120	111	143	175	4.1	71.1
168	115	145	177	3.8	69.3
336	108	140	174	3.3	55.1
504	108	140	175	3.1	54.2

Table 2.3 Characteristics of exothermic effect at 830–890 °C of synthesis products

Duration of hydrothermal treatment, h	T_{onset}, °C	T_{max}, °C	T_{end}, °C	Mass loss, %	Heat of reaction, J/g
32	850	862	875	0.3	25.2
48	855	866	876	0.2	31.5
72	856	866	876	0.2	48.3
120	857	868	878	0.2	49.9
168	855	866	876	0.2	49.9
336	857	869	880	0.2	49.1
504	858	870	882	0.2	47.8

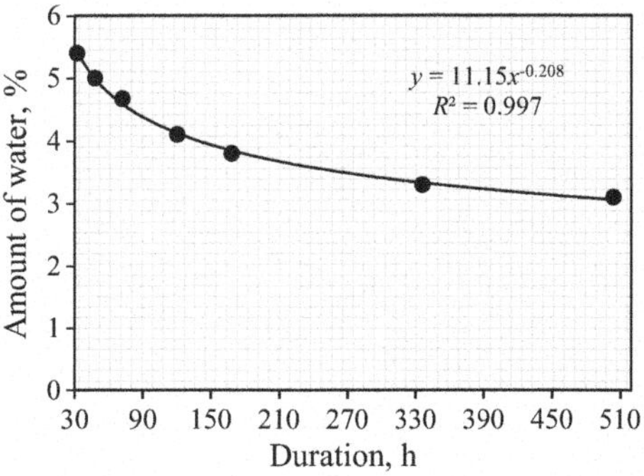

Fig. 2.9 Dependence of the adsorbed and interlayer water in the structure of synthesis products on the duration of hydrothermal treatment at 200 °C

similar, specifically, ~0.2%. Due to gyrolite recrystallization to truscottite, the heat characteristic to the exothermic effect slightly decreased after 336 and 504 h of synthesis. However, the mass loss remained the same (Table 2.3). Furthermore, by prolonging the synthesis duration, the typical temperatures of this effect shifted to higher temperatures: T_{onset} from 850 to 858 °C; T_{peak} from 862 to 870 °C; T_{end} from 875 to 882 °C.

By summarizing the STA and XRD results, it can be stated that the quantity of the adsorbed and interlayer water decreased by extending the duration of hydrothermal synthesis, and, at the end of the synthesis (336–504 h), it was equal to ~3.3 wt% (Fig. 2.9). Contrarily, while the amount of the adsorbed and interlayer water decreased, the crystallinity (the crystallite size and the intensity of the diffraction peaks) of synthetic gyrolite increased (Fig. 2.7 and Table 2.1). The obtained results essentially corroborate the data obtained by Gard et al. [66] and Lachowski et al. [77].

The FT-IR analysis data supports the above-discussed results that the hydrothermal treatment duration exerts influence on the quantity of the adsorbed and interlayer water in synthetic gyrolite (Figs. 2.9 and 2.10). The FT-IR spectra of the synthesized products showed that, by prolonging the synthesis duration to 504 h, a sharp peak at 3639 cm^{-1} becomes more prominent and shows that the OH^- group is strongly bounded in the crystal lattice of gyrolite (Fig. 2.10). In the frequency interval of 3444–3554 cm^{-1}, a broad absorption band can be assigned to molecular water vibration in the interlayers (as hydrogen bridges are being formed). The decrease of this band intensity confirms that, by prolonging the synthesis duration, the quantity of molecular H_2O in the gyrolite interlayer decreases (Fig. 2.9). After 336 and 504 h of synthesis, two absorption bands at 1252 and 807 cm^{-1}, which can be assigned to vibrations in Q^4 (formed due to the polymerization of the single silicate sheets to double sheets), were observed (Fig. 2.10). According to Garbev et al. [78], these bands should be assigned to truscottite.

Thus, gyrolite remains stable when the hydrothermal synthesis duration varies from 32 to 168 h at 200 °C, although its structure is altered when the synthesis duration is increased. Meanwhile, after 336–504 h of treatment, gyrolite became metastable and started to recrystallize to truscottite. Thus, in the next step of the

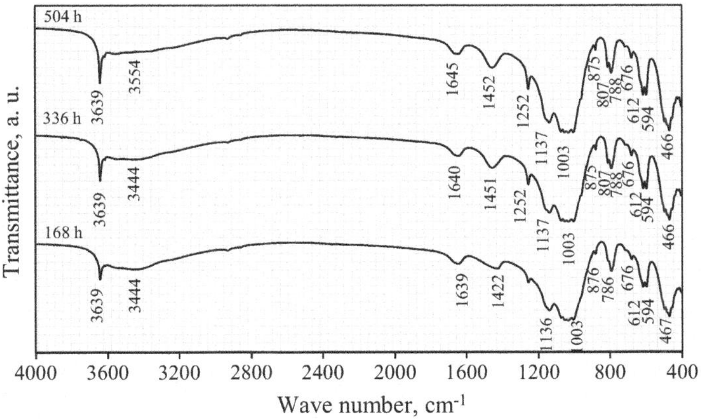

Fig. 2.10 FT-IR spectra of synthesis products after different durations of hydrothermal synthesis

research, gyrolite synthesized after 32, 48, 72, 120, and 168 h at 200 °C temperature was used for the determination of the impact of hydrothermal synthesis duration on its porosity.

2.4.2 Effect of Hydrothermal Synthesis Duration on the Porosity of Gyrolite

It is widely accepted that there is a correlation between the shape of the adsorption and desorption isotherms and the morphology (e.g. pore size distribution, shape of pores, connectivity) of porous materials. Typically, nitrogen adsorption and desorption isotherms are used to analyze the properties of porous materials [79–84]. However, the analysis of nitrogen gas isotherms itself is a complex process. The amount of the adsorbed nitrogen gas can be measured in any unit; however, this parameter is most commonly expressed in cm^3 per gram of the used adsorbent [85–88]. For a simple comparison of the adsorption data, the adsorption isotherms are presented in the graphical form with the adsorbed amount of nitrogen plotted against the equilibrium relative pressure (p/p_0).

Classification of adsorption isotherms and hysteresis loops

In 1985, according to the IUPAC guidelines, the adsorption isotherms were divided into six types, but, over the last decades, various new materials featuring unconventional isotherms have been synthesized [87]. Therefore, in 2015, the initial classification of the IUPAC adsorption isotherms and hysteresis loops was revised. The updated classifications of adsorption isotherms and hysteresis loops are given in Figs. 2.11 and 2.12, respectively.

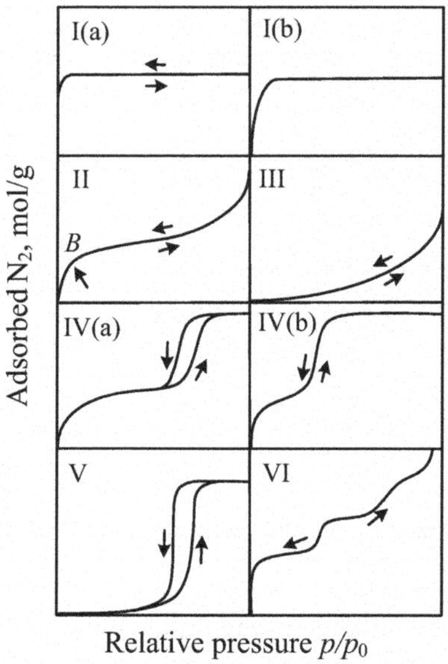

Fig. 2.11 Classification of adsorption isotherms

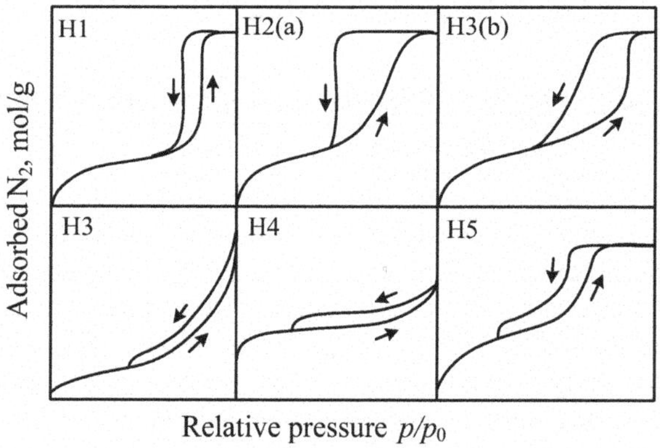

Fig. 2.12 Classification of hysteresis loops

Type I isotherms are reversible and characteristic to microporous solids with a relatively small external surface area [85–95]. These isotherms are concave with respect to the p/p_0 axis, and the limiting value is reached at a low p/p_0 ratio (filling of the micropores); thus, the amount of the adsorbed gas mostly depends on the micropore volume (Fig. 2.11). Type I isotherms are further classified into Type I(a) (characteristic to the materials with narrow micropores {width < ~1 nm}) and Type I (b) (characteristic to the materials with a wider pore size distribution {< ~2.5 nm}). For the description of Type I isotherms, the Langmuir equation can be used.

The reversible Type II isotherm is present in non-porous or macroporous materials [85–95]. At the beginning of adsorption ($p/p_0 < 0.35$), the formation of a monolayer on the surfaces of materials takes place (Fig. 2.11). Point B (the beginning of the linear part) shows the completion of the formation of a monolayer and the beginning of unlimited multilayer adsorption. This type of the isotherm can be described by using the BET equation.

Type III isotherm is convex (starting at the origin) with respect to the p/p_0 axis (Fig. 2.11) [85–95]. This type of the isotherm is usually present in non-porous or macroporous solid materials, in which the formation of a monolayer is not identified (i.e. point *B* does not exist). The interaction between the adsorbent and the adsorbate is relatively weak; thus, the adsorption of nitrogen gas proceeds on the active sites. It should be noted that the amount of gas adsorbed at $p/p_0 = 1$ is limited.

Type IV isotherm is present in mesoporous materials (the pore diameter varies within the 2–50 nm range), where, at first, a monolayer is formed, then a multilayer, and, finally, capillary condensation occurs [85–95]. At the beginning, the shape of the Type IV isotherm is the same as that of Type II; however, at a higher p/p_0 (> 0.7), gas condensation takes place in pores, and a plateau can be observed (Fig. 2.11). According to the literature [85–95], Type IV isotherms can be reversible {Type IV(b)} or have a hysteresis loop {Type IV(a)}.

The Type V isotherm is present in micro- and mesoporous as well as in flat and homogenous materials [86, 87]. In the range of low p/p_0 values (0–0.5), the shape of the Type V isotherm is the same as that of Type III; however, at higher values, the pores are filled (Fig. 2.11). This type of the adsorption isotherm has a hysteresis loop.

Type VI isotherm is present in highly uniform non-porous materials, in which, layer-by-layer adsorption takes place (Fig. 2.11) [87]. In this case, the step height shows the capacity for each adsorbed layer.

Hysteresis loops are formed in the multilayer region and are commonly related with adsorbate condensation in pores and adsorption metastability. Hysteresis loops are very closely related to the peculiarities of the pore structure and the adsorption mechanism. According to the IUPAC classification which has now been extended and updated, six distinct types of hysteresis loops are singled out. The types are presented in Fig. 2.12.

The Type H1 hysteresis loop is characteristic of mesoporous materials with cylindrical pores or which are composed of spherical particles of similar sizes. In this case, the effect of the pore network is minimal, and the upright, narrow loops are a clear sign of delayed condensation on the adsorption branch. Besides, the Type H1 hysteresis loop can also be found in networks of ink-bottle pores where the diameters of the channel (neck) and the pores are similar.

Type H2 hysteresis loops are characteristic to the materials with predominantly disordered ink-bottle neck-shaped pores (i.e. a different size and distribution); thus, the effect of the pore network is significant. This type of the hysteresis loop is classified as H2(a) (in the triangular shape with a sharp desorption branch) and H2(b). The Type H2(b) hysteresis loop is characteristic to adsorbents with a larger size distribution of the neck widths.

Type H3 hysteresis loops have two distinctive features: (1) a narrow hysteresis loop; (2) the adsorption and desorption isotherms coincide at p/p_0 equal to ~0.3. This type of the hysteresis loop is characteristic to the materials in which slit-shape pores are predominant between the parallel crystalline plates, or to macroporous materials, in which a network of pores is formed.

The Type H4 hysteresis loop is characteristic to micro- and mesopores with narrow slit pores. In this case, the adsorption and desorption isotherms are almost parallel and horizontal. The final type of hysteresis loops is Type H5 which is unusual and can be found in mesopores with both open and partially closed pores.

It must be taken into consideration that, in real life, materials may consist of more complex pore structures/networks (pores with different sizes and shapes); thus, combinations of hysteresis loops which are interpreted by mathematical calculations and other studies of instrumental analysis (e.g. microscopy) are most common.

Effect of hydrothermal synthesis duration on the shape of adsorption isotherms and hysteresis loops of gyrolite

The nitrogen adsorption by synthetic gyrolite experiments at 77 K temperature was performed with a BET surface area analyzer (Nitrogen adsorption porosimeter) *Nova 1200 E-Series* (*Quantachrome Instruments*, USA). Before conducting the nitrogen adsorption analysis, the samples were degassed under vacuum at 100 °C temperature.

The assumptions about the shape of the pores of the synthetic gyrolite samples were made based on the nature of nitrogen adsorption–desorption isotherms which are presented in Figs. 2.13 and 2.14. The shape of the adsorption isotherms is unconventional because, based on the IUPAC classification (Figs. 2.11 and 2.12), it can be classified either as Type II or as Type IV. This uncertainty occurs because the shape of the isotherms is more similar to reversible Type II isotherms (the absence of a plateau at higher p/p_0 values) which are present in non-porous or macroporous materials. However, all the samples were characterized by the occurrence of the hysteresis loop: adsorption and desorption isotherms do not coincide, the desorption isotherm is positioned to the left in comparison with the adsorption isotherm (Figs. 2.13 and 2.14). The existence of the hysteresis loop is a characteristic feature of the Type IV (a) isotherm (Fig. 2.11). This type of isotherms is characteristic to mesoporous adsorbents. Besides, all adsorption isotherms have an apparent bend in the lower p/p_0 area (point B). Microporous sorbents do not have a point B in the interval $0.05 \leq p/p_0 \leq 0.30$. Thus, it can be assumed that mesoporous pores are dominant in the structure of all synthetic gyrolite samples, although macroporous pores are also likely to be present (Figs. 2.13 and 2.14).

The hysteresis loops of the synthesis products synthesized for 32–72 h can be classified as Type H3 (Figs. 2.13 and 2.14a). This type of the hysteresis loop is obtained in the case of materials with mesoporous pores (where the pore diameter is within the range of 15–500 Å) formed between parallel crystal planes (slit-shape pores), or in macroporous materials in which a pore network is formed. However, according to the IUPAC classification (Fig. 2.12), the isotherms of adsorption and desorption must coincide within the range of relative pressures $p/p_0 \approx 0.30$; moreover, the curve of the desorption isotherm plummets extremely sharply (Figs. 2.13 and 2.14a). This indicates that, in the product structure, the pores formed between disordered plates were dominating. Meanwhile, by prolonging the synthesis duration to 120 and 168 h, the hysteresis loops become much narrower and overlap at a slightly higher pressure

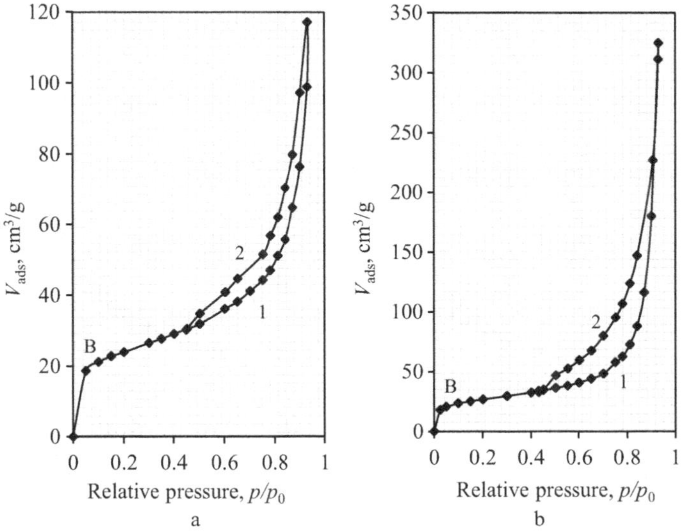

Fig. 2.13 Adsorption-desorption isotherms of synthetic gyrolite obtained at 200 °C temperature after 32 h **a** and 48 h **b**

p/p_0 (Fig. 2.14b, c). As a result, these hysteresis loops become more similar to Type H1 which is present in mesoporous materials with cylindrical pores or in materials consisting of spherical particles (Fig. 2.12). However, the adsorption and desorption isotherms did not coincide when the relative pressure $p/p_0 \approx 0.70$, which means that slit-shape pores may also be present in the sample. Thus, during the hydrothermal synthesis, changes both in the structure of the synthesized gyrolite and in the form of the pores were observed, specifically, cylindrical pores started to dominate.

Since the adsorption isotherms of synthetic gyrolite are characterized as Type IV and have a point B (the completion of the formation of a monolayer), the specific surface area can be calculated with the Brunauer–Emmett–Teller (BET) method.

Effect of hydrothermal synthesis duration on the specific surface area of gyrolite

Despite the weakness of some theoretical foundations, the Brunauer–Emmett–Teller (BET) method is the most commonly used method for the calculations of the specific surface area of porous and powder materials [85–95]. In some cases, the area obtained by using the BET method (when using Type II or Type IV (a) isotherms) can be considered as the area of the active adsorption centers. The specific surface area of the samples can be calculated by employing the numbers of the bottom segment of the N_2 adsorption isotherm ($0.05 < p/p_0 < 0.35$) and the BET equation [85–95]:

$$\frac{1}{X\left(\frac{p_0}{p}-1\right)} = \frac{C-1}{X_m \cdot C} \cdot \frac{p}{p_0} + \frac{1}{X_m \cdot C}, \tag{2.2}$$

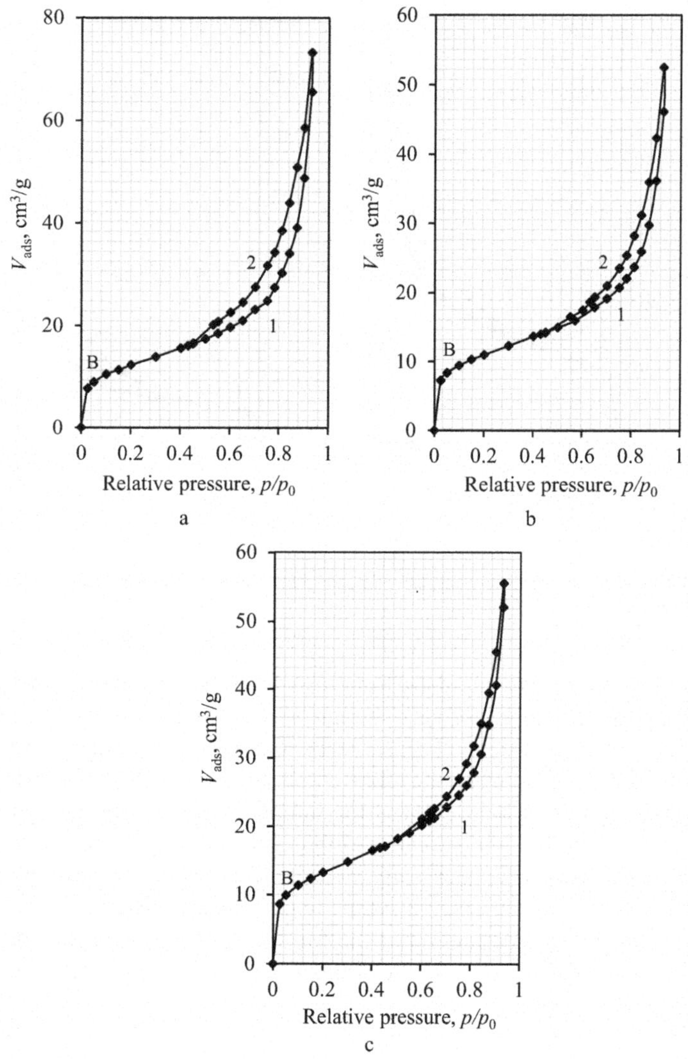

Fig. 2.14 Adsorption-desorption isotherms of synthetic gyrolite obtained at 200 °C temperature after 72 h **a**, 120 h **b** and 168 h **c**

where X is the mass of the adsorbate adsorbed on the sample at relative pressure p/p_0, p is the partial pressure of the adsorbate, p_0 is the saturated vapor pressure of the adsorbate, X_m is the mass of the adsorbate adsorbed at a coverage of one monolayer, C is a constant which is the function of the heat of adsorbate condensation and the heat of adsorption (C_{BET} is a constant).

The application of the BET method involves two stages:

An isotherm is displayed into the 'BET plot,' i.e. $\frac{1}{X\left(\frac{p_0}{p}-1\right)}$ is plotted versus $\frac{p}{p_0}$. The slope of $\frac{C-1}{X_m \cdot C}$ and the intercept of $\frac{1}{X_m \cdot C}$ are used to determine X_m and C: S = slope = $\frac{C-1}{X_m \cdot C}$ and I = intercept = $\frac{1}{X_m \cdot C}$. Solving X_m yields $X_m = 1/(S + I)$. For the calculations, the linear part of the BET plot should be used. The surface area is calculated by applying the obtained value of X_m and by using equation $S_t = \frac{X_m \cdot N \cdot A_{cs}}{M}$, where M is the molecular mass of the adsorbate, N is the Avogadro constant, A_{cs} is the cross-sectional area occupied by each nitrogen molecule (16.2×10^{-20} m^2). The specific surface area is given by the equation $S_{BET} = S_t/m$, where m is the mass of the sample.

Usually, for Type IV isotherms, X_m is calculated in the linear range, i.e. p/p_0 ~0.05–0.35. However, for the calculations using the latter interval, the negative intercept values were observed in the case of gyrolite samples synthesized after 32 and 48 h. Negative values may occur due to gas condensation in the narrow pores at relatively low p/p_0 values or due to the strong interaction between the adsorbent and the adsorptive which leads to the 'sooner' formation of the monolayer. For this reason, for the calculations, the linear BET interval was shifted toward lower relative pressure values (p/p_0 = 0.05–0.2).

The BET equation of all the investigated samples gives a linear plot in the range of relative pressures $0.05 \leq p/p_0 \leq 0.20$ (Fig. 2.15). Straight lines with the correlation coefficient $R^2 > 0.999$ were obtained for all the synthetic gyrolite samples in the BET coordination $\frac{1}{X\left(\frac{p_0}{p}-1\right)} - \frac{p}{p_0}$. The calculated parameters are given in Table 2.4.

According to the BET theory [85–95], constant C_{BET} is related to the monolayer adsorption energy, and its values give a useful indication of the shape of the isotherm within the BET range. If the value of the presently discussed constant is higher than 50–150 (depending on the literature), the knee of the isotherm curve is sharp, and point B (indicating the completion of the monolayer) is well-defined (Fig. 2.11). Meanwhile, a value of C_{BET} higher than 150 can be associated with the strong interaction between the adsorbent and the adsorptive, or with the filling of narrow pores.

The calculations revealed that, within the range of relative pressures of $0.05 \leq p/p_0 \leq 0.20$, all samples are characterized by the values of the C_{BET} constant higher than 50 (Table 2.4). It shows that the monolayer of absorbed N_2 was formed on the surface of synthetic gyrolite. It was thus established that the C_{BET} values of the gyrolite samples synthesized after 32 h and 48 h of hydrothermal synthesis were higher than 150, which shows strong interaction between the adsorbent and the adsorptive (Table 2.4). The strong interaction may be caused by the low crystallinity of synthetic gyrolite and/or by the high quantity of the adsorbed and interlayer water in the synthetic gyrolite structure (Figs. 2.7 and 2.9). Also, it is probable that, due to the high amount of water in the interlayer, the thickness of the interlayer increases by forming some micropores, and filling with liquid nitrogen is possible. Thus, structural elements of the synthetic gyrolite lattice were combined with an undefined composition (Ca^{2+}, H_2O) of the X interlayer (Fig. 2.3), and thus, it is not coincidental that C_{BET} constant values are higher than 150. Meanwhile, the C_{BET} constant values of the gyrolite

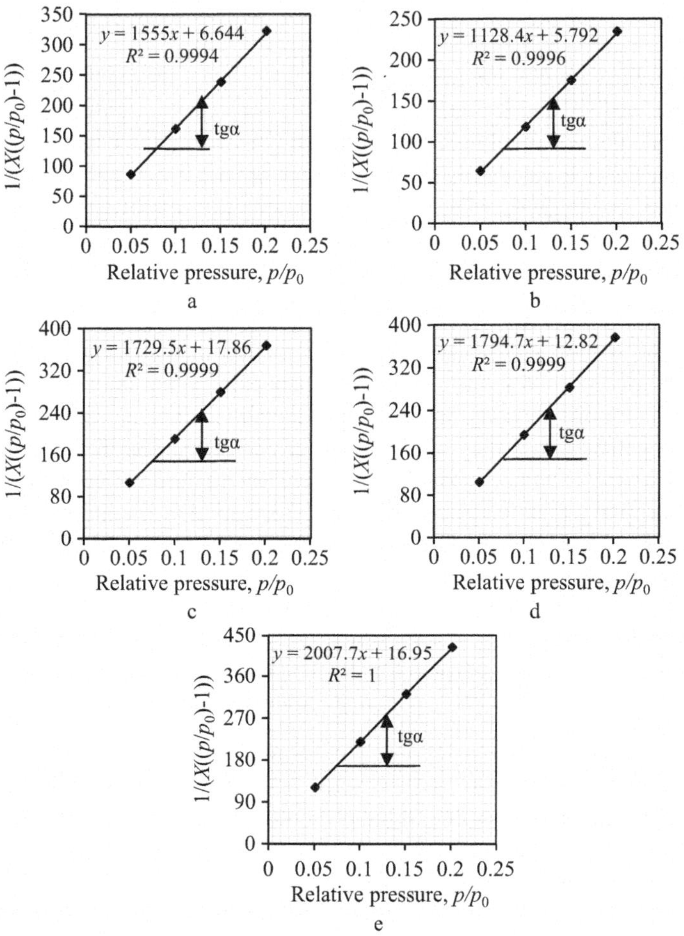

Fig. 2.15 Adsorption-desorption isotherms in the BET plot of synthetic gyrolite obtained at 200 °C temperature after 32 h **a**, 48 h **b**, 72 h **c**, 120 h **d**, and 168 h **e**

Table 2.4 Calculated parameters of synthetic gyrolite specific surface area (S_{BET})

Duration of synthesis, h	Sample mass, g	BET equation constants		Capacity of monolayer X_m, g ($\times 10^5$)	S_{BET}, m^2/g	C_{BET}	Reliability coefficient R^2
		Slope $S = tg\alpha$	Intercept, I				
32	0.0262	1555.04	6.64	6.40	81.17	235.06	0.9994
48	0.0323	1128.43	5.79	8.82	95.12	195.82	0.9996
72	0.0448	1729.50	17.86	5.72	44.52	97.85	0.9999
120	0.0490	1794.71	18.82	5.53	39.35	141.03	0.9999
168	0.0363	2007.68	16.95	4.94	47.42	119.43	1.0000

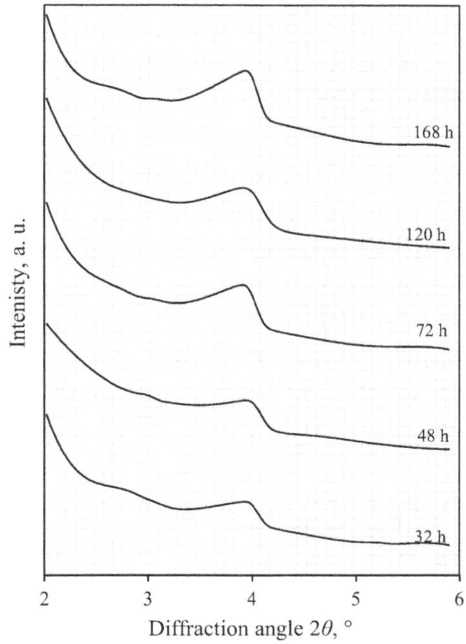

Fig. 2.16 Dependence of the intensity of diffraction peak *d*-spacing (2.273 nm) on the duration of hydrothermal treatment at 200 °C

obtained after 72, 120, and 168 h of synthesis decreased to 97–141 (Table 2.4). The decrement of this value is related to the improvement in the order of the crystal lattice, i.e. the decrement of water in the X interlayer (Fig. 2.9) and the increment of the intensity of the main diffraction peak *d*-spacing—2.273 nm (Fig. 2.16)—and the crystallite size (Fig. 2.7).

It was identified that the length of hydrothermal synthesis has an impact not only on the structure of gyrolite but also its specific surface area (Table 2.4). It was determined that, after 32 h of hydrothermal synthesis, the value of the specific surface area (S_{BET}) of gyrolite was equal to 81.17 m^2/g. When prolonging the hydrothermal synthesis duration to 48 h, the latter value also increased to 95.12 m^2/g (Table 2.4). As expected, a further prolongation of the duration of hydrothermal synthesis negatively influences S_{BET} values because, after 72 and 120 h of synthesis, it decreased by more than two times to 44.52 m^2/g and 39.35 m^2/g, respectively. The opposite results were obtained after 168 h of isothermal curing because the S_{BET} value of gyrolite slightly increased (47.42 m^2/g). Likely, one of the reasons is the changes in the synthetic gyrolite crystal structure where the water content varies from 12 to 20 wt% [31, 96].

Thus, the shape of isotherms along with the calculations by using the BET equation allows stating that, after a sufficiently long synthesis duration, the structural elements of the synthetic gyrolite lattice connect to a more orderly structure of interlayer X. This assumption is confirmed by XRD analysis: When extending the duration of isothermal curing, the intensity of the main diffraction peak *d*-spacing equals 2.273 nm, and the area increases as well (Figs. 2.7 and 2.16).

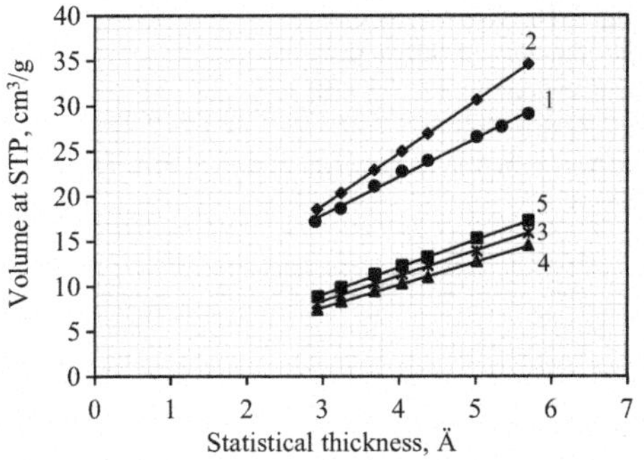

Fig. 2.17 t-plot of synthetic gyrolite when the duration of hydrothermal treatment at 200 °C, h equals: 1—32; 2—48; 3—72; 4—120; 5—168

Effect of hydrothermal synthesis duration on the pore volumes and size distributions of gyrolite

It is fairly complicated to evaluate the existence of micropores in mesoporous materials; thus, in order to determine the absence or presence of micropores in the samples, the t-plot model was used [85, 86, 90, 91]. By adopting this model, the experimental adsorbed volume was plotted versus the thickness (nm) of the adsorbed layer (Fig. 2.17). The Halsey equation was used in a low p/p_0 range of 0.02–0.4 to conduct the calculations of the thickness:

$$t = 3.54\left(\frac{5}{2.303lg\frac{p_0}{p}}\right)^{1/3}, \tag{2.3}$$

where the value of 3.54 represents the thickness of one adsorbed nitrogen layer.

In the t-plot graphs, straight lines with the correlation coefficient $R^2 > 0.99$ were established for all synthetic gyrolite samples (Fig. 2.17). This confirms the formation of a N_2 monolayer on the surfaces of synthetic gyrolite.

The micropore volume can be calculated by using the intercept of the obtained t-plot and the following equation:

$$V_{\text{micro}} = I \cdot 0.0015468\ \left(\text{cm}^3/\text{g}\right) \tag{2.4}$$

The total surface area S_t of all the pores can be calculated by using the intercept of the obtained t-plot and the following equation:

$$S_t = S \cdot 1.547\ \left(\text{m}^2/\text{g}\right) \tag{2.5}$$

The results of the calculations are given in Table 2.5. According to the literature, in the case of the absence of micropores, the value of the specific surface area determined by the BET method should be similar to the calculated S_t value. The calculations showed that gyrolite synthesized after 32 h of synthesis contains some micropores because the difference between the S_{BET} and S_t values was ~16 m^2/g (20%), while the micropore volume was equal to 8.202 × 10^3 mm^3/g (Table 2.5). This data is in good agreement with the high C_{BET} value (Table 2.4) and the presumption that, due to the high amount of water in the interlayer and a low crystallinity, some micropores were formed. It was thus established that the difference between the S_{BET} and S_t values was prominently decreased by prolonging the hydrothermal synthesis duration, and, after 72–168 h, it was lower than 5% (Table 2.5). Thus, it can be stated that, under these synthesis conditions, synthetic gyrolite does not have micropores. This data confirms the previously discussed results that, by prolonging the hydrothermal synthesis duration, the order of the crystal lattice of synthetic gyrolite increases.

As it was shown, synthetic gyrolite is a mesoporous compound. Thus, for the determination of the average diameter of such pores and the total pore volume, it is essential to find out and use the correct shape of the dominant pores. However, the classification of adsorption isotherms and hysteresis loops (Figs. 2.13 and 2.14) involves some ambiguities because both cylindrical-shape and slit-shape (pores between parallel plates) pores can be identified.

It is known that vapor in equilibrium with liquid contained in fine capillaries or pores will have depressed vapor pressures. In fact, if the pore is adequately small in diameter, vapor will condense at pressures far below normal. As the Kelvin equation (Eq. 2.6) indicates, nitrogen gas will condense into all pores with a radius of under 150 nm at a relative pressure of 0.99. By measuring the volume of nitrogen adsorbed at a relative pressure of 0.99 and with the prior knowledge of the surface area, the average pore radius can be calculated. It is worth noting that, due to the metastability of the adsorbed multilayer, delayed condensation occurs in an open-ended pore. Hence, the adsorption branch of the hysteresis loop is not in thermodynamic equilibrium, and it cannot be used for calculations. Meanwhile, the desorption stage is equivalent to reversible liquid–vapor transition because particles/atoms/molecules do not agglomerate during evaporation. As a result, the desorption branch is in thermodynamic equilibrium. Thus, all the calculations were done by using the entire N_2 desorption isotherm at 77 K. The total pore volume and the pore size distribution were determined according to the corrected Kelvin equation and the scheme developed by Orr et al. [85, 86, 90, 91].

Table 2.5 Data of micropores (calculation for synthetic samples)

Duration of synthesis, h	S_{BET}, m^2/g	Slope	Intercept	S_t, m^2g	S_{BET}—S_t, m^2g	S_{BET}—S_t, %	Volume of micropores, (×10^3) mm^3/g
32	81.17	42.07	5.303	65.09	16.08	19.81	8.202
48	95.12	57.79	1.600	89.40	5.72	6.02	2.475
72	44.52	27.89	0.004	43.15	1.37	3.08	0.007
120	39.35	25.17	0.091	38.93	0.42	1.08	0.141
168	47.42	30.11	0.069	46.58	0.84	1.77	0.107

The Kelvin equation relates the adsorbate vapor pressure depression to the radius of a capillary that is filled with the adsorbate:

$$\ln \frac{p}{p_0} = -2\frac{\gamma \cdot V_m \cdot \cos\theta}{R \cdot T \cdot r_k}, \tag{2.6}$$

where p is the saturated vapor pressure in equilibrium with the adsorbate condensed in a capillary or a pore, p_0 is the normal adsorbate saturated vapor pressure, γ is the surface tension of nitrogen at its boiling point, V_m is the molar volume of liquid nitrogen, θ is the wetting angle (usually taken at 0°; hence, $\cos\theta = 1$), R is the gas constant, T is the absolute temperature, and r_k is the Kelvin radius of the pore. When N_2 is used as the adsorbate, the Kelvin equation can be restructured as shown in the following Eq. 2.7:

$$r_k = \frac{4.146}{lg\frac{P_0}{P}} \tag{2.7}$$

Kelvin radius r_k is not the actual pore radius because some adsorption takes place on the wall of the pore prior to the occurrence of condensation inside the pore. During desorption, an adsorbed layer remains on the wall when evaporation is taking place. Therefore, the true pore radius r_p was calculated with the equation $r_p = r_k + t$. t; it was determined by employing Eq. 2.3.

The shape of natural pores is usually non-uniform. Selected geometrical models usually approximate their complicated build-up for calculation purposes. Based on the classification of hysteresis loops (Fig. 2.12), it is presumable that slit-shape or cylindrical-shape pores will dominate in the crystal lattice of synthetic gyrolite (Figs. 2.13 and 2.14). According to these models, if the relative pressure is changed from $(p/p_0)_2$ to $(p/p_0)_1$, then the pores between radii r_2 and r_1 will get emptied ($p_2 < p_1$, $r_2 < r_1$). When p_2 is lowered to p_1, the thickness of the adsorbed film on the previously emptied pores changes from t_2 to t_1. The volume of the liquid (V_L) evaporated due to the emptying of cylindrical pores, and the decrease in the thickness of the adsorbed layers in the previously empty pores was calculated with the following equation:

$$V_L = \pi\left(\overline{r_p} - t_2\right)^2 \cdot L + (t_2 - t_1) \cdot \sum A, \tag{2.8}$$

where $\overline{r_p}$ is the average pore radii in the interval r_2 to r_1, L is the total length of all the pores within the range r_2 to r_1, A is the area of the adsorbed film remaining in the pores after evaporation out of the pores has occurred. The mean volume of pores with $\overline{r_p}$ is

$$V_p = \pi\overline{r_p}^2 L \tag{2.9}$$

Then

$$V_p = \left(\frac{\overline{r_p}}{\overline{r_p} - t_2}\right)^2 \cdot V_L - \left(\frac{\overline{r_p}}{\overline{r_p} - t_2}\right)^2 \cdot (t_2 - t_1) \cdot \sum A \tag{2.10}$$

By using

$$r_p = r_K + t \tag{2.11}$$

$$V_p = \left(\frac{\overline{r_p}}{\overline{r_k}}\right)^2 \cdot \left(V_L - \Delta t \cdot \sum A\right) \tag{2.12}$$

The volume of the liquid desorbed in any interval of the desorption isotherm is related to the volume of gas from the equation: ΔV_L (cm^3) = ΔV_{gas} (1.54 × 10^{-3}). For cylinders, A was calculated as:

$$A\left(m^2\right) = \frac{2\Delta V_p}{\overline{r_p}} \times 10^4 \tag{2.13}$$

For the calculations of slit-shape (pores between parallel plates) pores, the following equations were used:

$$\overline{d_p} = r_k + 2t \tag{2.14}$$

$r_k = l$, where l is the distance between plates:

$$\overline{d_p} = l + 2t \tag{2.15}$$

$$V_p = \frac{\overline{d_p}}{\overline{r_k}} \cdot \left(\Delta V_L - 2\Delta t \sum A\right) \tag{2.16}$$

$$A\left(m^2\right) = \frac{2\Delta V_p}{\overline{d_p}} \times 10^4, \tag{2.17}$$

where $\overline{d_p}$ is the distance between two plates, Å.

The analysis of the desorption isotherms was finished when ($\Delta t \Sigma A$) exceeded ΔV_L because the desorbed gas was coming from the adsorbed layer rather than from the evaporation of the liquid out of the pore center. ΣA is the cumulative surface area which is obtained by summing A for each distance between the plates.

Theoretically, if the shape of pores is uniform, then the S_{BET} value equals the calculated value of ΣA. However, powder materials with various diameters of particles and different pores are obtained during hydrothermal synthesis. For this reason, only dominant pores can be determined in such systems. Therefore, the most suitable model (cylindrical-shape, and pores between parallel plates) is the one whose experimentally defined S_{BET} value is the closest to the calculated specific surface ΣA value. Later on, when using the suitable model, the dominant pore volume, size, and their differential distribution according to the radius can be calculated. The summarized results for ΣA calculations of synthetic gyrolite samples are presented in Table 2.6.

The obtained numerical values and their patterns showed that both models did not fit the dominant pore shape of the gyrolite sample synthesized after 32 h of synthesis (the obtained errors showed >40%) (Table 2.6). This data confirms that, at the beginning of synthesis, gyrolite is poorly crystallized, and the structure of the pores is not well-defined. However, further calculations showed that, after 48 h of synthesis, synthetic gyrolite is characterized by well-defined slit-shape pores formed between parallel planes because the difference between S_{BET} and ΣA is negligible (4.43%) (Table 2.6). This data plausibly suggests the attribution of the hysteresis loop to Type H3 (Fig. 2.13b), which is the characteristic feature of materials with mesoporous pores formed between parallel crystal planes.

It was found out that, by extending the duration of the hydrothermal synthesis of gyrolite from 48 to 168 h and using the model of pores between parallel plates, the difference between the values of S_{BET} and ΣA increased from 4.43 to 71.62% (Table 2.6). Meanwhile, a different tendency was observed when using the cylindrical pore model because the latter difference decreased from 58.45 to 33.99%. Thus, it can be stated that, by prolonging the synthesis duration, the shape of the dominant pores changes from well-defined slit-shape pores formed between parallel planes to cylindrical-shape pores. This data is in good agreement with the changes in hysteresis loops which become much narrower and overlap at a slightly higher pressure p/p_0; i.e. they become more similar to Type H1.

The total pore volume and the pore size distribution of gyrolite samples synthesized after 32 and 48 h were calculated by using the slit-shape pores model; meanwhile, the presently discussed parameters of the samples synthesized after 72 120, and 168 h were calculated by using the cylindrical pores model (Figs. 2.18 and 2.19).

It was also discovered that the total pore volume of gyrolite synthesized after 32 h of treatment was equal to 170 mm^3/g (Fig. 2.18, curve 1). By extending the hydrothermal synthesis duration to 48 h, the total pore volume increased by almost

Table 2.6 Data of ΣA calculations of synthetic samples

Duration of synthesis, h	S_{BET}, m^2/g	Results of cylindrical pores model			Results of slit-shape pores model		
		ΣA, m^2/g	$\lvert S_{BET}-\Sigma A\rvert$, m^2/g	$\lvert S_{BET}-\Sigma A\rvert$, %	ΣA, m^2/g	$\lvert S_{BET}-\Sigma A\rvert$, m^2/g	$\lvert S_{BET}-\Sigma A\rvert$, %
32	81.17	27.62	53.55	65.97	45.93	35.24	43.42
48	95.12	150.72	55.60	58.45	99.33	4.21	4.43
72	44.52	27.78	16.74	37.60	27.83	16.69	37.49
120	39.35	25.44	13.91	35.35	24.98	24.98	63.48
168	47.42	31.30	16.12	33.99	13.46	33.96	71.62

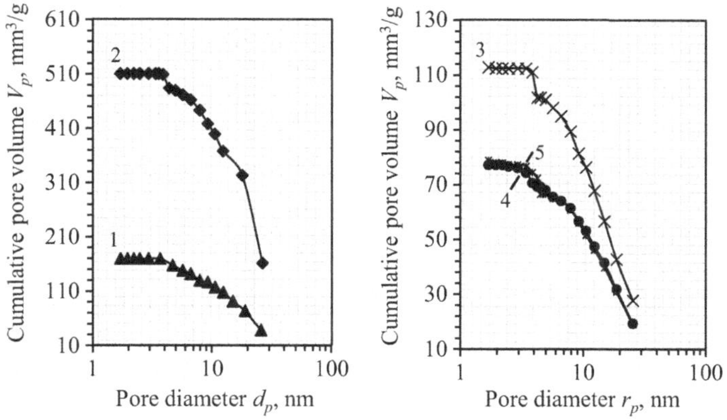

Fig. 2.18 Total pore volume of synthetic gyrolite when the duration of hydrothermal treatment at 200 °C, h equaled: 1—32; 2—48; 3—72; 4—120; 5—168

3 times, i.e. to 509 mm^3/g (Fig. 2.18, curve 1). This increment can be related to the increment of the surface area and the formation of well-defined slit-shape pores between plate particles (Table 2.6). Calculations also yielded the fact that the formation of cylindrical-shape pores leads to a decrement of specific surface area (Table 2.4) and the total pore volume (Fig. 2.18). It was calculated that the total pore volume of gyrolite synthesized for 72 h and 120–168 h decreased to 110 mm^3/g and 80 mm^3/g, respectively (Fig. 2.18, curves 3–5). Furthermore, the research data demonstrated that a stable gyrolite crystal lattice was formed only after 120 h of isothermal curing as the further prolongation of the synthesis duration did not influence the cumulative pore volume.

The differential distributions of the volume pore sizes (Fig. 2.19) highlighted that the pore diameter of synthetic gyrolite varied within the 2–30 nm range under all experimental conditions. It was established that the pores formed in gyrolite obtained after 32 h of synthesis had different sizes (Fig. 2.19a). Meanwhile, after 48 h of synthesis, the dominant distance between parallel plates was equal to 3.5–4.3 nm (Fig. 2.19b). Further calculations showed that the cylindrical-shape pores with a 2.0–3.0 nm radius were predominant in the structure of gyrolite obtained after 72 h of synthesis (Fig. 2.19c). Meanwhile, pores with a bigger radius (3.0–4.0 nm) were dominant after 120 h of hydrothermal treatment (Fig. 2.19d). The pore growth tendency remained when extending isothermal curing (up to 168 h), as the pores with a radius of 3.0–5.0 nm were dominant in the pore distribution by the radius curve (Fig. 2.19e).

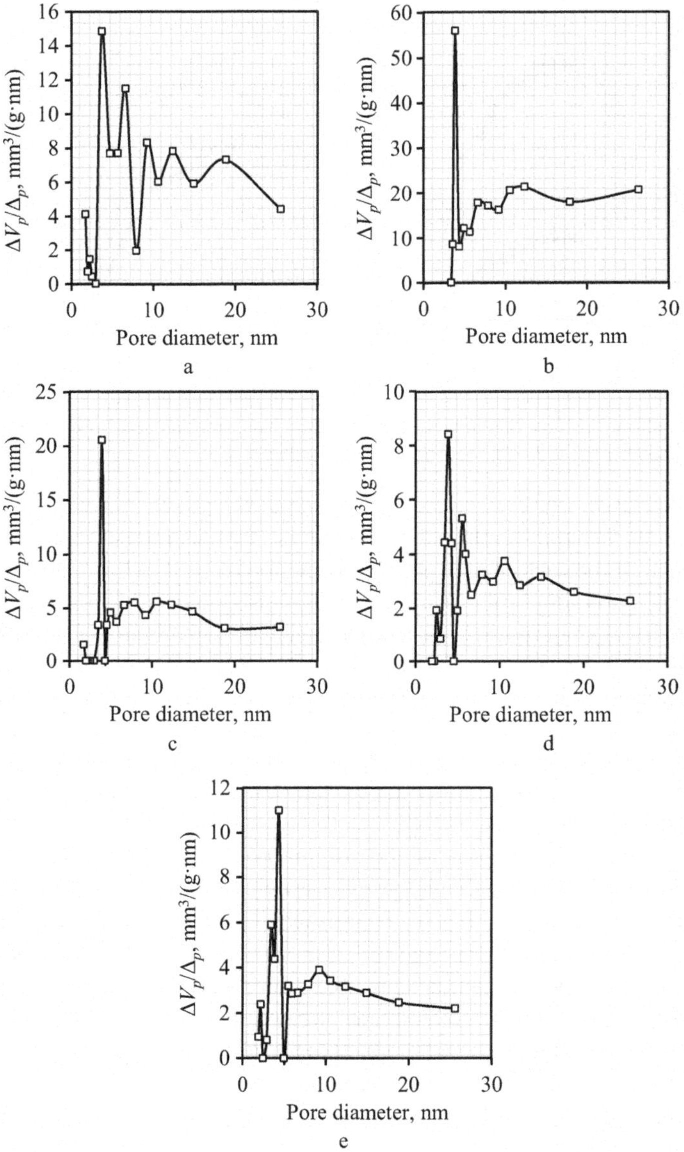

Fig. 2.19 Differential pore volume of synthetic gyrolite when the duration of hydrothermal treatment at 200 °C, h equaled: **a** 32; **b** 48; **c** 72; **d** 120; **e** 168

2.5 Summary

Thus, in the mixtures of calcium oxide and amorphous silicon dioxide, gyrolite is stable till 168 h of hydrothermal treatment at 200 °C because the increment of the intensity and the crystallite size of the main diffraction peak was observed. Meanwhile, by prolonging the duration of hydrothermal treatment to 336 h, gyrolite became metastable and started to recrystallize to another calcium silicate hydrate–truscottite. The results of the experiments were validated with the thermodynamic calculations of the hypothetical crystallization reactions of calcium silicate hydrates in the investigated system (Chap. 3). It was also established that the length of hydrothermal synthesis has an impact on the crystal structure and porosity of gyrolite.

References

1. Dambrauskas T, Baltakys K, Eisinas A, Kitrys S (2018) The specific surface area and porosity of synthetic and calcined α-C_2SH, kilchoanite and hydroxyledgrewite. Powder Technol 355:504–513
2. Anderson T (1851) XV. Description and analysis of gurolite, a new mineral species. Lond Edinb Dublin Philos Mag J Sci 1:111–113
3. Mackay AL, Taylor HFW (1953) Gyrolite. Mineral Mag J Mineral Soc 30:80–92
4. How H (1862) On gyrolite occurring with calcite in apophyllite in the trap of the Bay of Fundy. In: Proceedings of the royal society of Edinburgh 4:426–428
5. Heddle MF (1889) On the crystalline form of gyrolite. Mineral Mag 8:272–273
6. Heddle MF (1891) On the optic properties of gyrolite. Mineral Mag 9:391
7. Heddle MF (1889) On the occurrence of gyrolite in India. Mineral Mag 8:199
8. Clarke FW (1889) ART. XV. A new occurrence of gyrolite. Am J Sci 38:128
9. Currie J (1905) Note on some new localities for gyrolite and tobermorite. Mineral Mag J Mineral Soc 14:93–95
10. Fleischmann F (1910) On the occurrence of gyrolite in County Antrim. Mineral Mag J Mineral Soc 15:288–293
11. Cristie WAK (1925) Gyrolite and okenite from Bombay. J Geol Soc India 56:199–203
12. Flint EP, McMurdie HF, Wells LS (1938) Formation of hydrated calcium silicates at elevated temperatures and pressures. J Res Natl Bur Stand 21:617–638
13. Seager AF (1971) Mineralisation and paragenesis at Dean quarry, the Lizard, Cornwall. Trans Roy Geol Soc Corn 20:97–113
14. Elton NJ, Hooper JJ, Holyer VAD (1998) Al-free gyrolite from the Lizard, Cornwall, England. Mineral Mag 62(2):271–272
15. Ottens B (2003) Minerals of the Deccan traps, India. Mineral Rec 34:5–82
16. Merlino S (1988) Gyrolite: its crystal structure and crystal chemistry. Mineral Mag 52(366):377–387
17. Stevula L, Harman M, Horvath I, Putyera K (1990) The mineral gyrolite and its stability under hydrothermal conditions. Ceram Silik 34:315–330
18. Mamedov KS, Belov NV (1958) Crystal structure of mica-like hydrous calcium silicates: okenite, nekoite, truscottite, and gyrolite: new silicate radical [Si_6O_{15}]. Dokl Akad Nauk SSSR 121:713–716
19. Cann JR (1965) Gyrolite and reyerite from S Airde Beinn, northern Mull. Mineral Mag J Mineral Soc 35:1–4
20. Gottardi G, Passaglia E (1966) Tobermorite "non espandibile" e Gyrolite del Monte Biaena (Trento). Period Mineral 36:1079–1083

21. Garavelli CL, Vurro F (1984) Gyrolite from Ortano (Island of Elba). Rend Soc Ital Mineral Petrol 39:695–704
22. Koritnig S (1972) Gyrolite, okenite and tacharanite metasomatic products of reaction from the basalt occurrence of the Bramburg near Goettingen, Germany. Contrib Mineral Petrol 35:293–301
23. Meyer JW, Jaunarajs KL (1961) Synthesis and crystal chemistry of gyrolite and reyerite. Am Mineral 46:913–933
24. Różycka A, Kotwica Ł (2020) Effect of alkali on the synthesis of single phase gyrolite in the system CaO-quartz-H_2O. Constr Build Mater 239:117799
25. Okada Y, Masuda T, Ishida H (1995) Behavior of silicate anions on the formation processes of gyrolite followed by ^{29}Si NMR. J Ceram Soc Jpn 103:124–127
26. Eisinas A, Dambrauskas T (2023) Cu^{2+} ions adsorption by dense calcium silicate hydrate samples and their catalytic activity. Ceram Int 49:14886–14894
27. Eisinas A, Dambrauskas T, Eisinaite V, Sipailiene A (2022) Synthesis of antibacterially active wollastonite by using dense calcium silicate hydrate samples. J Therm Anal Calorim 147:14163–14174
28. Krakowiak KJ, Thomas JJ, James S, Abuhaikal M, Ulm F (2018) Development of silica-enriched cement-based materials with improved aging resistance for application in high-temperature environments. Cem Concr Res 105:91–110
29. Yamaguchi N, Kanda T, Fujimori H, Ioku K, Goto S (2001) Gyrolite fine particles synthesized from glass in the system of CaO-SiO_2 by hydrothermal reaction. J Mater Sci Lett 20:489–490
30. Gong WL, Wang LM, Ewing RC, Vernaz E, Bates JK, Ebert WL (1998) Analytical electron microscopy study of surface layers formed on the French SON68 nuclear waste glass during vapor hydration at 200 °C. J Nucl Mater 254:249–265
31. Shaw S, Henderson CMB, Clark SM (2002) In-situ synchrotron study of the kinetics, thermodynamics, and reaction mechanisms of the hydrothermal crystallization of gyrolite, $Ca_{16}Si_{24}O_{60}(OH)_8{\cdot}14H_2O$. Am Mineral 87:533–541
32. Garbev K, Black L, Stumm A, Stemmermann P, Gasharova B (2004) Polymerization reactions by thermal treatment of gyrolite-group minerals—an IR spectroscopic and X-ray diffraction study based on synchrotron radiation. In: Proceedings of the 8th international congress on applied mineralogy, vol 1. ICAM, pp 245–248
33. Dana JD, Dana ES, Gaines RV, Skinner HCW, Foord EE, Mason B, Rosenzweig A (1997) Dana's new mineralogy: the system of mineralogy of James Dwight Dana and Edward Salisbury Dana. Wiley-Interscience
34. Dana JD (2022) A system of mineralogy. BoD–Books on Demand
35. Baltušnikas A, Šiaučiūnas R, Lukošiūtė I, Baltakys K, Eisinas A, Kriūkienė R (2015) Crystal structure refinement of synthetic pure gyrolite. Medziagotyre 21:111–116
36. Strunz H, Micheelsen H (1958) Calcium-phyllosilikate. Naturwissenschaften 45:515
37. Eberhard E, Rahman SH (1982) The principle of the crystal structure of gyrolite $Ca_{13}[Si_8O_{20}]_3(OH)_2{\cdot}22H_2O$. Z Kristallogr 159:34–36
38. Baltušnikas A, Lukošiūtė I, Levinskas R, Grybėnas A, Baltakys K, Eisinas A (2012) Analysis of Rietveld method application for gyrolite crystal structure refinement. Medziagotyra 18:379–384
39. Levinskas R, Baltušnikas A, Lukošiūtė I, Baltakys K, Kalpokaitė-Dičkuvienė R, Grybėnas A (2013) Modification of structure of synthetic gyrolite. Mater Res Innov 17:495–500
40. Ferraris G, Pavese A, Soboleva SV (1995) Tungusite: new data, relationship with gyrolite and structural model. Mineral Mag 59:535–543
41. Garbev K, Black L, Beuchle G, Stemmermann P (2002) Inorganic polymers in cement based materials. Wasser-und Geotechnologie 1:19–30
42. Giacovazzo C (2002) Fundamentals of crystallography. Oxford University Press
43. Bailey SW (2018) Hydrous phyllosilicates:(exclusive of micas). Walter de Gruyter GmbH & Co KG
44. Różycka A, Kotwica Ł, Małolepszy J (2014) Synthesis of single phase gyrolite in the CaO-quartz-Na_2O-H_2O system. Mater Lett 120:166–169

45. Jander W, Franke B (1941) Die Bildung von Calciumhydrosilikaten aus Calciumoxyd und Kieselsäuregel bei 300° und 350° und hohen Drucken. III. Mitteilung über hydrothermale Reaktionen. Z Anorg Allg Chem 247:161–179
46. Speakman K (1968) The stability of tobermorite in the system CaO-SiO_2-H_2O at elevated temperatures and pressures. Mineral Mag J Mineral Soc 36:1090–1103
47. Merlino S (1972) New tetrahedral sheets in reyerite. Nat Phys Sci 238:124–125
48. Mitsuda T, Kobayakawa S, Toraya H (1986) Characterization of hydrothermally formed CSH. In: The 8th international congress on the chemistry of cement, vol 3. Rio de Janeiro, p 176
49. Babushkin VI, Matveev GM, Mchedlov-Petrosyan OP (1962) Thermodynamics of silicates. Gosstoiizdat, Moscow
50. Mchedlov-Petrosyan OP (1967) Thermodynamics and thermochemistry of cement. In: 6th international congress on the chemistry of cement 1967. Moscow, p 6–16
51. Blanc P, Bourbon X, Lassin A, Gaucher EC (2010) Chemical model for cement-based materials: temperature dependence of thermodynamic functions for nanocrystalline and crystalline C–S–H phases. Cem Concr Res 40:851–866
52. Roy DM, Harker RI (1960) Chemistry of cement: proceedings of the fourth international symposium, Washington, vol 1, pp 196–201
53. Harker RI (1964) Dehydration series in the system $CaSiO_3$–SiO_2–H_2O. J Am Ceram Soc 47:521–529
54. Taylor HFW (1964) The chemistry of cements. Academic Press
55. Kalousek GL, Nelson EB (1978) Hydrothermal reactions of dicalcium silicate and silica. Cem Concr Res 8:283–289
56. Števula L, Petrović J (1983) Formation of an intermediate C-S-H phase during the hydrothermal synthesis of gyrolite. Cem Concr Res 13:684–688
57. Chan CF, Mitsuda T (1978) Formation of 11 Å tobermorite from mixtures of lime and colloidal silica with quartz. Cem Concr Res 8:135–138
58. Hara N, Inoue N (1980) Thermal behaviour of 11Å tobermorite and its lattice parameters. Cem Concr Res 10:53–60
59. Moertel H (1980) Mineralbestand, Gefüge und physikalische Eigenschaften von Kalksandsteinen. Fortschr Mineral 58:37–67
60. Jauberthie R (1985) Etude comparée des tobermorites à 10 et à 11 Å obtenues par synthèse directe. Comptes-rendus des séances de l'Académie des sciences. Série 2, Mécanique-physique, chimie, sciences de l'univers, sciences de la terre 300:341–344
61. Jauberthie R, Temimi M, Laquerbe M (1996) Hydrothermal transformation of tobermorite gel to 10 Å tobermorite. Cem Concr Res 26:1335–1339
62. Esteoule J, Jauberthie R (1987) La genèse de la gyrolite. Comptes rendus de l'Académie des sciences. Série 2, Mécanique, Physique, Chimie, Sciences de l'univers, Sciences de la Terre 304:349–354
63. Esteoule J, Jauberthie R (1984) Synthèse directe de la tobermorite à 10Å par réaction de la chaux sur de la silice amorphe. Comptes-rendus des séances de l'Académie des sciences. Série 2, Mécanique-physique, chimie, sciences de l'univers, sciences de la terre 298:207–210
64. Shaw S, Clark SM, Henderson CMB (2000) Hydrothermal formation of the calcium silicate hydrates, tobermorite ($Ca_5Si_6O_{16}(OH)_2{\cdot}4H_2O$) and xonotlite ($Ca_6Si_6O_{17}(OH)_2$): an in-situ synchrotron study. Chem Geol 167:129–140
65. Shaw S, Henderson C, Komanschek BU (2000) Dehydration/recrystallization mechanisms, energetics, and kinetics of hydrated calcium silicate minerals: an in situ TGA/DSC and synchrotron radiation SAXS/WAXS study. Chem Geol 167:141–159
66. Gard JA, Mitsuda T, Taylor HFW (1975) Some observations on Assarsson's Z-phase and its structural relations to gyrolite, truscottite, and reyerite. Mineral Mag 40:325–333
67. Gundlach H (1973) Dampfgehärtete Baustoffe. Bauverl
68. Siauciunas R, Baltakys K (2004) Formation of gyrolite during hydrothermal synthesis in the mixtures of CaO and amorphous SiO_2 or quartz. Cem Concr Res 34:2029–2036
69. De Ferri L, Lottici PP, Vezzalini G (2014) Characterization of alteration phases on Potash–Lime–Silica glass. Corros Sci 80:434–441

70. Smalakys G, Siauciunas R (2020) Peculiarities of xonotlite synthesis from the raw materials with different SiO_2 activities. J Therm Anal Calorim 142:1671–1679
71. Meller N, Hall C, Kyritsis K, Giriat G (2007) Synthesis of cement based $CaO–Al_2O_3–SiO_2–H_2O$ (CASH) hydroceramics at 200 and 250 °C: ex-situ and in-situ diffraction. Cem Concr Res 37:823–833
72. Luke K (2004) Phase studies of pozzolanic stabilized calcium silicate hydrates at 180 °C. Cem Concr Res 34:1725–1732
73. Yang Z, Kang D, Zhang D, Yan C, Zhang J (2021) Crystal transformation of calcium silicate minerals synthesized by calcium silicate slag and silica fume with increase of C/S molar ratio. J Mater Res Technol 15:4185–4192
74. Kyritsis K, Hall C, Bentz DP, Meller N, Wilson MA (2009) Relationship between engineering properties, mineralogy, and microstructure in cement-based hydroceramic materials cured at 200°–350°C. J Am Ceram Soc 92:694–701
75. Meller N, Kyritsis K, Hall C (2009) The mineralogy of the $CaO–Al_2O_3–SiO_2–H_2O$ (CASH) hydroceramic system from 200 to 350 °C. Cem Concr Res 39:45–53
76. Meller N, Hall C, Phipps JS (2005) A new phase diagram for the CaO-Al_2O_3-SiO_2-H2O hydroceramic system at 200 °C. Mater Res Bull 40:715–723
77. Lachowski EE, Murray LW, Taylor HFW (1979) Truscottite: composition and ionic substitutions. Mineral Mag 43:333–336
78. Garbev K, Gasharova B, Stumm A, Black L, Mathis YL, Stemmermann P (2004) Phase transition of gyrolite into truscottite upon thermal treatment as studied at the ANKA infrared and XRD beamlines. Mineralogische und Technische Kristallographie 21:118
79. Choma J, Jagiello J, Jaroniec M (2021) Assessing the contribution of micropores and mesopores from nitrogen adsorption on nanoporous carbons: application to pore size analysis. Carbon 183:150–157
80. Rouquerol J, Avnir D, Fairbridge CW, Everett DH, Haynes JM, Pernicone N et al (1994) Recommendations for the characterization of porous solids (Technical Report). Pure Appl Chem 66:1739–1758
81. Khan HA, Kim S, Jung K (2020) Origin of high stability of Pt/anatase-TiO_2 catalyst in sulfuric acid decomposition for SI cycle to produce hydrogen. Catal Today 352:316–322
82. Saidian M, Godinez LJ, Prasad M (2016) Effect of clay and organic matter on nitrogen adsorption specific surface area and cation exchange capacity in shales (mudrocks). J Nat Gas Sci Eng 33:1095–1106
83. Lapham DP, Lapham JL (2017) Gas adsorption on commercial magnesium stearate: effects of degassing conditions on nitrogen BET surface area and isotherm characteristics. Int J Pharm 530:364–376
84. Arvaniti EC, Juenger MC, Bernal SA, Duchesne J, Courard L, Leroy S et al (2015) Determination of particle size, surface area, and shape of supplementary cementitious materials by different techniques. Mater Struct 48:3687–3701
85. Rouquerol J, Rouquerol F, Llewellyn P, Maurin G, Sing KS (2013) Adsorption by powders and porous solids: principles, methodology and applications. Academic Press
86. Naderi M (2015) Surface area: brunauer–emmett–teller (BET). In: Progress in filtration and separation. Academic Press, pp 585–608
87. Thommes M, Kaneko K, Neimark AV, Olivier JP, Rodriguez-Reinoso F, Rouquerol J et al (2015) Physisorption of gases, with special reference to the evaluation of surface area and pore size distribution (IUPAC Technical Report). Pure Appl Chem 87:1051–1069
88. Rudzinski W, Everett DH (2012) Adsorption of gases on heterogeneous surfaces. Academic Press
89. Keller JU, Staudt R (2005) Gas adsorption equilibria: experimental methods and adsorptive isotherms. Springer Science & Business Media
90. Webb PA, Orr C (1997) Analytical methods in fine particle technology. Micromeritics Instrument Corporation
91. Lowell S, Shields JE, Thomas MA, Thommes M (2006) Characterization of porous solids and powders: surface area, pore size and density. Springer Science & Business Media

92. Gregg SJ, Sing KSW (1982) Adsorption, surface area and porosity. Academic Press
93. García-Tejeda YV, Barrera-Figueroa V (2019) Least squares fitting-polynomials for determining inflection points in adsorption isotherms of spray-dried açaí juice (Euterpe oleracea Mart.) and soy sauce powders. Powder Technol 342:829–839
94. Zhou L, Bai S, Zhou Y, Yang B (2002) Adsorption of nitrogen on silica gel over a large range of temperatures. Adsorption 8:79–87
95. Storck S, Bretinger H, Maier WF (1998) Characterization of micro- and mesoporous solids by physisorption methods and pore-size analysis. Appl Catal A Gen 174:137–146
96. Stumm A, Garbev K, Beuchle G, Black L, Stemmermann P, Nüesch R (2005) Incorporation of zinc into calcium silicate hydrates, Part I: formation of C-S-H(I) XE "C-S-H(I)" with C/S=2/3 and its isochemical counterpart gyrolite. Cem Concr Res 35:1665–1675

Chapter 3
Solid Waste. Application of AlF_3 Production Waste to Gyrolite Synthesis

The third chapter of the monograph begins with a literature review about the influence of foreign ions on the formation of synthetic gyrolite. As the next step, a detailed description of solid waste—silica gel contaminated with aluminum and fluorine ions—is presented. This waste is usually landfilled, which may lead to serious environmental issues. The application of silica gel waste is limited because it contains ~10 wt% of F^- ions; however, the high concentration of SiO_2 (78.9 wt%) makes this waste a potential source of silica for the synthesis of silicates. Finally, the synthesis of gyrolite was performed by changing the molar ratio of the initial mixture and raw materials. The results of the experimental part and the thermodynamic calculations of hypothetical crystallization reactions of calcium silicate hydrates in the investigated system showed that the formation of gyrolite family compounds under hydrothermal synthesis conditions proceeds according to the following sequence: $0.66CaO + SiO_2 + H_2O \rightarrow$ C–S–H(I) $\rightarrow$ *Z*-phase $\rightarrow$ gyrolite gel $\rightarrow$ gyrolite $\rightarrow$ truscottite. Also, it was determined that silica gel waste is a prospective SiO_2 source for the synthesis of low-base calcium silicate hydrates because the formation of gyrolite is two times shorter in comparison with pure mixtures. Meanwhile, calcium nitrate is not recommended as a source of calcium because, during synthesis, nitrate anions intercalate into the structure of gyrolite. As a result, the porosity of gyrolite decreases, which can negatively affect the properties of synthetic gyrolite.

3.1 Influence of Additives on the Synthesis and Structure Properties of Synthetic Gyrolite

As outlined in Chap. 2, both sodium and aluminum ions are usually present in natural gyrolite; however, there is no sufficient data about the impact of Al_2O_3 and Na_2O additives on the synthesis of gyrolite (compared to 1.13 nm tobermorite). It is known that the formation sequence, stability, and properties of calcium silicate hydrates are

K. Baltakys and T. Dambrauskas, *Synthetic Gyrolite*,
https://doi.org/10.1007/978-3-031-49259-4_3

changed by the presence of these additives [1–6]. The effectiveness of each additive is caused by the composition of the primary mixture, the nature and properties of the raw materials, the synthesis conditions, as well as some other factors. According to the literature, additives containing alkali metals Na^+, K^+ [2, 3], aluminum [1, 3–6], and other elements [3, 7] exert the most significant effect on the formation and properties of CSH. The amount of Na^+ ions that can intercalate into the structure of calcium silicate hydrates is higher in lower basicity CSH than in dibasic CSH [2, 3, 8, 9].

Therefore, the C/S molar ratio of the formed calcium silicate hydrates is affected more by alkaline additives than by temperature changes. Even a modest amount of NaOH noticeably impacts the interaction between CaO and SiO_2 under hydrothermal conditions because, whenever employed, this additive increases the solubility of SiO_2, whereas CaO decreases its solubility [10, 11]. Moreover, the duration of the hydrothermal synthesis can be considerably reduced by changing the concentration of Na^+ ions. In 1974, Blakeman et al. determined that, in mixtures with a constant C/S molar ratio of 0.83, the mineral composition of the formed compounds [C–S–H(I), 1.13 nm tobermorite, pectolite, xonotlite] depends on the quantity of NaOH [12].

The distance between $[SiO_4]^{4-}$ tetrahedra chains in the gyrolite structure is equal to 0.03 nm, which is sufficient to intercalate other ions by maintaining the neutral charge of the crystal lattice (Fig. 3.1) [13–15]. For example, when Si^{4+} ions are replaced by Na^+ with Al^{3+} ions, the crystal lattice of gyrolite remains neutral [16]. Therefore, after the incorporation of Na^+ ions, they can easily participate in ion exchange reactions with other monovalent and divalent cations which are present in the liquid medium.

Miyake et al. [16] successfully synthesized gyrolite with intercalated aluminum and sodium ions [$Ca_8Si_{11.32}Al_{0.68}Na_{0.44}O_{30}(OH)_4{\cdot}6.6H_2O$] and applied it for ion exchange reactions. Since synthetic gyrolite features a smaller amount of Na^+ ions than substituted Al^{3+} ions, it was presumed that Ca^{2+} ions would compensate for the deficiency in order to keep the charge neutrality. The authors determined that the main diffraction peak of synthetic gyrolite shifts toward larger angles when Al^{3+} ions

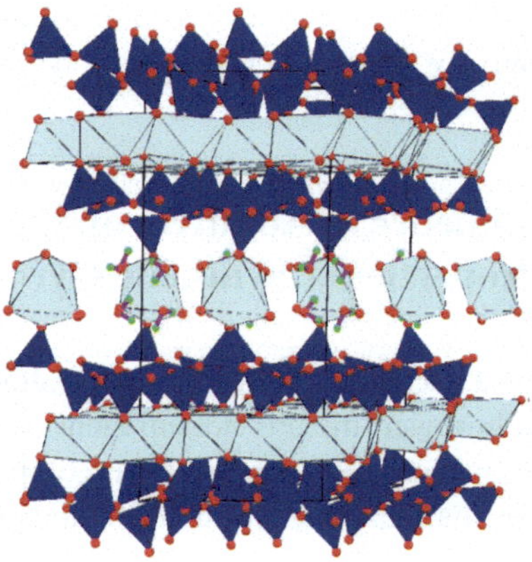

Fig. 3.1 Crystal structure of pure synthetic gyrolite determined by Baltušnikas et al. [13]. Reprinted from *Medziagotyra*, 2015, Vol. 21, Baltušnikas et al., Crystal Structure Refinement of Synthetic Pure Gyrolite, pp. 111–116. Open access under a CC BY 4.0 license, https://creativecommons.org/licenses/by/4.0/

intercalate into the structure of synthetic gyrolite, but it remains constant during ion exchange reactions.

Nocun-Wczelik et al. [10, 17, 18] investigated the impact of Na^+ and Al^{3+} ions containing additives on the formation of CSH under hydrothermal conditions and its properties. These scholars discovered that, at the beginning of the synthesis process, Na^+ ions accelerate the formation of semicrystalline CSH; however, further crystallization reactions depend on the amount of the sodium additive. After hydrothermal treatment of CaO and SiO_2 mixtures with C/S = 1.0, xonotlite usually dominates in the products. However, if over 20 wt% Na_2O is added to the mixture, gyrolite is formed after 24 h of hydrothermal treatment at 200 °C. Meanwhile, the crystallization of gyrolite in mixtures with lower C/S ratios (0.67 or less) proceeds more rapidly in pure mixtures than in the mixtures containing Na_2O additive (5–10 wt%).

Baltakys and Siauciunas [19] also investigated the influence of various parameters of hydrothermal synthesis on the formation of gyrolite in mixtures with the molar ratio C/(S + Al) = 0.66 and Al/(S + A) = 0.025. The authors determined that in the cases when calcium oxide and amorphous silica are used as raw materials, at 150 °C temperature, the aluminum oxide additive inhibits the formation of Z-phase, but, instead, promotes the formation of 1.13 nm tobermorite with substituted Al^{3+} ions. The latter compound was identified under all experimental conditions. Further research showed that, in pure mixtures with quartz, gyrolite does not form even at 200 °C (Table 3.1) [20]. These researchers established that Na_2O additive positively impacts the formation of gyrolite because the duration of hydrothermal synthesis can be decreased by 50–67%. However, the formed gyrolite is not stable, and it recrystallizes into pectolite [$Ca_2NaH(SiO_3)_3$]. Synthetic gyrolite can be stabilized by using a complex additive (Na_2O + γ-Al_2O_3) which prevents its transformation to pectolite.

Shaw [21] used high-energy synchrotron X-ray radiation to investigate the early stage formation of gyrolite at 200 °C temperature. This scientist determined that, in mixtures with NaOH additive, C–S–H gel formed after 4 min of isothermal curing. When hydrothermal synthesis was extended to 140 and 190 min, diffraction peaks characteristic of gyrolite and pectolite, respectively, were identified (Fig. 3.2). It is worth noting that gyrolite formation proceeds not through the intermediate compound—Z-phase—but directly from C–S–H gel.

Finally, Różycka et al. [22, 23] successfully synthesized single-phase gyrolite by using quartz as a siliceous material. This team of researchers established that the amount of NaOH in the CaO–SiO_2–Na_2O–H_2O system exerts crucial influence on the formation of gyrolite. The optimum amount of NaOH additive varied between 0.75 and 1.00 wt%. In other cases, unreacted quartz was observed in the final products.

It should be noted that not only the above-discussed aluminum and sodium ions but also some other ions affect the formation and properties of gyrolite. Stumm et al. [24] indicated that, during mechanochemical-hydrothermal synthesis at 210 °C temperature, it is possible to incorporate up to Zn/(Zn + Ca) = 1/6 (roughly 6 wt%) zinc ions into the structure of gyrolite. It was determined that gyrolite was formed, although quartz did not react completely when the Zn/(Zn + Ca) molar ratio was

Table 3.1 Formation of calcium silicate hydrates at 200 °C according to Baltakys and Siauciunas [20]

Initial mixture	Duration of synthesis, h	p	Q	α	C	k	T	Xo	P	G	Z
CaO–quartz–H_2O	4–32		●		●		●	●			
	48–168		●				●	●			
CaO–quartz–Al_2O_3–H_2O	4–16	●	●	●	●	●	●				
	24		●	●	●	●	●	●			
	32		●		●		●	●			
	48–72		●		●		●	●			
	168		●		●		●				
CaO–quartz–Na_2O–H_2O	4–8		●		●						
	16		●		●					●	●
	24		●						●	●	●
	32–168								●	●	
CaO–quartz–Al_2O_3–Na_2O–H_2O	4–8		●		●		●				
	16–24		●		●		●			●	
	32				●		●			●	
	48–168						●			●	

Indexes p—portlandite; Q—quartz; α—α-C_2SH; C—C–S–H(I); k—calcium carbonate; T—tobermorite; Xo—xonotlite; P—pectolite; G—gyrolite; Z—Z-phase

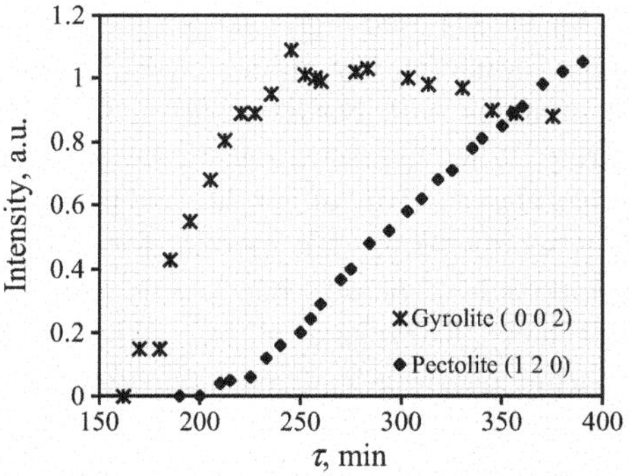

Fig. 3.2 Dependence of diffraction peaks intensity of gyrolite (0 0 2) and pectolite (1 2 0) on the duration of hydrothermal synthesis

higher than 0.02. What concerns nanocrystalline phases, the increment of the zinc content leads to gradual lowering of the main diffraction peak of synthetic gyrolite.

Baltakys and Siauciunas [25] investigated the impact of a gypsum additive on the formation and structure of gyrolite. Sulfate ions promote the formation of C–S–H(I) and Z-phase in a low-basicity mixture, and, only after 72 h at 200 °C, high crystallinity gyrolite is formed. The amount of sulfur intercalating into the crystal lattice of synthetic gyrolite depends on the synthesis conditions as well as on the composition of the primary mixture; when the molar concentration of $SO_4^{2-}/(SiO_2 + SO_4^{2-})$ in the initial mixture increases from 0.025 to 0.1, the amount of sulfur intercalated into synthetic gyrolite increases from 0.1 to 0.35 wt%. These authors proposed calculating the composition of the initial mixtures according to the molar ratios of the primary compounds. In other scenarios, upon enhancing the amount of sulfate ions, the basicity level of the mixture is lowered, and gyrolite does not form that easily. Also, it was determined that intercalated sulfur ions do not affect the thermal conversion of C–S–H(I), *Z*-phase, and gyrolite into wollastonite.

Nocun-Wczelik [26] investigated the influence of $CaCl_2$, $AlCl_3$, $CrCl_3$, Na_2CrO_4, NaOH, and $Al(OH)_3$ additives on the formation of gyrolite at 200 °C temperature. The obtained results showed that, regardless of the used additives, gyrolite did not form, and the main product of the synthesis was semicrystalline CSH. According to the author, the main reason for such results was the low activity of raw materials containing CaO and SiO_2 components.

Siauciunas and Baltakys [27] determined the influence of chromium (VI) ions (1, 2, and 5 wt%) on the formation of gyrolite at 200 °C temperature. They established that the used additive promotes the crystallization process of CSH under all experimental conditions. In the mixtures with Cr^{6+} ions and under the molar ratio of C/S = 0.66, the formation of gyrolite proceeds two times faster than in a pure mixture. Meanwhile, in the higher basicity mixtures (C/S = 0.83), the formation of CSH strongly depends on the content of additives. For example, when the content of CrO_3 in the mixture was equal to 1 or 2 wt%, 1.13 nm tobermorite and xonotlite remained stable till 72 h of hydrothermal treatment; however, when the content of the additive was increased to 5 wt%, tobermorite started to get recrystallized into gyrolite.

Many researchers have contributed to the determination of the synthesis conditions of gyrolite as well as the structure, stability, and properties of synthetic gyrolite [13, 16, 23, 28–34]. These researches have shown that the formation process of gyrolite is determined by a variety of factors, such as the type of lime and silica which is used, the molar ratio of CaO and SiO_2, the curing conditions, the surface area of materials, the w/s ratio, the nature of the used additives, etc. Even though the hydrothermal synthesis of gyrolite is a lengthy and intricate process, it offers a great feasibility to be successfully used in many industrial areas. It should be highlighted that various solid waste containing SiO_2 can be used to synthesize calcium silicate hydrates. By using waste for the synthesis, natural sources are saved, and the landfill problems can be reduced. Thus, in the next chapter of this work, the formation of AlF_3 production waste (a potential source of SiO_2) and its properties shall be discussed.

3.2 Formation of AlF_3 Production Waste, Its Application, and Properties

The growing needs of people lead to an increased use of natural resources and the generation of waste which causes air, water, and soil pollution (Fig. 3.3) [35–37]. Waste is generated at all stages of the production process (raw material extraction, preparation, production, and even packaging of the final product), which consequently makes recycling or reuse a huge problem worldwide.

Every year, in EU alone, more than 2.15 billion tons of waste is generated by economic activities and in the households [38]. The most significant part of waste consists of mineral waste (construction, mining and quarrying), which accounts for ~50.5 wt%, followed by production waste (~11 wt%), waste management and water treatment (~11 wt%), and household waste (~9.5 wt%) (Fig. 3.4). It is estimated that around 2.03 billion tons of the waste are processed: ~60 wt% of it is recovered, whereas ~40 wt% of the waste is disposed of in landfill sites or in some other ways. It should be noted that the outcome of recycling, the environmental impact of the waste, and the storage of specific waste groups in landfills cannot be assessed unequivocally. For example, mineral waste can be used for landscaping [39, 40], wood or tire waste is suitable for energy production [41, 42], etc.; however, the reuse of industrial (production) and environmentally hazardous waste is limited by the variability in its chemical and mineral composition as well as by its physical and chemical properties [43–46]. In addition, the reuse of the presently discussed waste usually complicates the manufacturing process, the properties of the products deteriorate, or the improvement is negligible, and the product price is subsequently elevated. Finally, it is important to avoid the formation of third-generation waste. Therefore, as environmental issues must be mitigated and/or completely resolved while still pursuing economic benefits, technologies for turning waste materials into commercial products are in great demand.

Fig. 3.3 Soil and water pollution (Photo ID: 1398960038, *Source* iStock by Getty Images)

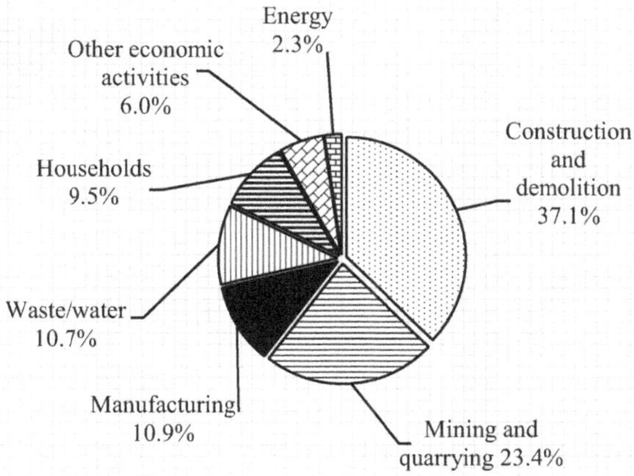

Fig. 3.4 Waste generation by economic activities and households in the EU in 2020

3.2.1 Solid Wastes and Their Utilization

Solid waste is usually perceived as all the unwanted or useless solid materials generated by all human activities (commercial and industrial sectors, households, etc.). These types of waste can be classified according to their origin (industrial, commercial, recycling, construction, institutional, or households); according to their content (inorganic or organic materials, polymers, glass, ceramic, metal, plastic, paper, etc.); or according to its hazard potential (toxic or non-toxic, chemical, radioactive, flammable, biomedical, etc.). From the ecological point of view and based on the waste management hierarchy, it is most favorable when the generation of solid waste is avoided completely (Fig. 3.5) [47, 48]. Still, most of the materials or products sooner or later turn into waste that should be recycled [39–46]. The next alternative is to reuse the solid waste in other production processes as a raw material, or to incinerate it to obtain energy [47, 48]. It should be noted that incineration should be avoided because products are released into the environment during their combustion, and ash containing hazardous elements is formed. Landfilling is the most avoidable and the cheapest solution; thus, it is widely used for waste storage [38, 49]. However, landfills extend over vast territories where, due to the high concentration of noxious materials, air, water (groundwater, surface water), and soil pollution originates, thus leading to a wide array of environmental issues [49–51]. That is why the long-term goal of waste management is to move to the circular economy in which one of the main objectives is converting waste into raw materials or commercially valuable products (Fig. 3.5).

Unfortunately, a large part of solid waste is still landfilled or discharged into the water without sufficient treatment. Thus, one of the EU's priorities is the transition toward a circular economy (Fig. 3.6), while the *EU Waste Framework Directive* (2008/98/EC) sets the basic concepts and definitions of waste management [47]. The

Fig. 3.5 Waste hierarchy

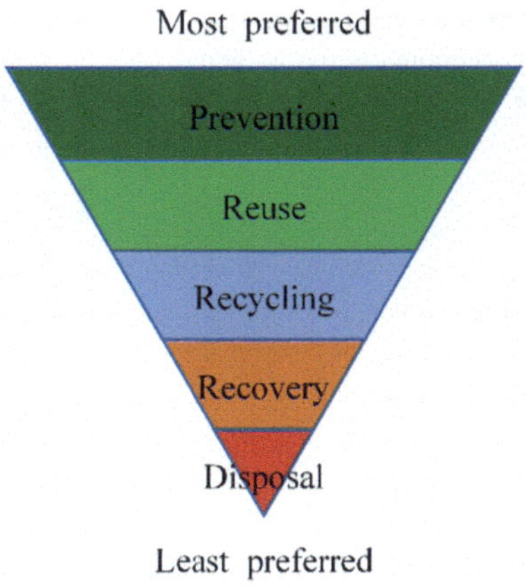

objective of the Directive is that waste must be managed without a negative impact on the health of humans and other living organisms, and that the environment cannot be harmed. The directive proclaims that producers are responsible for their waste, and the polluter should pay for it. More recently, in 2015, the EU Commission launched the first Circular Economy Package which has been constantly updated since then [52, 53]. The revised legislative proposal on waste includes some elements of waste reduction: 60 wt% of municipal and 70 wt% of packaging waste should be recycled by 2030; in 2030, landfilling of municipal waste should be less than 10 wt%; the reduction of landfills is promoted, and the reuse of by-products along with the introduction of greener products in the market is encouraged [47, 52–54]. Furthermore, by emphasizing legislation and considering the negative impact on the environment, it is recommended to change the manufacturing process by reducing by-products, reusing the waste in other productions as a raw material, and by adopting efficient waste treatment technologies. Finally, the EU has issued itself a challenge of achieving climate neutrality by 2050.

Therefore, there is a great possibility for setting up new factories or manufacturing lines for recycling and reusing solid waste. It is estimated that the application of the circular economy principles across the EU economy has the potential to increase EU GDP by an additional 0.5% by 2030 [55] and reduce greenhouse gas emissions by up to 4% [56]. In addition, recycling and reusing would avoid other environmental challenges, such as landscape degradation due to landfills and water and air pollution. The production of environmentally friendly construction materials is one of the most prominent fields where huge quantities of solid wastes could be reused [57–59]. Such construction materials offer a good potential to fulfill people's needs in urban and rural areas. The next great area where solid waste can be utilized is the synthesis of adsorbents, zeolites, and other inorganic materials which can be used later on as additives to cementitious materials [60–63]. However, in order to use solid waste in manufacturing products of commercial value, the mineralogical and chemical composition, the physical and chemical properties, as well as the morphology

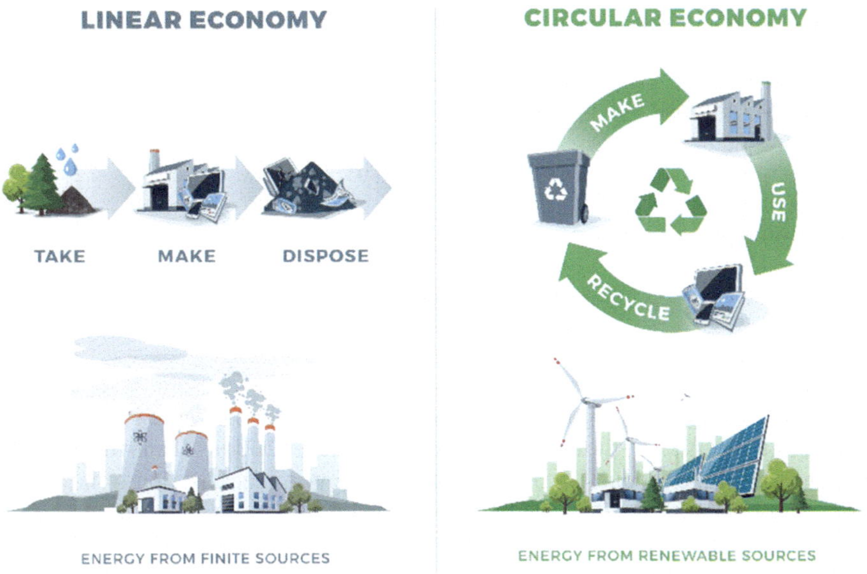

Fig. 3.6 Difference between linear and circular economy (Photo ID: 904493694, *Source* iStock by Getty Images)

of the waste should be evaluated in detail. The application of solid waste for production is impossible if it negatively impacts the durability and stability of the obtained products. Thus, the durability performance of products based on solid waste must be evaluated before commercialization. Therefore, rigorous quality control and assurance should be practiced when solid waste is used for the production. In order to facilitate effective commercialization and to maximize the use of products based on solid waste, technology centers are needed to make lab production feasible in the real world. The usage of industrial solid waste can reduce the costs of products to some extent. Thus, not only researchers, but also the entire industry are looking for new economically attractive and viable ways so that sustainable technologies of solid waste application for the production of commercially valuable products could be developed.

3.2.2 *Formation and Properties of By-Product of AlF_3 Production*

Worldwide, about 1 million tons of aluminum fluoride (AlF_3) per year is produced, most of which is used to reduce the melting point of aluminum in the aluminum industry [64, 65]. There are many other applications of this compound, including the glass industry, ceramics, in the production of white glazes, catalysts, etc. [64, 66,

67]. It should be noted that, in the industry, aluminum fluoride can be produced in two ways:

- by the dry fluorite (calcium fluoride, CaF_2) method:

$$CaF_2 + H_2SO_4 \rightarrow 2HF + CaSO_4 \tag{3.1}$$

followed by

$$Al_2O_3 + 6HF \rightarrow 2AlF_3 + 3H_2O \tag{3.2}$$

- from hexafluorosilicic acid.

Both of these industrial processes generate large amounts of waste: the manufacturing of 1 ton of AlF_3 by the dry fluorite method produces 2.7 tons of anhydrite waste. In comparison, the production of 1 ton of AlF_3 by wet neutralization of hexafluorosilicic acid generates ~0.7 tons of silica gel with incorporated F^- and Al^{3+} ions (SGW-F) [64, 67, 68]. The presently discussed by-products are contaminated with fluorine ions; hence, their recycling and application in other processes is a challenging task. For this reason, companies producing AlF_3 are looking for ways to diminish the concentration of fluoride ions in their waste by changing the manufacturing process or treating the resulting waste.

Dry AlF_3 production method

About 60% of the global AlF_3 production is based on a two-stage dry process [64, 67, 69, 70]:

1) production of gaseous HF from fluorite and sulfuric acid;

2) manufacturing of AlF_3 from gaseous HF and Al_2O_3.

The dry AlF_3 manufacturing process is characterized by the high consumption of natural raw materials and energy; i.e. ~4.4 tons of raw materials and ~1.5 MWh (~5 MJ) of energy are required to produce one ton of AlF_3 [68]. This process is not only demanding in terms of natural resources, but it also produces a large amount of waste contaminated with fluorine ions which is usually stored in landfill sites. Calcium sulfate, which is generated in a rotary kiln during the production of HF acid, is the main component of such pollution (Fig. 3.7). Usually, this by-product is landfilled without any further processing or neutralization [68, 71, 72]. Therefore, it is difficult to lower to a major extent the harmful effect on the environment during the production because the technologies can clean harmful gases with an efficiency of > 99%, while calcium sulfate processing technologies are complex and expensive. For this reason, the search for cheaper alternative raw materials and a higher cost-efficiency of the production technologies, such as hexafluorosilicic acid processing, has become more appealing than ever before. Well-grounded estimates suggest that about 20% of worldwide AlF_3 production proceeds this way [64, 68, 69, 73, 74].

Fig. 3.7 Landfill of calcium sulfate (with permission from *JSC Lifosa*)

Production of AlF_3 from hexafluorosilicic acid

The rapid growth of the human population has increased the demand for fertilizers. According to the *International Fertilizer Association* (IFA) [75] and the *European Phosphate Fertilizer Alliance* (AEEP) [76], the global annual consumption of phosphate rock has increased from 184 (2010) to 207 million tons (2019), and the worldwide manufacturing of phosphoric acid went up from 74 (2010) to 86 million tons (2019) during the 2010s. It is known that phosphate rocks contain 2–5 wt% (6.24 million tons) of fluorine (approximately 3 wt% F^- on average) [77–80]. This amount of fluorine ions is formed as a by-product in the production of phosphoric acid, which is usually obtained by decomposing the phosphate raw material with sulfuric acid:

$$Ca_5(PO_4)_3F + 5H_2SO_4 + nH_2O \rightarrow 3H_3PO_4 + 5CaSO_4 \cdot nH_2O + HF + Q. \quad (3.3)$$

Since phosphate rocks contain impurities, together with the main reaction (Eq. 3.3), additional reactions take place; i.e. various silicates react with sulfuric acid:

$$\begin{aligned} &Na_2O \cdot K_2O \cdot Al_2O_3 \cdot SiO_2 + 5H_2SO_4 \\ &\quad \rightarrow Na_2SO_4 + K_2SO_4 + Al_2(SO_4)_3 + 2SiO_2 + 5H_2O \end{aligned} \quad (3.4)$$

The formed silica interacts with fluoric acid, which is released in the course of Reaction 3.3:

$$SiO_2 + 4HF \rightarrow SiF_4 + 2H_2O \quad (3.5)$$

A part of SiF_4 is removed as gas, whereas the remaining part further reacts with fluoric acid to form hexafluorosilicic acid, which remains in the solution:

$$SiF_4 + 2HF \rightarrow H_2SiF_6 \quad (3.6)$$

During the production, 10–50 wt% of fluorine (from its primary content) is converted into gaseous SiF_4 which is absorbed together with steam by using water in the absorption system [69, 78, 81]. Thus, absorption produces hexafluorosilicic acid:

$$(n + 2)H_2O + 3SiF_4 \rightarrow 2H_2SiF_6 + SiO_2 \cdot nH_2O \quad (3.7)$$

According to various literature sources, an estimated 1.5–3 million tons of hexafluorosilicic acid is generated annually worldwide, of which, only about 0.5 million tons are used for the fluoridation of potable water, the synthesis of silicates, and the production of low-density aluminum fluoride (Table 3.2) [64, 67, 82]. Moreover, this acid is also applied in the production of silicon, ceramics and glass, tanning of animal skins, etc. Finally, it should be noted that the increasing fluorite prices resulted in the development of new economically attractive aluminum fluoride production technologies which allow the use of hexafluorosilicic acid and reduce the environmental pollution [70].

Currently, there are two leading technologies for the manufacturing of AlF_3 by using hexafluorosilicic acid. The first technology is based on the production of anhydrous hydrofluoric acid from hexafluorosilicic acid and HF conversion to aluminum fluoride. This technology is often referred to as the dry conversion of hexafluorosilicic acid. The second technology involves direct neutralization of hexafluorosilicic acid with aluminum hydroxide. This technology will likely be rejected in the future because of a low product density, high production costs, and, worst of all, the negative impact on the environment. Nevertheless, it is constantly updated due to its technological simplicity.

Table 3.2 Application of hexafluorosilicic acid

Application area	Quantity, tons per year
Pure H_2SiF_6 (for the fluoridation of drinking water)	~100,000
Synthesis of silicates	> 100,000
Production of AlF_3	~200,000
Production of SiF_4	30,000
Other applications	30,000

Production of AlF_3 by neutralizing hexafluorosilicic acid with aluminum hydroxide

The first known method of producing aluminum fluoride from hexafluorosilicic acid and aluminum hydroxide was patented more than 60 years ago [64, 83]. Academic literature suggests that, worldwide, more than one-sixth of aluminum fluoride is prepared this way [84]. In the wet method, aluminum fluoride is obtained when a direct reaction between hexafluorosilicic acid and aluminum hydroxide takes place:

$$H_2SiF_6 + 2Al(OH)_3 \rightarrow 2AlF_3 + SiO_2 + 4H_2O. \tag{3.8}$$

In the course of this reaction, low-density aluminum fluoride (700–800 kg/ton) is formed, which ensures a higher yield in aluminum production compared to high-density aluminum fluoride, but the production technology is much simpler. The production of aluminum fluoride by neutralizing hexafluorosilicic acid with aluminum hydroxide can be divided into five main stages: the storage of raw materials, the neutralization reaction, crystallization, calcination, and packaging of the final product [64, 67, 68].

According to the literature, producing 1 ton of aluminum fluoride by the wet process requires about 1,200 kg of hexafluorosilicic acid (a by-product of phosphoric acid) and ~1 ton of aluminum hydroxide (Table 3.3) [68]. Due to the economic aspects and the peculiarities of the technological process, an excess of ~10% hexafluorosilicic acid is used during this production. Therefore, the wet process reduces the total amount of raw substances needed to produce 1 ton of aluminum fluoride approximately by half in comparison with the requirements of the dry process.

In the course of the neutralization reaction (Eq. 3.8), approximately 670 kg of silica gel contaminated with fluorine ions (up to 10 wt% of dry material) is produced, which ultimately gets landfilled. Because of the dissociation, solubility, w/s ratio and adsorption properties of SiO_2, the decrease of the fluorine ion concentration in silica gel represents major technological challenges. Therefore, industries are looking for efficient ways to process and use silica gel in other areas.

Possibilities of purification and use of AlF_3 production waste

Fluoride is a highly aggressive and non-biodegradable anionic pollutant which, even at low concentrations, causes serious health problems, such as skeletal and tissue fluorosis, neurological disorders, renal effects, genotoxic effects, and excitotoxicity

Table 3.3 Consumption of raw materials and energy for the production of aluminum fluoride

Consumption of raw materials, t/t AlF_3	
100% H_2SiF_6	1.2
$Al(OH)_3$	1.03
$Ca(OH)_2$	0.09
Consumption of energy, MWh/t AlF_3	
Fuel (liquefied petroleum gas and natural gas)	1.37
Electricity	0.34

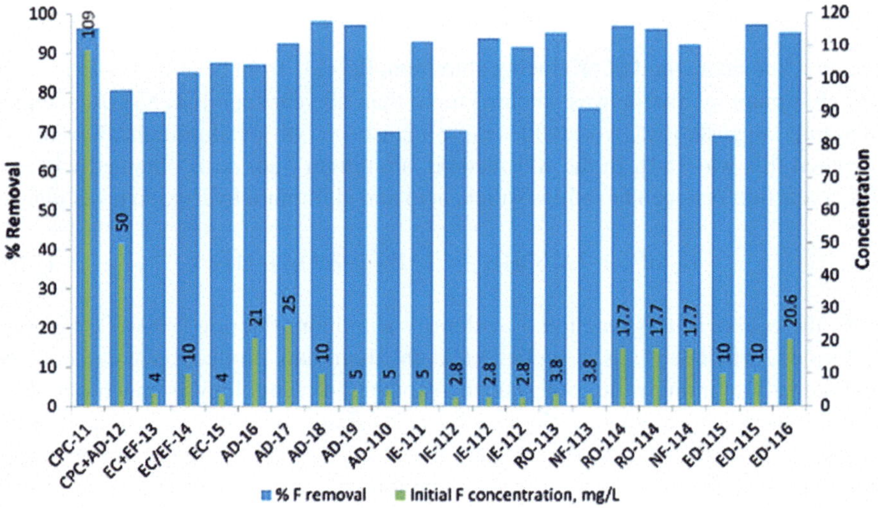

Fig. 3.8 Fluoride removal performance of various technologies [90]. *Indexes* CPC—chemical precipitation/coagulation; AD—adsorption; EC—electrocoagulation; EF—electrocoagulation/flotation; IE—ion exchange; RO—reverse osmosis; NF—nanofiltration; ED—electrodialysis. Reused with permission from Ref. [30] Copyright 2015 Elsevier

[85]. The pollution of surface water and groundwater can occur naturally by the dissolution of fluoride-rich minerals and artificially by human activities (mining and quarrying, metal and plating industries, aluminum smelters, etc.) [86, 87]. A variety of technologies for the removal of fluoride ions from wastewater have been created: precipitation, ion exchange, electrodialysis, osmosis, nanofiltration, electrocoagulation, adsorption, etc. (Fig. 3.8) [88–91]. The choice of the method depends on the area, the concentration of fluoride, the availability of resources, etc.

Reverse osmosis is a membrane filtration process in which water is passed through a semi-permeable membrane by using a high pressure, while pollutants are detained. According to the literature, the removal efficiency of F^- ions from drinking water is above 95% [90]. *Nanofiltration* is similar to both reverse osmosis and ultrafiltration. It works at a low pressure; however, the removal efficiency is under 60% for monovalent ions [88–91].

The *chemical precipitation* technique is based on fluoride reaction with cations and the formation of insoluble salts [92, 93]:

$$Ca^{2+}{}_{(aq)} + 2F^{-}{}_{(aq)} \rightarrow CaF_{2(solid)} \tag{3.9}$$

Also, aluminum salt which acts as a coagulant is usually used in this method [94]. As a result, the concentration of fluoride in water can be reduced by up to 96% [92–94]. Similar findings on fluoride removal by calcium salts have been confirmed by Jadhav et al. [95].

Another method that can be effectively employed to treat wastewater containing fluoride is *electrocoagulation* [88–91, 94]. The key part of an electrocoagulation reactor is an electrolytic cell with an anode and a cathode, in which the following reactions take place:

$$Al \rightarrow Al^{3+} + 3e \text{ (at the anode)}. \tag{3.10}$$

$$Al^{3+} + 3H_2O \rightarrow Al(OH)_3 + 3H^+. \tag{3.11}$$

$$Al(OH)_3 + xF^- \rightarrow Al(OH)_{3-x}F_x + xOH^- \tag{3.12}$$

$$2H_2O + 2e^- \rightarrow H_2 + 2OH^- \text{ (at the cathode)}. \tag{3.13}$$

When $Al(OH)_{3-x}F_x$ particles coagulate and produce a mass, they can be removed by flotation and sedimentation.

Electrodialysis can also be applied for water purification. During purification, an ion exchange membrane (responsible for the pollutant and water separation) is placed between the two electrodes [88–91, 94]. The electric field between the electrodes is the driving force to move fluorine ions through the membrane to the anode.

Ion exchange or adsorption has been highlighted in several scientific works of the literature due to its efficacy for wastewater purification [88–91, 94]. This method can be applied for the removal of both cations and anions. Among all the presently listed techniques, adsorption is denoted by the highest number of advantages: greater accessibility, acceptable price, ease of operation, and effectiveness. Unfortunately, it also involves such challenges as the volatility of the properties of adsorbents, the regeneration of adsorbents, and the utilization of the used adsorbents (most of them are landfilled, which causes secondary pollution).

Each of the abovementioned processes is suitable for fluoride removal from ground or wastewater; however, they cannot be applied to solid waste containing fluoride. In the scientific literature, only theoretical data on the application and purification of silica gel with incorporated F^- and Al^{3+} ions is provided. One of the earliest researches on the utilization of silica gel with impurities in cement was performed by Mandeikyte et al. [96]. The authors investigated the production of silicate cement from silica gel with impurities, lime, and phosphogypsum. It has been determined that, by using these wastes, we can produce a binder that hardens at the ambient temperature or under water vapor pressure, and it can reach a compressive strength value as high as 18 MPa. Investigations suggest that such strength values are determined by the ratio of the formed ettringite and calcium silicate hydrates in the hydrated sample. Further studies have shown that, when the binder is compressed (36 × 36 mm samples) and cured at 90 °C, the samples show a compressive strength of 21.5 MPa, and their frost resistance and water absorption values meet the requirements of the silicate brick standard.

Vaičiukynienė et al. [97, 98] researched the feasibility of utilizing silica gel with impurities in cement stone. These scholars determined that the compressive strength of the standard concrete samples got reduced by augmenting the amount of dried silica gel with impurities (up to 15 wt%). This research concluded that a decrease in the strength can be explained by the formation of calcium fluoride which was obtained during the interaction of aluminum fluoride trihydrate and calcium hydroxide. Different results were obtained when using calcined (800 °C, 1 h) silica gel with impurities. It was found that, during cement hydration, this additive reacted with calcium hydroxide and formed semicrystalline calcium silicate hydrates. Moreover, it was measured that, after 28 days of hydration, an increase of the share of calcined silica gel with impurities by 10 wt% had a positive effect on the compressive strength values of the samples with the values enhanced by ~5–7 MPa. However, it is worth noting that these studies did not provide any data on the stability of fluoride ions in cement stone.

One of the earliest attempts to use silica gel with impurities for synthesizing geopolymer concrete was unsuccessful because, even by changing the composition of the mixtures, the values of the compressive strength of the best-obtained samples did not even reach 3 MPa [99]. In further research, Vaičiukynienė et al. [100] used silica gel with impurities to produce alkali-activated additives for cement. Different combinations of silica gel and clay (0:1; 1:9; 1:3; 1:1) were prepared for the experiments, which were then activated with sodium hydroxide and thermally treated at 600 °C (3 h). It was found that, when the mixture of silica gel and clay was treated with sodium hydroxide, intense reactions took place between the initial components, during which, sodium aluminum silicate hydrates were formed, and fluorine ions were bound to calcium fluoride. The addition of silica gel was observed to enhance the mechanical properties of the samples: The one containing 25 wt% silica gel had the highest compressive strength values (17.50 MPa). It should be noted that this value was almost 3 times higher than that of the control clay sample.

Krivenko et al. [101] researched the impact of sodium silicate (Na_2SiO_4) activated silica gel with impurities on the hydration characteristics of cement samples. At the beginning of the experiments, silica gel was mixed with liquid sodium silicate glass 1:1 (by weight). These materials were observed to react with each other to form crystalline Na_3AlF_6 and amorphous $SiO_2{\cdot}nH_2O$. Further, the obtained additive was mixed with water (used to produce cement dough) to reach the amount of silica gel with impurities in the cement samples equal to 0.5, 1.0, 1.5, and 2.0 wt%. It was subsequently established that the additive modifies the mechanism of early hydration: (1) It increases the amount of heat released during the main hydration reaction; (2) it prolongs the length of the induction period; (3) it promotes crystallization of both crystalline and amorphous calcium silicates hydrates and calcium aluminate hydrates. Moreover, after 28 days of hydration, the compressive strength values of the sample with 1.5 wt% silica gel were about 38% higher than those of the sample without an additive. However, the discussed studies [96–101] determined neither the durability of the samples nor the stability of fluorine ions.

Other researchers have shown in their investigations that silica gel with impurities can be used to produce pozzolanic materials without releasing F^- ion-containing compounds into the environment. Kaminskas et al. [102–104] discovered that the calcination of a mixture containing 20 wt% of clay and 80 wt% of silica gel with impurities at 600 °C doubled its pozzolanic activity, and the ultimate value was 230–260 mg CaO/g. Moreover, the pozzolanic activity of the mixture was found to decrease with a further increase in the calcination temperature. Further theoretical and experimental research showed that, for binding fluorine ions to stable compounds (CaF_2 and MgF_2), 10–20 wt% of limestone must be used in the production of the pozzolanic additive. Notably, the activity of the pozzolanic additive with limestone (calcination temperature – 700 °C) is 250 mg CaO/g. The authors investigated the influence of a synthetic pozzolanic additive on cement hydration and found that this additive promotes not only the crystallization of ettringite, but also the hydration reactions of calcium silicates. It was discovered that the replacement of a part of cement with a pozzolanic additive (5–15 wt%) in concrete products led to an increase of the compressive strength values of the samples by 5%. It should be noted that, during cement hydration (up to 3 months), most of the fluorine ions remained bound to calcium fluoride, and less than 1 ppm was released into the liquid medium.

Data of previous researchers suggests that silica gel with impurities can be used to synthesize Na zeolite [105, 106]. These papers state that this zeolite can be synthesized under hydrothermal conditions or by sonication of suspensions of silica gel, aluminum hydroxide, sodium hydroxide, and water. The more efficient synthesis method is sonification because, after 24 h of treatment, the synthesis product has a higher degree of crystallinity (25.81%).

Sarkar et al. [107] discovered that the burning temperature of products can be reduced by using silica gel with impurities in the production of white porcelain without compromising the performance levels.

Krysztafkiewicz et al. [108] investigated the use of silica gel when manufacturing rubber fillers, sodium or potassium silicate, and in other fields. It is stated that, in order to use silica gel in practice, fluorine ions must be removed. Therefore, in research [109], the authors investigated the use of silica gel treated with hot water for the manufacturing of rubber fillers. The scientists found that, by treating a silica gel sample with water at a temperature of 90–100 °C, the concentration of fluorine ions can be reduced by 1.7–0.2 wt%. However, it is worth noting that a fraction of fluorine evaporates into the environment during the treatment of silica gel with hot water. Moreover, another issue with this paper is that the authors did not provide the w/s ratio used in this work.

Dambrauskas et al. [110] investigated the influence of various hydroxide and salt additives on the removal of F^- ions from silica gel waste. The scientists applied various leaching of F^- ions from silica gel waste to the liquid medium techniques: (1) leaching under static conditions; (2) leaching under dynamic conditions by using continuous liquid medium flow; and (3) leaching in cycles under dynamic conditions. It was determined that the efficiency of fluoride removal from this waste depends on w/s ratio, leaching conditions, and used additives. The authors proved that it is possible to reduce the concentration of fluorine ions from 10 to < 5 wt% by changing the treatment conditions and by adding alkaline compounds. It was determined that the most efficient technique to remove fluoride from silica gel waste is the treatment with NH_4OH solution under dynamic conditions (w/s = 10).

To summarize, the share of fluorine ions can be reduced via two alternative options: by removing them or by reducing their mobility-immobilization/adsorption. However, as Sect. 3.2.1 and 3.2.2 showed, currently, there are no appropriate SGW-F purification or immobilization technologies available. It is known that the use of such a material involves the initial step of the determination of its chemical and physical properties. Thus, in the following part of this monograph, SGW-F was characterized by instrumental analysis.

Properties of silica gel contaminated with F^- and Al^{3+} ions

According to the conducted chemical analysis, dried SGW-F consists of 79.0 wt% SiO_2 and 10.0 wt% F^- ions (Fig. 3.9a). A roughly identical pattern was noted when applying XRF: It was established that SGW-F contains 36.2 wt% of silica, which is equivalent to 78.9 wt% of SiO_2, and 2.5 wt% of Al^{3+} ions (Fig. 3.9a).

The data of chemical and XRF analysis was confirmed by the results of XRD analysis. In the XRD pattern, the intensive diffraction peaks characteristic of $AlF_3{\cdot}3H_2O$ (PDF No. 00-035-0827; *d*-spacing: 0.546, 0.386, 0.330, and 0.244 nm) and a broad basal reflection (within the 18–37° diffraction angle range) which correspond to amorphous silicon dioxide were observed (Fig. 3.9b).

Particle size distribution analysis showed that the diameter of SGW-F particles spread across a wide range (0.03–170 μm), while the particles of a size of 34–72 μm were predominant (Fig. 3.9c). Furthermore, the density and the surface area were equal to 2141 kg/m^3 and 281.01 m^2/kg, respectively.

DSC analysis confirmed the above-discussed results. In the DSC curve (Fig. 3.10), a two-step decomposition of $AlF_3{\cdot}3H_2O$ was observed at 146 and 165 °C temperatures [111, 112]:

$$AlF_3 \cdot 3H_2O \rightarrow AlF_3 \cdot 0.5H_2O + 2.5H_2O \tag{3.14}$$

$$AlF_3 \cdot 0.5H_2O \rightarrow AlF_3 + 0.5H_2O \tag{3.15}$$

The third endothermic effect at 188 °C can be attributed to the dehydration of silica gel:

$$SiO_2 \cdot nH_2O \rightarrow SiO_2 + nH_2O. \tag{3.16}$$

The experiment of water vapor adsorption proved the nature of the third effect. A SGW-F sample was placed above water (the ratio of the relative water vapor pressure (p/p_0) was equal to 1.0) and kept for 24 h. It was found that, after the experiment, the dehydration heat of amorphous silicon dioxide increased by more than three times (59 to 189 J/g). This process was concurrent with the endothermic effect attributed to the decomposition of $AlF_3{\cdot}3H_2O$ (Fig. 3.10).

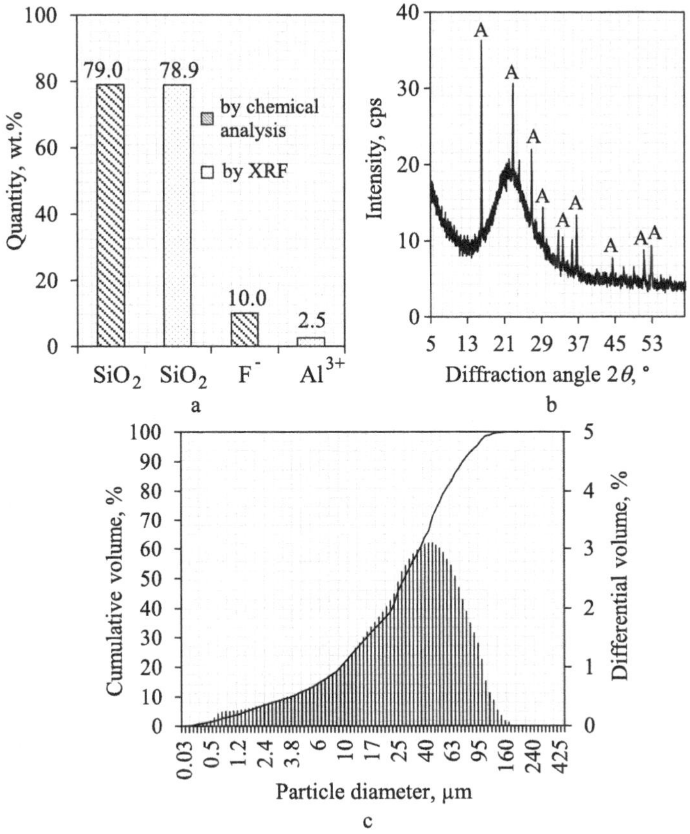

Fig. 3.9 Chemical composition (**a**), XRD pattern (**b**), and particle size distribution (**c**) of SGW-F. *Indexes* A—$AlF_3 \cdot 3H_2O$

With the objective to maximize the perception of the thermal behavior of SGW-F, the calcination of this waste was implemented throughout the temperature range of 50–1000 °C (Fig. 3.11). XRD results showed that the diffraction peaks commonly demonstrated by $AlF_3 \cdot 3H_2O$ somewhat decreased with the rising temperature, and it fully decomposed at 188 °C (Fig. 3.11). It should be noted that, after the decomposition of $AlF_3 \cdot 3H_2O$, the peaks characteristic of crystalline compounds containing aluminum or fluoride were not detected until the temperature of calcination was elevated to 1000 °C. At this temperature, the reaction between aluminum containing components and SiO_2 proceeded, and, as a result, mullite ($Al_{4.95} \cdot Si_{1.05} \cdot O_{9.52}$; PDF 00-015-0776; *d*-spacing: 0.539, 0.342, 0.339, and 0.221 nm) was formed (Fig. 3.11). It should be noted that F^- ions are firmly bound in the structure of SGW-F as the conducted chemical analysis yielded data of the calcined product (at 188 and 550 °C) which demonstrated that the share of fluoride ions in the solid equals to ~9.8 wt%.

The results of instrumental and chemical analysis showed that AlF_3 production waste is predominantly amorphous silicon dioxide and fluoride (~10 wt%).

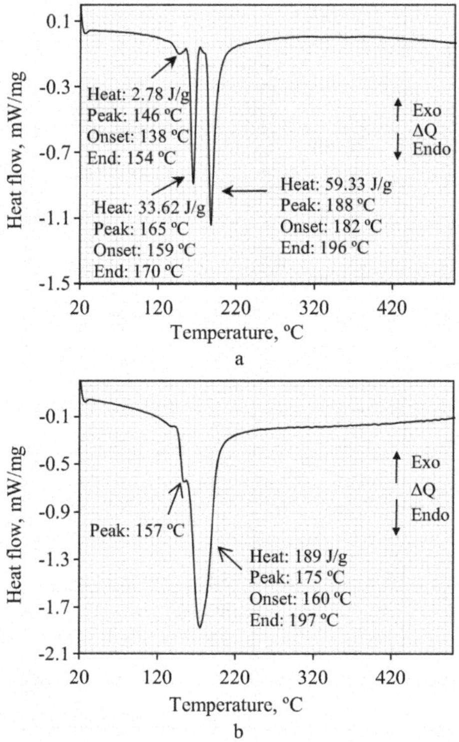

Fig. 3.10 DSC curves of SGW-F samples before (**a**) and after (**b**) water vapor adsorption ($p/p_0 = 1.0$)

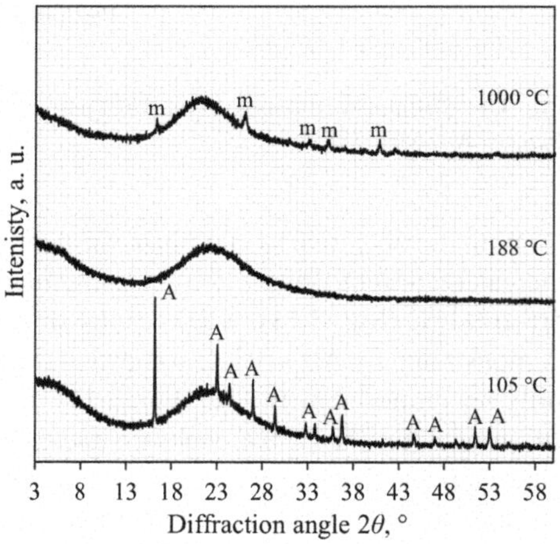

Fig. 3.11 XRD patterns of calcined in the temperature range of 105–1000 °C for 1 h SGW-F samples. *Indexes* A—$AlF_3{\cdot}3H_2O$; m—mullite

According to the literature, these ions can be leached by treating AlF_3 production waste with a high volume of water [108, 110, 113]. Thus, when this waste is stored in landfills and gets in contact with rain water, it can pollute groundwater with fluoride. On top of that, it contains ~2.5 wt% aluminum ions which positively affect the formation and stabilization of some calcium silicate hydrates [1, 3–6, 16]. Whereas AlF_3 production waste contains a high quantity of amorphous silica and some aluminum, it could be used as a raw material for the synthesis of calcium silicate hydrates. It is known that some Ca^{2+} and Si^{4+} ions in the crystal lattice of calcium silicate hydrates can be replaced with various cations (Na^+, K^+, Cs^+, Cu^{2+}, Fe^{2+}, Zn^{2+}, Cd^{2+}, Pb^{2+}, Mg^{2+}, etc.) under the conditions of hydrothermal synthesis. The impact of each additive on the formation and properties of CSH is conditioned by the nature of the raw materials, the parameters of hydrothermal reactions (time, temperature, w/s ratio), and some other factors. Also, these calcium silicate hydrates can be used as adsorbents for various cations and anions. Thus, it is likely that the use of AlF_3 production waste in the synthesis of CSH promotes the formation of calcium silicate hydrates, and fluoride will be combined by the formed compounds.

It should be noted that amorphous/semicrystalline calcium silicate hydrates are the main hydration products of ordinary Portland cement (OPC) [3, 114, 115]. Moreover, synthetic crystalline CSH is chemically and structurally similar to those formed in the course of OPC hydration. Therefore, it is likely that gyrolite family compounds containing fluorine and/or other pollutants can be used as an additive to OPC. These pollutants are chemically bound into the structure of CSH; thus, they are not released into the environment.

3.3 Influence of AlF_3 Production Waste on the Formation of Gyrolite Family Compounds

In this part of the monograph, a non-standard method for the synthesis of lower basicity calcium silicate hydrates combining the utilization of a by-product of AlF_3 production (silica gel with incorporated F^- and Al^{3+} ions {SGW-F}) and the immobilization of F^- ions in the derived products during hydrothermal synthesis is outlined. As a matter of fact, SGW-F mainly consists of amorphous SiO_2; it could thus be used for the synthesis of gyrolite family compounds which have a great adsorption capacity for cations and anions. It is plausibly considered that, during hydrothermal synthesis, F^- ions will be chemically adsorbed by CSH or combined into a stable compound, such as calcium fluoride. Afterward, the obtained products could be used as an additive in binders and in the industry of construction materials.

It was established that, in a pure calcium oxide, amorphous silica dioxide, and water system, when the molar ratio of $CaO/SiO_2 = 0.55$, after 8 h of hydrothermal treatment at 200 °C, the primary compounds were fully reacted, and semicrystalline calcium silicate hydrate C–S–H(I) (*d*-spacing: 0.304, 0.278, and 0.183 nm) was formed (Fig. 3.12, curve 1). By prolonging the duration of isothermal curing

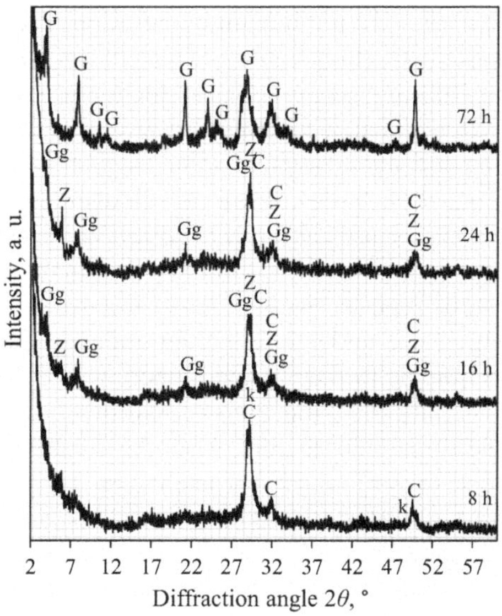

Fig. 3.12 XRD patterns of synthesis products when the molar ratio of CaO and amorphous silica dioxide mixture was equal to CaO/SiO_2 = 0.55. *Indexes* G—gyrolite; C—C–S–H(I); Z—*Z*-phase; Gg—gyrolite gel; k—calcite

to 16–24 h, together with the latter compound, a low crystallinity calcium silicate hydrates *Z*-phase (PDF No. 01-074-7587; *d*-spacing: 1.526, 0.755, 0.501, and 0.441 nm) and gyrolite gel (*d*-spacing: 2.312, 1.138, 0.839, 0.421, and 0.280 nm) were formed (Fig. 3.12, curve 2). Meanwhile, after 48 h of synthesis, the above-listed compounds became metastable and recrystallized into gyrolite (PDF No. 00-042-1452; *d*-spacing: 2.273, 1.126, 0.844, 0.420, and 0.365 nm) which remained stable till 72 h (Fig. 3.12, curve 4). The identification of the diffraction peaks characteristic of C–S–H(I) and gyrolite in their mixtures is complicated; however, earlier works indicated that C–S–H(I) is not stable under the conditions of lengthy synthesis. For this reason, it can be stated that gyrolite is the dominant product formed after 48 and 72 h of synthesis.

It should be noted that the synthesis products were carbonized while being dried in an air-conditioned chamber (50 °C, 24 h) because small intensity diffraction peaks characteristic of calcium carbonate (PDF No. 01–083-1762; *d*-spacing: 0.306, 0.250, 0.209, and 0.187 nm) were detected in the XRD pattern after 8 h of synthesis (Fig. 3.12, curve 1). The data on the presence of calcium carbonate was also corroborated by the small endothermic effect at ~706 °C in the DSC curve (Fig. 3.13a, curve 2).

The XRD results were confirmed by STA analysis data. In the DSC curve, the first endothermic effect within the temperature range of 50–300 °C can be attributed to the dehydration of calcium silicate hydrates (Fig. 3.13a, b, curve 2). The exothermic effects at 850 and 852 °C were assigned to C–S–H(I) and gyrolite recrystallization into wollastonite, respectively (Fig. 3.13a, b, curve 2). The heat of the latter process increases from 27.36 to 39.77 J/g by increasing the synthesis duration.

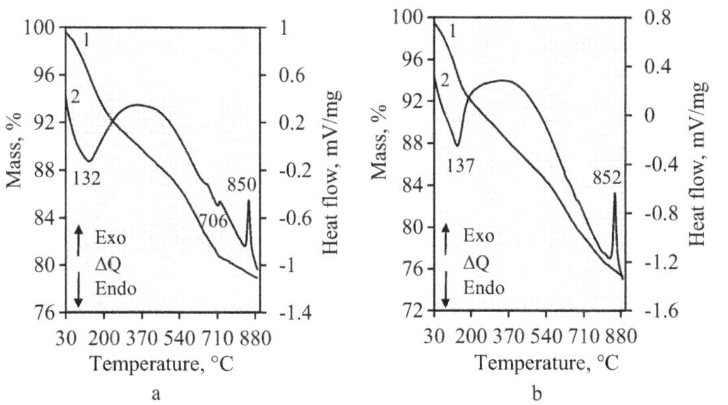

Fig. 3.13 STA (1—TGA curve; 2—DSC curve) curves of the synthesis products when the molar ratio of CaO and amorphous silica dioxide mixture was equal to $CaO/SiO_2 = 0.55$. Duration of hydrothermal synthesis at 200 °C temperature, h: **a** 8; **b** 72

It can be plausibly assumed that, when curing the mixture of CaO and amorphous SiO_2 with $C/S = 0.55$ at 200 °C temperature under hydrothermal conditions, the following sequence of the formation of new compounds is observed:

$$0.55CaO + SiO_2 \cdot nH_2O + nH_2O \xrightarrow{8\,h} \begin{vmatrix} CaCO_3^* \\ C\text{–}S\text{–}H(I) \end{vmatrix} \xrightarrow{16\,h} \begin{vmatrix} Z\text{-phase} \\ C\text{–}S\text{–}H(I) \\ \text{gyrolite gel} \end{vmatrix} \xrightarrow{48\text{–}72\,h} \text{gyrolite}$$

*—calcite is formed during the drying process of synthesis products.

It was established that silica gel with incorporated F^- and Al^{3+} ions promotes the formation of crystalline calcium silicate hydrates because, in the CaO-SGW-F-H_2O system with $CaO/SiO_2 = 0.55$, as early as after 8 h of isothermal curing at 200 °C, low-base calcium silicate hydrate *Z*-phase was formed (Fig. 3.12, curve 1 and Fig. 3.14, curve 1). It should be noted that the stoichiometric composition of this compound conforms to the molar ratio $CaO/SiO_2 = 0.55$ of the initial mixture. Also, together with this compound, C–S–H(I) was detected (Fig. 3.14, curve 1). It was established that, during synthesis, $AlF_3{\cdot}3H_2O$, which is present in SGW-F, decomposed, and, as a result, F^- and Al^{3+} ions were released to the liquid medium. These ions actively interacted with calcium ions; thus, calcium fluoride (PDF No. 00-035-0816; *d*-spacing: 0.315, 0.193, and 0.165 nm) and katoite [$Ca_3Al_2O_6(H_2O)_6$, PDF No. 04-007-2666; *d*-spacing: 0.513, 0.281, 0.230, and 0.204 nm] were formed (Fig. 3.14, curves 1–4). It should be noted that both compounds remained stable under all the researched synthesis conditions (Fig. 3.14, curves 1–4). It was thus discovered that, after 16 h of synthesis, *Z*-phase becomes metastable and starts recrystallizing to another intermediate phase–gyrolite gel. Finally, after 24 h of synthesis, both compounds fully transform into gyrolite (Fig. 3.14, curves 2 and 3). It should be noted that, in the mixture with SGW-F, the synthesis duration of gyrolite is halved; i.e. it goes down from 48 to 24 h (Figs. 3.12 and 3.14, curve 3).

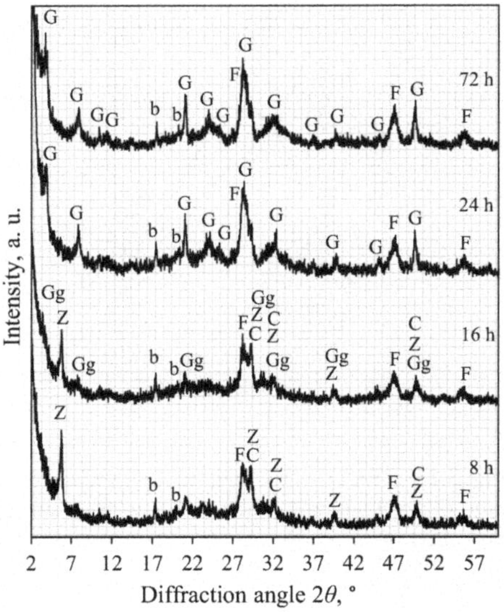

Fig. 3.14 XRD patterns of synthesis products with the molar ratio of CaO and SGW-F mixture equal to $CaO/SiO_2 = 0.55$. *Indexes* G—gyrolite; C—C–S–H(I); Z—*Z*-phase; Gg—gyrolite gel; k—calcite; b—katoite; F—CaF_2

In the DSC curves, as in the previous case (Fig. 3.15), the first broad endothermic effect is associated with the loss of the adsorbed and interlayer water from the crystal structure of C–S–H (Fig. 3.15). The second endothermic effect at 326 °C temperature corresponds to the dehydration process of katoite. It is thus evident that the nature of the SiO_2 source exerts influence on C–S–H(I) and gyrolite recrystallization into wollastonite as, in the mixtures containing SGW-F, this process is accelerated and identified at 812 and 797 °C temperatures. However, in the pure system, the above-outlined reaction proceeds at more elevated temperatures (~850 °C) (Fig. 3.15 a, b, curve 2). Also, the heat released during this process is lower in the mixtures with SGW-F.

The reactions of the formation of the new compounds in the CaO-SGW-F-H_2O system when C/S = 0.55 occur in the following sequence:

$$0.55\mathrm{CaO} + \mathrm{SGW} - \mathrm{F} + n\mathrm{H_2O} \xrightarrow{8\,\mathrm{h}} \left|\begin{array}{l} Z\text{-phase} \\ \mathrm{C{-}S{-}H}(I) \\ \mathrm{CaF_2} \\ \text{katoite} \end{array}\right. \xrightarrow{16\,\mathrm{h}} \left|\begin{array}{l} Z\text{-phase} \\ \mathrm{C{-}S{-}H}(I) \\ \mathrm{CaF_2} \\ \text{gyrolite gel} \\ \text{katoite} \end{array}\right. \xrightarrow{24-72\,\mathrm{h}} \left|\begin{array}{l} \text{gyrolite} \\ \mathrm{CaF_2} \\ \text{katoite} \end{array}\right. .$$

It was established that the increment of the C/S molar ratio to 0.66 does not have any significant influence on the crystallization of calcium silicate hydrates. Only the beginning of gyrolite gel crystallization was prolonged from 16 h in the mixtures with C/S = 0.55 to 24 h in the mixtures with C/S = 0.66 (Fig. 3.12, curve 2 and Fig. 3.16, curve 3). When prolonging the duration of isothermal curing to 48 h, Z-phase, C–S–H(I) and gyrolite gel fully recrystallize into gyrolite after 48 h of synthesis. Gyrolite remains stable until 72 h (Fig. 3.16, curve 4).

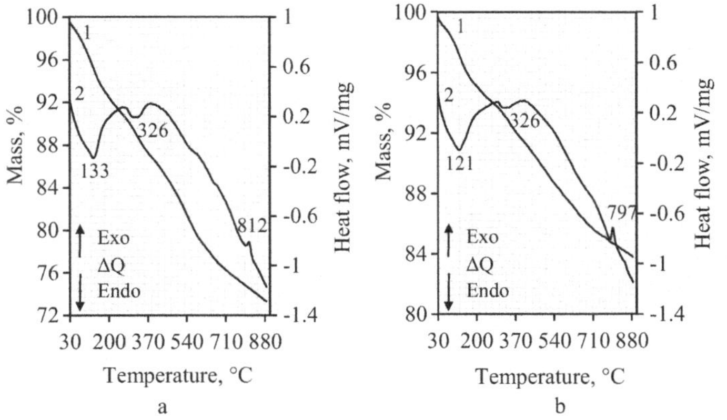

Fig. 3.15 STA (1—TGA curve; 2—DSC curve) curves of synthesis products with the molar ratio of CaO and SGW-F mixture equal to $CaO/SiO_2 = 0.55$. Duration of hydrothermal synthesis at 200 °C temperature, h: 1—8; 2—72

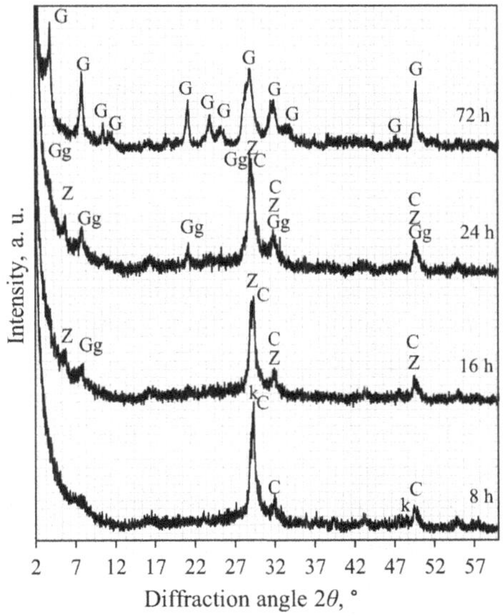

Fig. 3.16 XRD patterns of synthesis products when the molar ratio of CaO and amorphous silica dioxide mixture was equal to $CaO/SiO_2 = 0.66$. *Indexes* G—gyrolite; C—C-S-H(I); Z—Z-phase; Gg—gyrolite gel; k—calcite

The XRD data was in line with the results of STA analysis (Fig. 3.17). It was determined that the STA curves of the synthesis products change negligibly in comparison with the STA curves of the products synthesized in mixtures with C/S = 0.55 (Figs. 3.13 and 3.17). The main difference is the increased heat of the exothermic effect at ~ 850 °C. It could relate to the higher crystallinity of calcium silicate hydrates.

Thus, the sequence of the compounds formed in the $CaO–SiO_2–H_2O$ system with the C/S = 0.66 molar ratio during the synthesis is as follows:

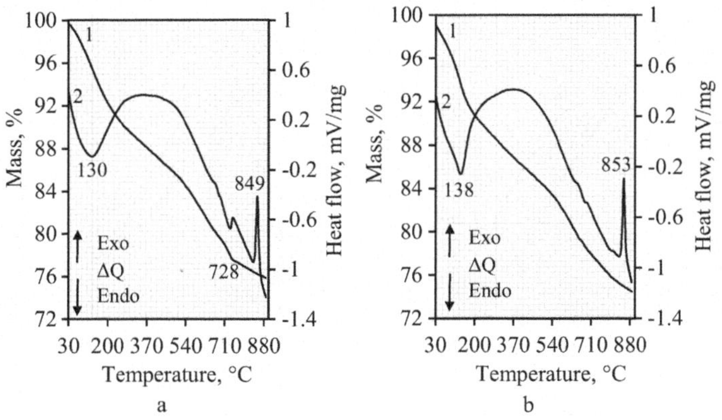

Fig. 3.17 STA (1—TGA curve; 2—DSC curve) curves of hydrothermal synthesis products when the molar ratio of CaO and amorphous silica dioxide mixture was equal to $CaO/SiO_2 = 0.66$. Duration of hydrothermal synthesis at 200 °C temperature, h: 1—8; 2—72

$$0.66CaO + SiO_2 \cdot nH_2O + nH_2O$$

$$\xrightarrow{8\,h} \begin{vmatrix} CaCO_3^* \\ C{-}S{-}H(I) \end{vmatrix} \xrightarrow{16\,h} \begin{vmatrix} Z\text{-phase} \\ C{-}S{-}H(I) \end{vmatrix} \xrightarrow{24\,h} \begin{vmatrix} Z\text{-phase} \\ C{-}S{-}H(I) \\ \text{gyrolite gel} \end{vmatrix} \xrightarrow{48-72\,h} \text{gyrolite.}$$

The above-discussed results overlap fairly well with the literature data highlighting that the most suitable temperature of gyrolite synthesis is relatively high (circa 200 °C), and the duration is relatively extensive (about 32 h) when amorphous silica is employed as a SiO_2 component. Also, the sequence of the formation of intermediate phases (C–S–H(I) → *Z*-phase → gyrolite gel → gyrolite) is similar to the one established previously.

Another important point is that the mechanism of hydrothermal reactions and the sequence of the compounds formed in the CaO-SGW-F-H_2O mixture are fairly different when C/S = 0.66. As established, after 8 h of isothermal curing, all the intermediate phases formed in the synthesis product (Fig. 3.18, curve 1). Also, as in the mixtures with C/S = 0.55, calcium fluoride and katoite were formed due to the interaction between calcium and fluorine or aluminum ions. Gyrolite became the dominant compound after 16 h of hydrothermal treatment and maintained stability until 72 h (Fig. 3.18, curves 2–4). It should be underlined that the intensity of the main diffraction peak (*d*-spacing – 2.273 nm) characteristic of synthetic gyrolite is 2–3 times lower in the mixtures with SGW-F in comparison with the pure system. This data conforms fairly well to the plots of the DSC curves: the heat of the exothermic effect corresponding to synthetic gyrolite recrystallization into wollastonite is also several times lower (Fig. 3.19, curve 2). These facts confirm the assumption that gyrolite of a lower crystallinity gets formed in the mixtures with SGW-F.

It could be summarized that SGW-F has a positive effect on the beginning of gyrolite formation, but it acts negatively on its crystallinity. The sequence of the

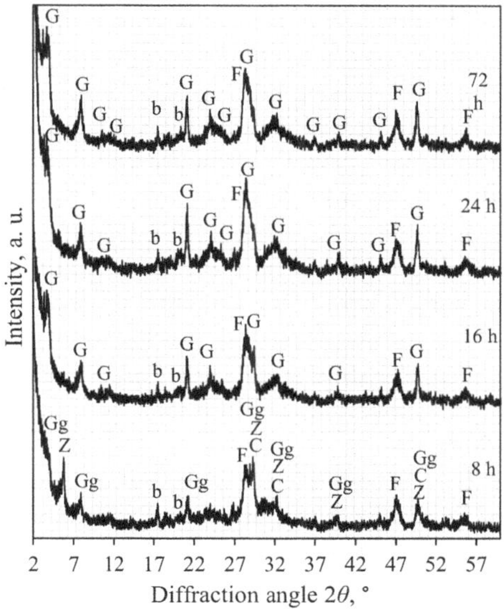

Fig. 3.18 XRD patterns of synthesis products with the molar ratio of CaO and SGW-F mixture equal to $CaO/SiO_2 = 0.66$. *Indexes* G—gyrolite; C—C–S–H(I); Z—*Z*-phase; Gg—gyrolite gel; b—katoite; F—CaF_2

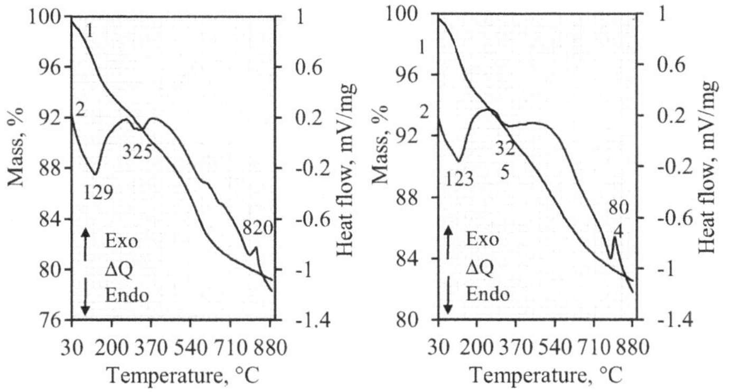

Fig. 3.19 STA (1—TGA curve; 2—DSC curve) curves of synthesis products with the molar ratio of CaO and SGW-F mixture equal to $CaO/SiO_2 = 0.66$. Duration of hydrothermal synthesis at 200 °C temperature, h: 1—8; 2—72

formation of new compounds in the mixture of CaO and SGW-F with the C/S = 0.66 at 200 °C temperature under hydrothermal conditions can be presented as follows:

$$0.66\mathrm{CaO} + \text{SGW-F} + n\mathrm{H_2O} \xrightarrow{8\,\mathrm{h}} \begin{array}{|l} Z\text{-phase} \\ \text{C–S–H}(I) \\ \mathrm{CaF_2} \\ \text{gyrolite gel} \\ \text{katoite} \end{array} \xrightarrow{16-72\,\mathrm{h}} \begin{array}{|l} \text{gyrolite} \\ \mathrm{CaF_2} \\ \text{katoite} \end{array}.$$

The duration of the synthesis was determined to influence the particle size of the synthesis products obtained in the CaO-SGW-F-H_2O system. The results showed that, by increasing the duration of the synthesis from 8 to 72 h, the amount (wt%) of < 80 μm fraction decreased: When $CaO/SiO_2 = 0.55$, it decreased by 1.5 times (Fig. 3.20a, curve 1); when $CaO/SiO_2 = 0.66$, it decreased by 1.2 times (Fig. 3.20a, curve 2). Meanwhile, the quantity of the 80–106 μm fraction increased by about 1.2 times in both cases (Fig. 3.20b). Thus the length of the hydrothermal treatment makes an impact not only on the mineralogical composition but also on the distribution of the particle size of the synthesis products.

Furthermore, data of the liquid medium analysis demonstrated that, in the course of hydrothermal synthesis, F^- ions were adsorbed by CSH or combined into stable compounds because their concentration in the liquid medium was very low (Table 3.4).

It is known that the temporal extension of hydrothermal treatment yields that calcium silicate hydrates which are stable at short synthesis durations recrystallize to thermodynamically stable compounds at high temperatures and long durations.

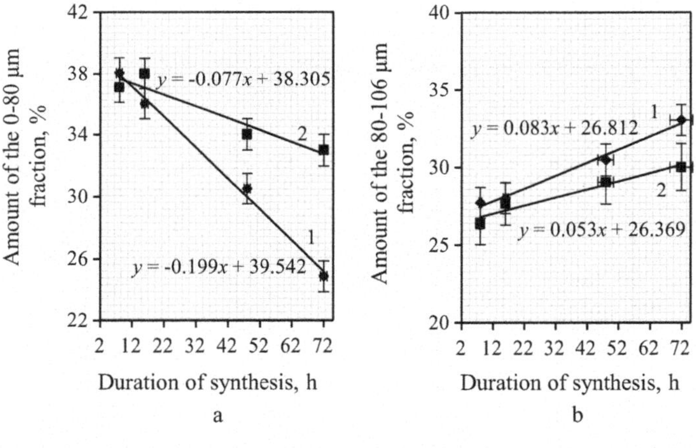

Fig. 3.20 Dependence of the amount of < 80 μm (**a**) and 80–106 μm (**b**) particle fractions of the synthesis products synthesized in CaO and SGW-F system on the synthesis duration and the molar ratio of the primary mixture: 1—0.55; 2—0.66

Table 3.4 Concentration (wt%) of F^- ions in the liquid medium after hydrothermal synthesis

Duration of synthesis, h	Molar ratio of CaO/SiO_2	
	0.55	0.66
	Concentration of F^- ions, wt%	
8	0.057	0.060
16	0.031	0.045
24	0.031	0.032
72	0.037	0.060

The crystallization process may last from a few hours to weeks. However, in this monograph, the effect of a longer duration of hydrothermal treatment (> 504 h) on the gyrolite stability was not investigated. Thus, in order to confirm the gyrolite formation mechanism, the sequence of the formation of intermediate compounds, their stability temperature, and duration intervals, as well as to predict the stability of gyrolite, the thermodynamic calculations of the theoretical CSH formation reactions were performed.

In order to thermodynamically calculate the parameters of CSH formation and gyrolite stability reactions, the method of absolute entropies was employed. According to the method [116–120], the alterations of the reaction standard free Gibbs energy $\Delta_r G_T^0$ can be calculated with the following equation:

$$\Delta_r G_T^0 = \Delta_r H_T^0 - T\Delta_r S_T^0, \tag{3.17}$$

where $\Delta_r H_T^0$ and $\Delta_r S_T^0$ are the standard change of the reaction enthalpy and entropy (J/mol) in temperature T, respectively. The enthalpy change of the reactions is defined by the following equations:

$$\Delta_r H_T^0 = \Delta_r H_{298}^0 + \int_{298}^{T} \Delta_r c_p \mathrm{d}T, \tag{3.18}$$

$$\Delta_r H_{298}^0 = \sum \left(n_i \Delta H_{298}^{0,f} \right)_{\text{products}} - \sum \left(n_i \Delta H_{298}^{0,f} \right)_{\text{reagents}}, \tag{3.19}$$

where

$\Delta H_{298}^{0,f}$ is the standard enthalpy change of the formation of the reactants, J/mol;
$\Delta_r c_p$ is the heat capacity change of the reactions, J/mol K.

The entropy change of the reactions at temperature T was calculated as follows:

$$\Delta_r S_T^0 = \Delta_r S_{298}^0 + \int_{298}^{T} \frac{\Delta_r c_p}{T} \mathrm{d}T \tag{3.20}$$

$$\Delta_r S_{298}^0 = \sum \left(n_i \Delta S_{298}^{0,f} \right)_{\text{products}} - \sum \left(n_i \Delta S_{298}^{0,f} \right)_{\text{reagents}}, \tag{3.21}$$

where $\Delta H_{298}^{0,f}$ is the standard entropy changes of the formation of the reactants, J/mol.

The values of thermodynamic constants which were used for the calculations are given in Table 3.5.

In order to evaluate the influence of the main factors of synthesis—the temperature and the molar ratio of the initial mixture—on the formation of gyrolite, thermodynamic calculations of the gyrolite formation reactions (Gibbs energy dependence on the molar C/S ratio and temperature) were performed. We should highlight that, in

Table 3.5 Standard molar thermodynamic properties at 25 °C and 1 bar [116–120]

Compound	$\Delta H_{298}^{0,f}$, kJ/mol	$\Delta S_{298}^{0,f}$, J/(mol K)	c_p, J/(mol K)
$H_2O_{(vapor)}$	– 241.8	188.7	33.6
SiO_2	– 910.9	41.8	44.5
$Ca(OH)_2$	– 985.1	83.3	87.5
Gyrolite	– 4914.4	271.2	325.9
1.4 nm tobermorite	– 12,175.1	874.5	973.5
1.1 nm tobermorite	– 10,680.9	692.5	764.9
Xonotlite	– 10,022.1	507.5	628.64
Okenite	– 3135.7	208.5	210.0
Truscottite	– 16,854.6	927.6	1,034.1
Foshagite	– 6032.4	295.1	309.4
C–S–H(I)	– 1916.0	80.0	160.7
Z-phase	– 3135.7	170.9	161.9

the course of hydrothermal synthesis in the $Ca(OH)_2$–SiO_2–H_2O system, depending on the temperature and the length of the synthesis, the crystallization of the most probable CSH can take place according to Reactions 4.6–4.14 (assuming a molar CaO/SiO_2 ratio of 0.66 and excess of water):

Foshagite:

$$0.66Ca(OH)_2 + SiO_2 + 10H_2O \rightarrow 0.17(4CaO \cdot 3SiO_2 \cdot 1.5H_2O) + 0.5SiO_2 + 10.41H_2O. \quad (3.22)$$

Xonotlite:

$$0.66Ca(OH)_2 + SiO_2 + 10H_2O \rightarrow 0.11(6CaO \cdot 6SiO_2 \cdot H_2O) + 0.34SiO_2 + 10.55H_2O. \quad (3.23)$$

C–S–H(I):

$$0.66Ca(OH)_2 + SiO_2 + 10H_2O \rightarrow 0.8(0.83CaO \cdot SiO_2 \cdot 1.3H_2O) + 0.21SiO_2 + 9.63H_2O. \quad (3.24)$$

1.1 nm tobermorite:

$$0.66Ca(OH)_2 + SiO_2 + 10H_2O \rightarrow 0.13(5CaO \cdot 6SiO_2 \cdot 5.5H_2O) + 0.21SiO_2 + 9.93H_2O. \quad (3.25)$$

1.4 nm tobermorite:

$$0.66Ca(OH)_2 + SiO_2 + 10H_2O \rightarrow 0.13(5CaO \cdot 6SiO_2 \cdot 10.5H_2O) + 0.21SiO_2 + 9.27H_2O. \quad (3.26)$$

Truscottite:

$$0.66Ca(OH)_2 + SiO_2 + 10H_2O \rightarrow 0.08(7CaO \cdot 12SiO_2 \cdot 3H_2O) + 0.08Ca(OH)_2 + 10.33H_2O. \quad (3.27)$$

Gyrolite:

$$0.66Ca(OH)_2 + SiO_2 + 10H_2O \rightarrow 0.33(2CaO \cdot 3SiO_2 \cdot 2.5H_2O) + 0.01SiO_2 + 9.83H_2O. \quad (3.28)$$

Okenite:

$$0.66Ca(OH)_2 + SiO_2 + 10H_2O \rightarrow 0.5(CaO \cdot 2SiO_2 \cdot 2H_2O) + 0.16Ca(OH)_2 + 9.5H_2O. \quad (3.29)$$

Z-phase:

$$0.66Ca(OH)_2 + SiO_2 + 10H_2O \rightarrow 0.06(9CaO \cdot 16SiO_2 \cdot 15H_2O) + 0.10Ca(OH)_2 + 9.6H_2O. \quad (3.30)$$

The theoretical possibilities of the formation reactions of calcium silicate hydrates (Eqs. 3.22–3.30) are given in Fig. 3.21 ($\Delta_r G_T^0 = f(CaO/SiO_2)$ plot). It should be highlighted that the obtained values for the formation reaction of Z-phase (Eq. 3.30) could contain some errors because the formula for Z-phase and the thermodynamic values as presented in the literature somewhat differ.

The calculations show that the formation of C–S–H(I) (Eq. 3.24) is most probable when the temperature of the hydrothermal synthesis is lower than 225 °C because the determined Gibbs free energy value of this compound formation is the lowest (Fig. 3.21). Furthermore, these calculations align fairly well with the experimental results because calcium oxide and amorphous silica at the beginning of synthesis (8 h) were only identified to contain C–S–H(I) (Figs. 3.12 and 3.16). Meanwhile, in the mixtures with SGW-F, together with C–S–H(I), other calcium silicate hydrates were identified (Figs. 3.14 and 3.18). Therefore, it could be related to the recrystallization of C–S–H(I) to stable compounds or to the direct formation of crystalline calcium silicate hydrates because the obtained Gibbs free energy values at all given temperatures and molar ratios are negative for all reactions (Fig. 3.21).

The formed C–S–H(I) recrystallization to the most probable calcium silicate hydrates was evaluated in the following step of thermodynamic calculation. The calculations were performed while assuming that the molar ratio of the reacting components is equal to $CaO/SiO_2 = 0.55$ and $CaO/SiO_2 = 0.66$, i.e. similar to the used molar ratio. C–S–H(I) recrystallization can take place according to Reactions 3.31–3.37 (while assuming a molar CaO/SiO_2 ratio of 0.66 and excess of water):

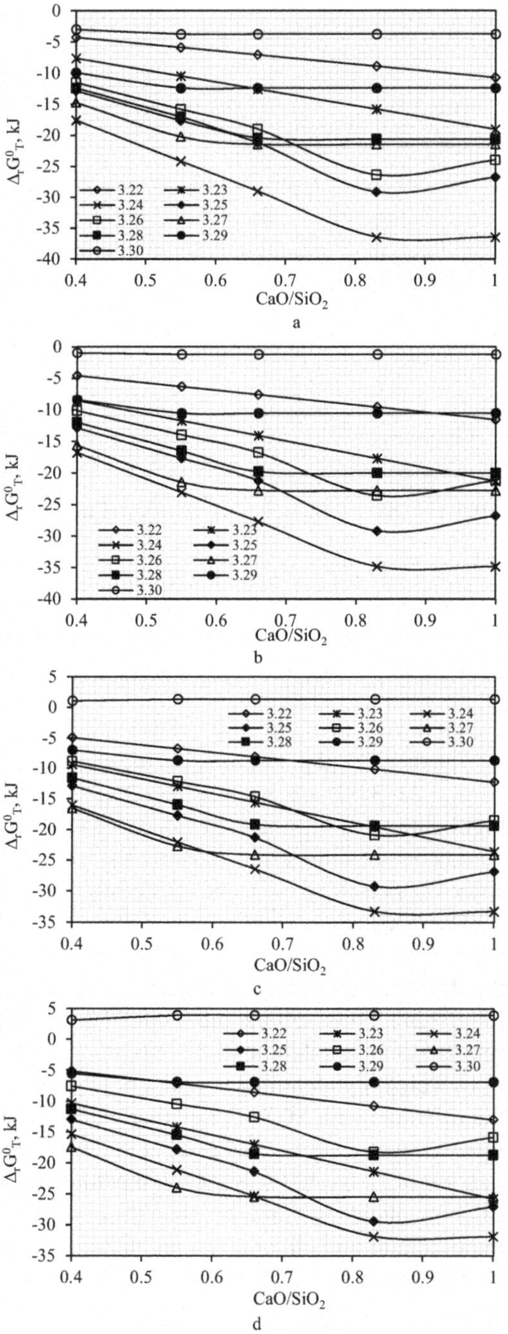

Fig. 3.21 Gibbs free energy dependence of Reactions 3.22–3.30 on the molar ratios of the initial mixtures CaO/SiO_2 at 150 °C (**a**), 175 °C (**b**), 200 °C (**c**), and 225 °C (**d**) temperatures

C–S–H(I) to foshagite

$$0.8(0.83CaO \cdot SiO_2 \cdot 1.3H_2O) + 0.21SiO_2 + 10H_2O \rightarrow 0.17(4CaO \cdot 3SiO_2 \cdot 1.5H_2O) + 0.5SiO_2 + 10.8H_2O. \tag{3.31}$$

C–S–H(I) to xonotlite

$$0.8(0.83CaO \cdot SiO_2 \cdot 1.3H_2O) + 0.21SiO_2 + 10H_2O \rightarrow 0.11(6CaO \cdot 6SiO_2 \cdot H_2O) + 0.34SiO_2 + 10.9H_2O. \tag{3.32}$$

C–S–H(I) to 1.1 nm tobermorite:

$$0.8(0.83CaO \cdot SiO_2 \cdot 1.3H_2O) + 0.21SiO_2 + 10H_2O \rightarrow 0.13(5CaO \cdot 6SiO_2 \cdot 5.5H_2O) + 0.21SiO_2 + 10.3H_2O. \tag{3.33}$$

C–S–H(I) to 1.4 nm tobermorite:

$$0.8(0.83CaO \cdot SiO_2 \cdot 1.3H_2O) + 0.21SiO_2 + 10H_2O \rightarrow 0.13(5CaO \cdot 6SiO_2 \cdot 10.5H_2O) + 0.21SiO_2 + 9.6H_2O. \tag{3.34}$$

C–S–H(I) to truscottite:

$$0.8(0.83CaO \cdot SiO_2 \cdot 1.3H_2O) + 0.21SiO_2 + 10H_2O \rightarrow 0.08(7CaO \cdot 12SiO_2 \cdot 3H_2O) + 0.08Ca(OH)_2 + 10.7H_2O. \tag{3.35}$$

C–S–H(I) to gyrolite:

$$0.8(0.83CaO \cdot SiO_2 \cdot 1.3H_2O) + 0.21SiO_2 + 10H_2O \rightarrow 0.33(2CaO \cdot 3SiO_2 \cdot 2.5H_2O) + 0.01SiO_2 + 10.2H_2O. \tag{3.36}$$

C–S–H(I) to *Z*-phase:

$$0.8(0.83CaO \cdot SiO_2 \cdot 1.3H_2O) + 0.21SiO_2 + 10H_2O \rightarrow 0.06(9CaO \cdot 16SiO_2 \cdot 15H_2O) + 0.10Ca(OH)_2 + 10H_2O. \tag{3.37}$$

The thermodynamic properties of the C–S–H(I) recrystallization reactions are presented in the $\Delta_r G^0_T = f(T)$ plot (Fig. 3.22). The calculations revealed that the C–S–H(I) recrystallization reaction to *Z*-phase (Eq. 3.37) is the most probable in both mixtures. However, by increasing the molar ratio of the reacting compounds from 0.55 to 0.66, the $\Delta_r G^0_T$ of the presently discussed reaction moves toward positive values. These results conform relatively well to the literature data since, in mixtures with a higher molar ratio (>0.83), C–S–H(I) does not recrystallize to *Z*-phase. Formation of other CSH is also possible, but the $\Delta_r G^0_T$ values of their formation reactions are higher (specifically, closer to 0) (Fig. 3.22).

In the next step of the synthesis of calcium silicate hydrates, *Z*-phase should recrystallize to the most thermodynamically stable compound according to the presented reactions (when assuming a molar CaO/SiO_2 ratio of 0.66 and excess of water):

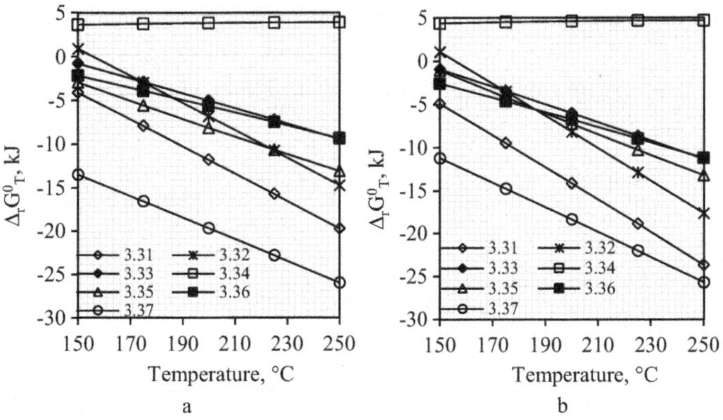

Fig. 3.22 Gibbs free energy dependence of Reactions 3.31–3.37 on the temperature when CaO/SiO_2 molar ratio of the reacting components equals to 0.55 (**a**) and 0.66 (**b**)

Z-phase to foshagite:

$$0.06(9CaO \cdot 16SiO_2 \cdot 15H_2O) + 0.10Ca(OH)_2 + 10H_2O \rightarrow 0.17(4CaO \cdot 3SiO_2 \cdot 1.5H_2O) + 0.5SiO_2 + 10.8H_2O. \quad (3.38)$$

Z-phase to xonotlite:

$$0.06(9CaO \cdot 16SiO_2 \cdot 15H_2O) + 0.10Ca(OH)_2 + 10H_2O \rightarrow 0.11(6CaO \cdot 6SiO_2 \cdot H_2O) + 0.34SiO_2 + 10.9H_2O. \quad (3.39)$$

Z-phase to 1.1 nm tobermorite:

$$0.06(9CaO \cdot 16SiO_2 \cdot 15H_2O) + 0.10Ca(OH)_2 + 10H_2O \rightarrow 0.13(5CaO \cdot 6SiO_2 \cdot 5.5H_2O) + 0.21SiO_2 + 10.3H_2O. \quad (3.40)$$

Z-phase to 1.4 nm tobermorite:

$$0.06(9CaO \cdot 16SiO_2 \cdot 15H_2O) + 0.10Ca(OH)_2 + 10H_2O \rightarrow 0.13(5CaO \cdot 6SiO_2 \cdot 10.5H_2O) + 0.21SiO_2 + 9.6H_2O. \quad (3.41)$$

Z-phase to truscottite:

$$0.06(9CaO \cdot 16SiO_2 \cdot 15H_2O) + 0.10Ca(OH)_2 + 10H_2O \rightarrow 0.08(7CaO \cdot 12SiO_2 \cdot 3H_2O) + 0.08Ca(OH)_2 + 10.7H_2O. \quad (3.42)$$

Z-phase to gyrolite:

$$0.06(9CaO \cdot 16SiO_2 \cdot 15H_2O) + 0.10Ca(OH)_2 + 10H_2O \rightarrow 0.33(2CaO \cdot 3SiO_2 \cdot 2.5H_2O) + 0.01SiO_2 + 10.2H_2O. \quad (3.43)$$

The results showed that, in mixtures with CaO/SiO_2 molar ratios equal to 0.55 and 0.66, Z-phase is metastable, and that it should recrystallize to a thermodynamically stable compound—gyrolite—because the obtained $\Delta_r G_T^0$ values of this compound formation (Eq. 3.43) are the lowest (Fig. 3.23). Furthermore, the obtained data can be well aligned with the experimental results because, under all investigated conditions, gyrolite was the only product obtained after 48–72 h of hydrothermal treatment.

As the experimental and theoretical results showed, the molar ratio, temperature, and duration exert decisive influence on the formation of the compounds of the gyrolite group. However, in the experimental part, the effect of the duration of the hydrothermal treatment longer than 504 h (Chap. 2) on the gyrolite stability was not investigated. Therefore, in order to predict the final products in the system, thermodynamic calculations of the gyrolite recrystallization reaction to other calcium silicate hydrates were performed. The calculations were performed while assuming that the molar ratio of the reacting components is equal to $CaO/SiO_2 = 0.55$ and $CaO/SiO_2 = 0.66$, i.e. equal to the molar ratio that was being used. Gyrolite recrystallization can take place according to Reactions 3.44–3.48:

Gyrolite to foshagite:

$$0.33(2CaO \cdot 3SiO_2 \cdot 2.5H_2O) + 0.01SiO_2 + 10H_2O \rightarrow 0.17(4CaO \cdot 3SiO_2 \cdot 1.5H_2O) + 0.5SiO_2 + 10.8H_2O. \quad (3.44)$$

Gyrolite to xonotlite:

$$0.33(2CaO \cdot 3SiO_2 \cdot 2.5H_2O) + 0.01SiO_2 + 10H_2O \rightarrow 0.11(6CaO \cdot 6SiO_2 \cdot H_2O) + 0.34SiO_2 + 10.9H_2O. \quad (3.45)$$

Gyrolite to 1.1 nm tobermorite:

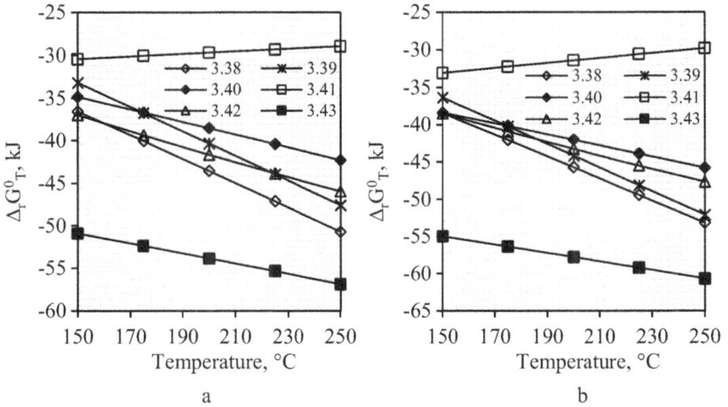

Fig. 3.23 Gibbs free energy dependence of Reactions 3.38–3.43 on the temperature when CaO/SiO_2 molar ratio of the reacting components equals to 0.55 (**a**) and 0.66 (**b**)

$$0.33(2CaO \cdot 3SiO_2 \cdot 2.5H_2O) + 0.01SiO_2 + 10H_2O \rightarrow 0.13(5CaO \cdot 6SiO_2 \cdot 5.5H_2O) + 0.21SiO_2 + 10.3H_2O. \quad (3.46)$$

Gyrolite to 1.4 nm tobermorite:

$$0.33(2CaO \cdot 3SiO_2 \cdot 2.5H_2O) + 0.01SiO_2 + 10H_2O \rightarrow 0.13(5CaO \cdot 6SiO_2 \cdot 10.5H_2O) + 0.21SiO_2 + 9.6H_2O. \quad (3.47)$$

Gyrolite to truscottite:

$$0.33(2CaO \cdot 3SiO_2 \cdot 2.5H_2O) + 0.01SiO_2 + 10H_2O \rightarrow 0.08(7CaO \cdot 12SiO_2 \cdot 3H_2O) + 0.08Ca(OH)_2 + 10.7H_2O. \quad (3.48)$$

The theoretical possibilities of foshagite, xonotlite, 1.1 nm tobermorite, 1.4 nm tobermorite, and truscottite formation reactions are given in Fig. 3.24.

The thermodynamic parameters of Reactions 3.44–3.48 showed that, when hydrothermally treating mixtures with CaO/SiO_2 (molar ratio 0.55–0.66) within the temperature range of 150–250 °C, the most significant possibility is that gyrolite will transform into truscottite (Eq. 3.48) because the Gibbs free energy values determined for this reaction are the lowest (Fig. 3.24). Gyrolite recrystallization to xonotlite (Eq. 3.45) might also have an impact on this process; however, its thermodynamic possibility is lower than the first option. It should be emphasized that the obtained $\Delta_r G_T^0$ values of the other compound formation reactions are positive; thus, these reactions should not proceed.

By summarizing the experimental and theoretical results of Sect. 3.3, it can be stated that the formation of the gyrolite family compounds under hydrothermal synthesis conditions proceeds according to the following sequence:

$$Ca(OH)_2 + SiO_2 + H_2O \rightarrow C{-}S{-}H(I) \rightarrow \text{Z-phase} \rightarrow \text{gyrolite gel} \rightarrow \text{gyrolite} \rightarrow \text{truscottite}$$

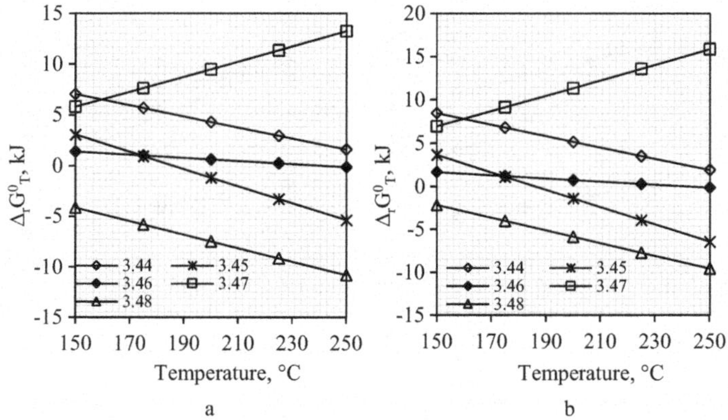

Fig. 3.24 Gibbs free energy dependence of Reactions 3.44–3.48 on the temperature when CaO/SiO_2 molar ratio of the reacting components is equal to 0.55 (**a**) and 0.66 (**b**)

Also, it has been established that AlF_3 production waste—silica gel with incorporated F^- and Al^{3+} ions—is a prospective SiO_2 source for the synthesis of low-base calcium silicate hydrates because the formation of gyrolite is shortened by half in comparison with pure $Ca(OH)_2$–SiO_2–H_2O mixtures. Besides, in the course of the hydrothermal synthesis, F^- ions were not released into the liquid medium, yet they were adsorbed by calcium silicate hydrates instead and/or combined into a stable compound, calcium fluoride.

3.4 Stability of Synthetic Calcium Silicate Hydrates with CaF_2 at 600–1000 °C Temperature

As established previously, one of the most important parameters of materials, which leads to their successful application, is their thermal stability. Various chemical and physical processes can occur during calcination due to which the mineralogical composition and morphology of samples can change. Thus, with the objective to determine the influence of CaF_2, which was formed during hydrothermal synthesis in the CaO-SGW-F-H_2O system, on the thermal behavior of calcium silicate hydrates, in the following section of this monograph, the obtained products were calcined in the temperature range of 600–1000 °C.

For the investigation, samples with different mineralogical compositions were chosen: 1—mainly consisting of synthetic gyrolite gel and synthetic *Z*-phase (CaO/SiO_2 = 0.55, 16 h, 200 °C); 2—mainly consisting of synthetic gyrolite (CaO/SiO_2 = 0.66, 48 h, 200 °C) (Fig. 3.25). It should be noted that, during synthesis, fluoride ions reacted and/or were adsorbed by-products of synthesis as the amount of F^- ions in the liquid medium was lower than 0.1 wt% (Table 3.4). Therefore, calcium fluoride was the only crystalline compound containing F^- ions after the synthesis.

It was ascertained that, after the calcination of the investigated mixtures at 600 °C temperature, synthetic gyrolite became metastable and recrystallized into related compounds: truscottite (PDF No. 00-017-0761; *d*-spacing: 2.047, 0.471, 0.314, and 0.282) and semicrystalline calcium silicate hydrate (Fig. 3.25, curve 2). Also, under these calcination conditions, katoite decomposed; however, the diffraction peaks characteristic of other compounds containing aluminum ions were not determined in the XRD pattern. The amount of the aluminate phase is probably small, and its peaks overlap with the diffraction maximums characteristic of calcium silicates. After incrementally increasing the calcination temperature to 800 °C, calcium fluoride silicate—cuspidine ($Ca_4Si_2O_7F_2$, PDF No. 00-011-0075; *d*-spacing: 0.336, 0.306, 0.287, and 0.188 nm)—and traces of wollastonite (PDF No. 00-043-1460; *d*-spacing: 0.352, 0.332, 0.298, and 0.231 nm) were formed (Fig. 3.25, curve 3). Finally, after calcination at 1000 °C temperature, wollastonite became the dominant compound in the calcination products (Fig. 3.25, curve 4). The interesting point is that CaF_2 remained stable in both mixtures after calcination at 1000 °C, although cuspidine, which contains fluoride, was formed after calcination at 800 °C and remained stable

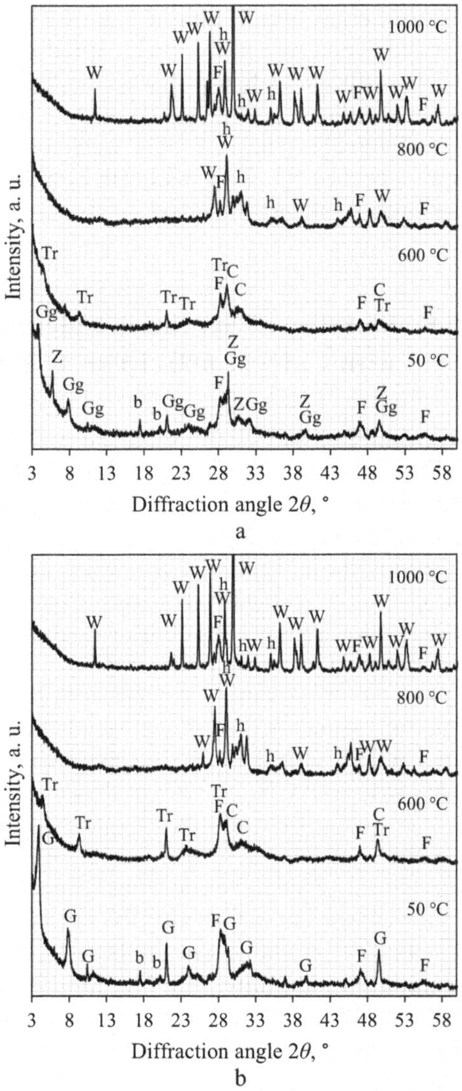

Fig. 3.25 XRD patterns of synthetic samples (**a**—CaO/SiO$_2$ = 0.55, 16 h, 200 °C; **b**—CaO/SiO$_2$ = 0.66, 48 h, 200 °C) calcined at different temperatures. *Indexes* G—gyrolite; C—C-S-H(I); Z—Z-phase; Gg—gyrolite gel; b—katoite; F—CaF_2; Tr—truscottite; h—cuspidine; W—wollastonite

till 1000 °C. It is possible that part of F^- ions reacted with calcium and formed calcium fluoride, while another part of fluoride was adsorbed by calcium silicate hydrates during the hydrothermal synthesis, and, finally, during calcination, it reacted with CSH to form cuspidine.

To determine the effect of the calcination temperature on the morphology of the products and to prove that F^- ions are strongly combined, SEM/EDX analysis of the synthetic and calcined samples was performed. In the SEM micrographs of the synthetic samples, crystals/particles of two different shapes can be seen: plate-shaped crystals characteristic of synthetic gyrolite [23] and cube-shaped crystals characteristic of CaF_2 [121] (Fig. 3.26). The data obtained in the course of EDX analysis proved that the latter particles consist of (in wt%) 20.18 F, 40.91 O, 5.21 Si, 20.18 Ca, and 13.52 Al.

It was determined that calcination influenced the morphology of the products. After calcination at 1000 °C temperature, plates with a smooth surface characteristic of wollastonite were dominant. Also, globule-shaped particles typical of fluorine-containing compounds were detected. The EDX analysis also proved that fluorine is strongly combined, and it remained in the sample structure even after calcination at 1000 °C (Fig. 3.27).

Data of FT-IR analysis (Fig. 3.28) confirmed the results of XRD and SEM. It was established that a small amount of katoite and calcium fluoride formed during the synthesis ($CaO/SiO_2 = 0.66$, 200 °C, 48 h) did not affect the FT-IR spectra of synthetic gyrolite because all the absorption bands characteristic of this compound were observed (Fig. 3.28, curve 1, Table 3.6).

It was subsequently established that, by increasing the temperature of calcination, the intensity of the absorption bands characteristic of H_2O and OH^- vibrations continuously decreased until they finally disappeared at 1000 °C (Fig. 3.28). Interesting results were obtained after calcination at 800 °C temperature because two absorption bands at 715 and 854 cm^{-1} were identified (Fig. 3.28, curve 3). These absorption bands disappeared after calcination at 1000 °C (Fig. 3.28, curve 4). The presently discussed absorption bands are probably related to the formation of truscottite or poorly crystalline calcium silicates during the partial decomposition of C–S–H(I). The FT-IR spectrum of the calcined sample at 1000 °C is complex because more than 11 absorption bands can be identified within the range of 800–1200 cm^{-1} (Fig. 3.28, curve 4). The five absorption bands positioned within the range of 1100–800 cm^{-1} could be assigned to the asymmetric and symmetric stretching vibrations of Si–O–Si bonds in SiO_4 tetrahedra. Meanwhile, the absorption bands at 647 cm^{-1} and

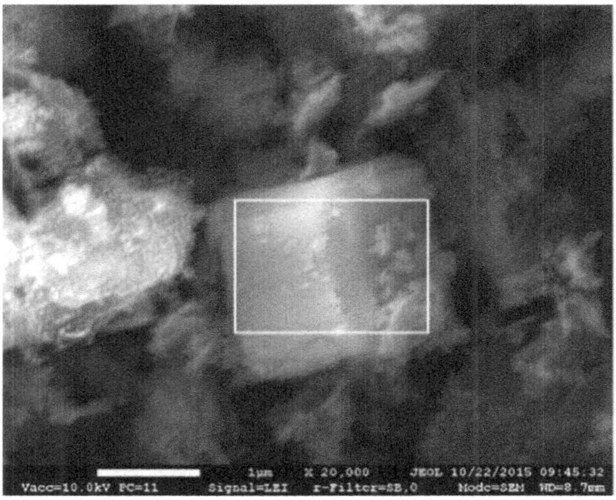

Fig. 3.26 Scanning electron micrograph of synthetic gyrolite with calcium fluoride. Chemical composition of spectrum 1, in wt%, was determined as follows: F—20.18; O—40.91; Si—5.21; Ca—20.18; Al—13.52

Fig. 3.27 Scanning electron micrographs of calcined gyrolite with calcium fluoride at 1000 °C temperature. Chemical composition of spectrum 2, in wt%, was determined as follows: F—9.05; O—47.02; Si—16.88; Ca—25.42; Al—1.63

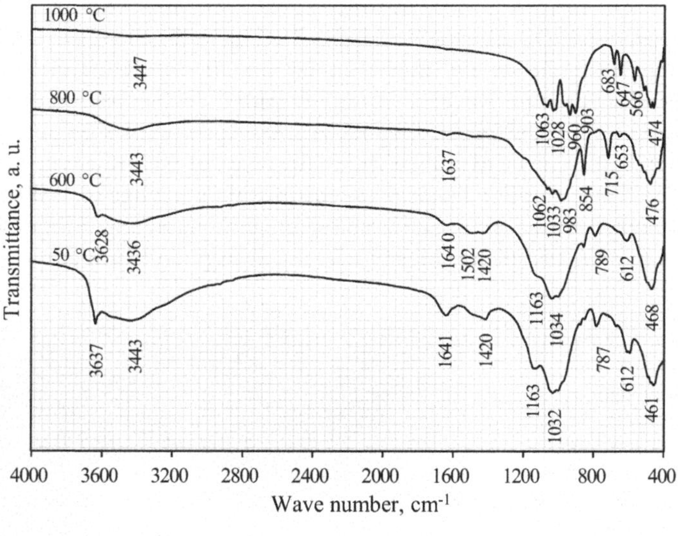

Fig. 3.28 FT-IR spectra of synthesized (curve 1) and calcined products (curve 2 at 600 °C; curve 3 at 800 °C; curve 4 at 1000 °C), when CaO/SiO_2 molar ratio was 0.66, and hydrothermal treatment duration was 48 h

683 cm^{-1} can be assigned to Si–O–Si symmetric stretching vibration in the wollastonite structure [122, 123] and to $[Si_2O_7]^{6-}$-dimer, $[SiF_6]^{2-}$-octahedral complexes of cuspidine [122, 124] (Fig. 3.28, curve 4). Within the lower frequency interval (400–800 cm^{-1}), the absorption maximums can be attributed to Si–O bending vibrations; in this specific region, a quintet is determined.

Table 3.6 FT-IR band assignments of synthetic gyrolite

Band assignment	Band position in spectra, cm^{-1}
δ(O–Si–O)	461
$\nu(OH^-)$	612
ν_s(O–Si–O)	787
ν_{as}(O–Si–O), from Q^3	1032, 1140
δH_2O (bending)	1641
$\nu(OH^-)$ (in gyrolite interlayer)	3443
$\nu(OH^-)$ (situated in lattice)	3637

Having summarized the data derived from the experiment, the formation sequence of compounds during the calcination of hydrothermal synthesis products at various temperatures is outlined in Table 3.7. The mineralogical composition of synthesized C–S–H was found out not to affect the stability of the formed CaF_2 as it remained stable under all calcination conditions. Also, during hydrothermal synthesis, part of F^- ions were adsorbed by calcium silicate hydrates, and, during calcination, they reacted with CSH to form cuspidine.

In addition, the obtained data showed that fluorine ions accelerate the recrystallization of calcium silicate hydrates into wollastonite as the formation of this compound was initiated at lower temperatures (at ~800 °C, Fig. 3.25) compared with the above-mentioned reaction in the pure systems.

Thus, AlF_3 production waste not only influences positively the formation of the gyrolite group compounds and the crystallization of wollastonite during calcination, but it also combines F^- ions into chemically and thermally stable compounds.

Table 3.7 Formation sequence of compounds during calcination of hydrothermal synthesis products under different conditions

Synthesis conditions	Products	Calcination temperature, °C				
		600	700	800	900	1000
$CaO/SiO_2 = 0.55$, 16 h, 200 °C	Gyrolite gel, Z-phase, CaF_2, katoite	Truscottite, C–S–H(I), CaF_2		Cuspidine, wollastonite, CaF_2		CaF_2, wollastonite, cuspidine
$CaO/SiO_2 = 0.66$, 16 h, 200 °C	Gyrolite, CaF_2, katoite					

3.5 Influence of $Ca(NO_3)_2$ Additive on the Formation and Porosity of Gyrolite Family Compounds

The previously outlined results and the scientific literature discussed in Sect. 3.1 *Influence of additives on the synthesis and structure properties of gyrolite* showed that the formation and properties of gyrolite depend on the conditions of hydrothermal synthesis as well as on the chemical composition of the primary mixtures. Unfortunately, a number of the possible mixture compositions that can reduce the production costs and improve specific properties have not been studied yet. The influence of the nature of silicon-containing compounds on the formation of gyrolite was described in the previous chapter; thus, in the next stage of this research, calcium oxide was partially replaced with calcium nitrate. The reduction of the pH value of the primary mixtures may change the formation and properties of gyrolite.

3.5.1 *Influence of $Ca(NO_3)_2$ on the Formation of Gyrolite Family Compounds*

To determine the impact of the calcium nitrate solution on the formation of gyrolite, five mixtures with different concentrations of $Ca(NO)_3$ were produced (Table 3.8). The samples were labeled as S0 (pure sample), S5, S10, S15, and S20 (Table 3.8).

In order to prepare the initial mixtures, the necessary amount of the solid components was weighed and poured into sealed plastic containers with 5 porcelain grinding bodies (so that to ensure the quality and consistency of mixing). The primary mixtures were homogenized for 1 h at 30 rpm speed. The prepared mixtures were mixed with $Ca(NO_3)_2{\cdot}4H_2O$ solutions to reach w/s = 20 and $CaO/SiO_2 = 0.66$. Afterward, the prepared suspensions were treated hydrothermally at 200 °C for 4, 16, 24, 32, 48, and 72 h.

It was discovered that the nature of a calcium containing compound strongly affects the formation of calcium silicate hydrates. At the initial stage of the synthesis, only semicrystalline calcium silicate hydrate C–S–H(I) was formed in samples S0 and

Table 3.8 Chemical composition of primary mixtures

Amount (mol%) of Ca^{2+} ions replaced by calcium from $Ca(NO_3)_2$ solution	0	5	10	15	20
Molar ratio of dry components	0.66	0.627	0.594	0.561	0.528
Ca^{2+} concentration in $Ca(NO_3)_2$ solution, g/dm^3	0.00	0.0069	0.0137	0.0206	0.0275
pH value of $Ca(NO_3)_2$ solution	6.51	5.42	5.03	4.97	4.95
w/s and CaO/SiO_2 ratio of final mixture	w/s = 10; $CaO/SiO_2 = 0.66$				
Abbreviation	S0	S5	S10	S15	S20

S5 (Fig. 3.29). Meanwhile, the increment of the $Ca(NO_3)_2{\cdot}4H_2O$ solution concentration leads to the formation of a compound of a higher crystallinity—*Z*-phase—in samples S10, S15, and S20. It is worth noting that the intensity of the diffraction peaks characteristic of the *Z*-phase was the highest in sample S20. It was established that carbonization proceeds under all synthesis conditions because traces of calcium carbonate were identified in all products.

The results of XRD were approved by FT-IR data (Fig. 3.30). In the FT-IR spectrum, all the absorption bands characteristic of C–S–H(I) were identified [125]:

- ~454 cm^{-1} characteristic of δ(Si–O–Si) vibrations;
- ~673 cm^{-1} characteristic of ν_s(Si–O–Si) vibrations;
- ~978 cm^{-1} characteristic of ν(Si–O) vibrations from Q^2.

Also, the intensive absorption bands characteristic of δH_2O (~1640 cm^{-1}) and $\nu(OH^-)$ (3460 cm^{-1}) vibrations were identified. The carbonization of the products was confirmed by the identification of the absorption bands at ~875 cm^{-1} and ~1460 cm^{-1} which can be assigned to the vibrations of the $(CO_3)^{2-}$ group. It is worth noting that, during the synthesis, $(NO_3)^-$ anions intercalate into the structure of the synthesis products because the absorption band at ~1384 cm^{-1} was identified in all the samples with calcium nitrate solutions [122, 126, 127] (Fig. 3.30). The intensity of the latter band increased when increasing the concentration of the calcium nitrate solution.

By extending the duration of the hydrothermal treatment to 16 h, *Z*-phase together with C–S–H(I) was obtained in all the samples (Fig. 3.31). Furthermore, it was established that the crystallinity of *Z*-phase depends on the composition of the initial mixture, and it increases by increasing the concentration of the calcium nitrate solution.

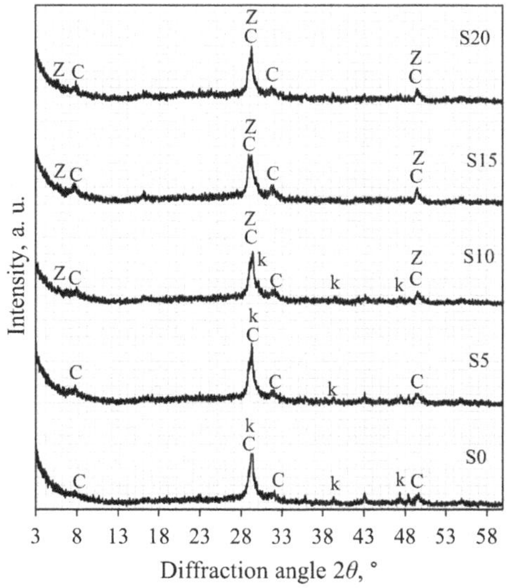

Fig. 3.29 XRD patterns of synthesis products with hydrothermal synthesis duration at 200 °C temperature of 4 h. *Indexes* C—C-S-H(I); Z—Z-phase; k—calcite

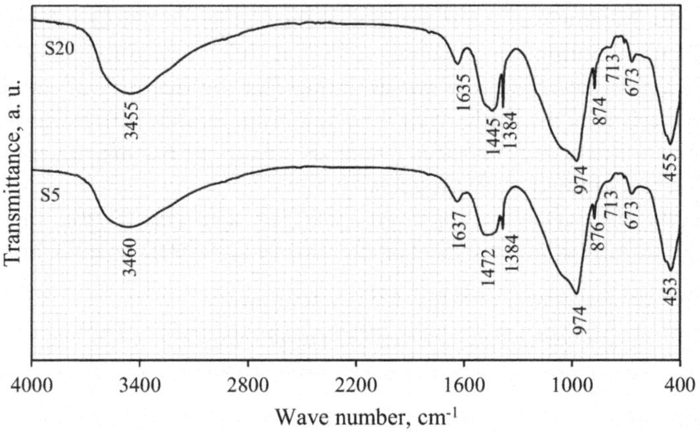

Fig. 3.30 FT-IR spectra of synthesis products with hydrothermal treatment duration at 200 °C equal to 4 h

Fig. 3.31 XRD patterns of the synthesis products with hydrothermal synthesis duration at 200 °C temperature equal to 16 h. *Indexes* C—C-S-H(I); Z—Z-phase; C—calcite

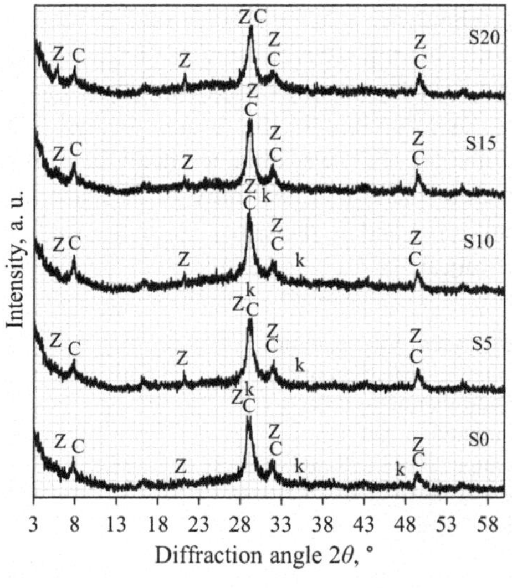

The duration of hydrothermal synthesis was established not to affect the FT-IR spectrum of the synthesis products significantly (Figs. 3.30 and 3.32a). The main differences are related to the changes of the absorption band's intensity. Due to the higher crystallinity of Z-phase, the intensity of the absorption bands characteristic of this compound increased in the S20 sample (Fig. 3.32). Meanwhile, the intensity of the absorption bands characteristic of $(CO_3)^{2-}$ decreased in all the samples, which shows that the synthesis products are more resistant to carbonization.

The DSC analysis showed that the synthesis products are characterized by three thermal effects (Fig. 3.32b). The first endothermic effect within the temperature range

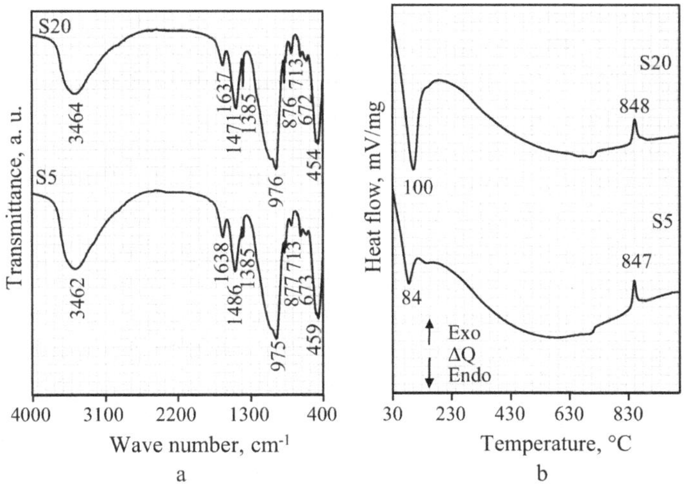

Fig. 3.32 FT-IR spectra (**a**) and DSC curves (**b**) of synthesis products with hydrothermal treatment duration at 200 °C equal to 16 h

of 30–200 °C can be assigned to the removal of adsorbed and structural water from calcium silicate hydrates. It was established that the heat of this effect is more than 10 times higher in sample S20 than in sample S5, which confirms that, in sample S20, *Z*-phase with a higher crystallinity was formed. The second minor endothermic effect at ~700 °C temperature corresponds to the decomposition of calcium carbonate, while the exothermic effect at ~848 °C may be assigned to calcium silicate hydrate recrystallization to wollastonite.

The increment of the length of hydrothermal synthesis to 24 h leads to the crystallization of compounds of a higher crystallinity (Fig. 3.33a). It was found out that *Z*-phase and C–S–H(I) become metastable and start recrystallizing to gyrolite gel in the pure sample (Fig. 3.33a). The same tendency was observed in the samples with 5–15 mol% of Ca^{2+} ions from the initial mixture replaced by calcium from the $Ca(NO_3)_2$ solution. Different results were observed in sample S20 because diffraction peaks characteristic of gyrolite were identified.

The formation of gyrolite in sample S20 was proven by FT-IR analysis (Fig. 3.34a). A new absorption band at 613 cm^{-1} can be attributed to $\nu(OH^-)$ vibrations, while the absorption bands within the range of 900–1200 cm^{-1} can be attributed to ν_s(O–Si–O) and ν_{as}(O–Si–O) vibrations. The most characteristic absorption band of synthetic gyrolite is observed at ~3640 cm^{-1} (Fig. 3.34a). This absorption band shows that the interlayer characteristic of synthetic gyrolite was formed, and water molecules are strongly bounded in the crystal lattice of synthetic gyrolite.

It was established that gyrolite gel became metastable and recrystallized to other products of the synthesis after 32 h of synthesizing (Fig. 3.33b). Furthermore, *Z*-phase was almost completely recrystallized to gyrolite in the pure sample, while, in the other samples, intensive diffraction peaks characteristic of *Z*-phase were still

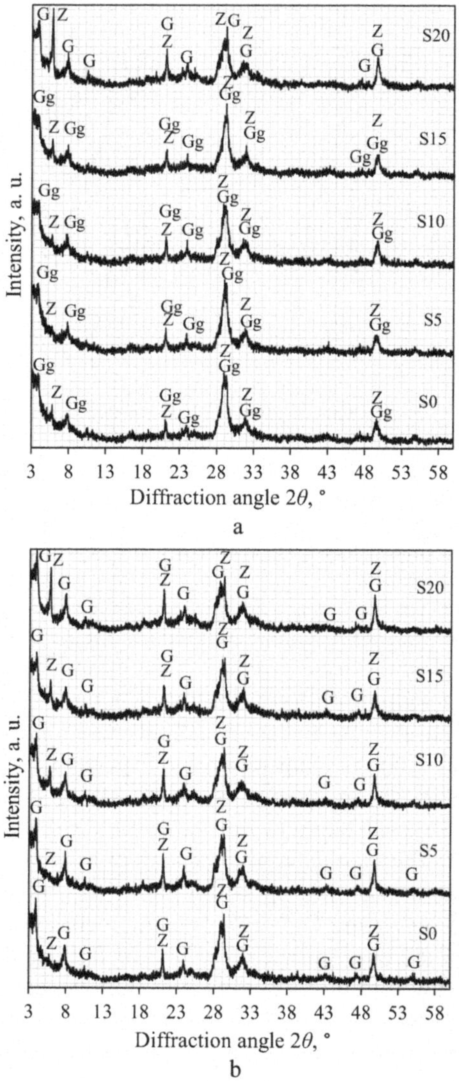

Fig. 3.33 XRD patterns of the synthesis products with hydrothermal synthesis duration at 200 °C temperature equal to 24 h (**a**) and 32 h (**b**). *Indexes* Z—Z-phase; k—calcite; Gg—gyrolite gel; G—gyrolite

observed. These results show that the calcium nitrate solution used for the synthesis of calcium silicate hydrates extends the duration of the existence of the intermediate phases.

XRD results were reiterated by the FT-IR data because, in the FT-IR spectra of the synthesis products, all absorption bands characteristic of gyrolite were observed (Fig. 3.34b). In addition, it is worth noting that there are no significant differences between the FT-IR spectra of samples S5 and S20.

The results of scanning electron microscopy also confirmed that both gyrolite and *Z*-phase are dominant compounds in the synthesis products (Fig. 3.35). Furthermore,

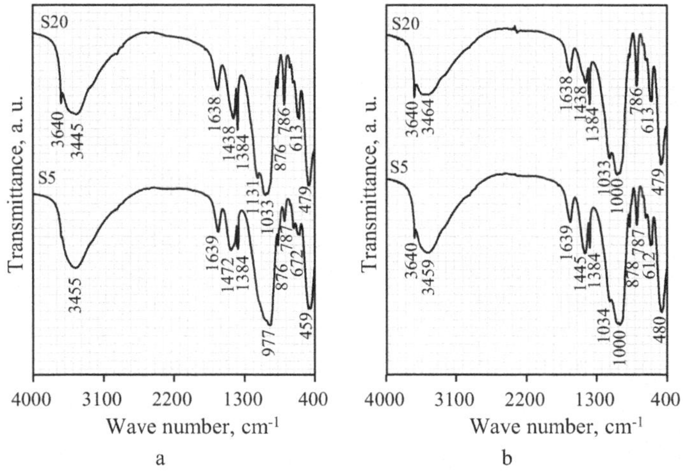

Fig. 3.34 FT-IR spectra of the synthesis products with hydrothermal treatment duration at 200 °C equal to 24 h (**a**) and 32 h (**b**)

it was observed that, in sample S0, a lower quantity of amorphous mass and small plate-shaped crystal aggregates characteristic of *Z*-phase were present than in sample S20. Also, large particles (1–5 μm) of synthetic gyrolite were dominant in the pure sample.

It was determined that gyrolite was the only product formed after 48–72 h of hydrothermal synthesis (Fig. 3.36a). However, in sample S20, the main diffraction peak characteristic of synthetic gyrolite (*d*—2.2 nm) became wider, and its intensity decreased. It can be related to the beginning of gyrolite recrystallization to truscottite. It was determined that, in sample S20, the intensity of the absorption band at 3640 cm^{-1} is significantly higher than in sample S5. Probably, more water molecules intercalate into the interlayer, and thus the intensity of the absorption band increased, whereas the intensity of the main diffraction peak decreased. It is worth highlighting that gyrolite adsorbed nitrate anions because the intensive absorption band at ~1384 cm^{-1} characteristic of $(NO_3)^-$ group vibrations was determined under all experimental conditions. Hence, synthetic gyrolite is applicable not only for cation adsorption, but also for anions.

It was determined that the pH value of the liquid medium derived after synthesis depends on the concentration of the calcium nitrate solution used for synthesis (Table 3.9). It was measured that the highest value of pH was reached in sample S0, while the lowest value was reached in sample S20. Such differences in pH values can be related to the different mineralogical compositions of the samples during hydrothermal treatment for 4–72 h at 200 °C.

The obtained data brings us to the conclusion that the calcium nitrate solution used for synthesizing calcium silicate hydrates changes the sequence of the compound formation. It was determined that, at the beginning of the synthesis, the calcium nitrate solution promotes the formation of *Z*-phase and later its recrystallization to

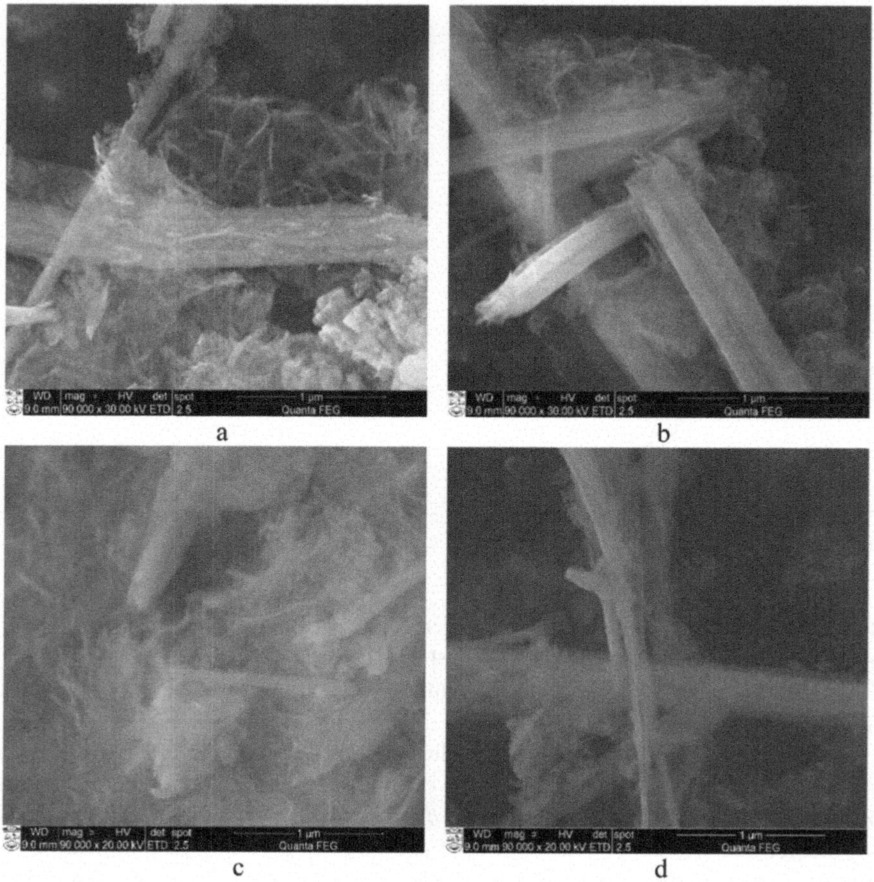

Fig. 3.35 Scanning electron micrograph of S0 (**a**, **b**) and S20 (**c**, **d**) samples after 32 h of synthesis

gyrolite. Also, when 20 mol% of calcium ions in the initial mixture were replaced by calcium from the calcium nitrate solution, the formation of gyrolite proceeded directly from Z-phase, while, in the pure system, it happened through an intermediate compound—gyrolite gel. The final product of hydrothermal synthesis is gyrolite with intercalated nitrate anions. Since the intercalated anion group can change the properties of synthetic gyrolite, the synthetic product was described by nitrogen adsorption at 77 K analysis.

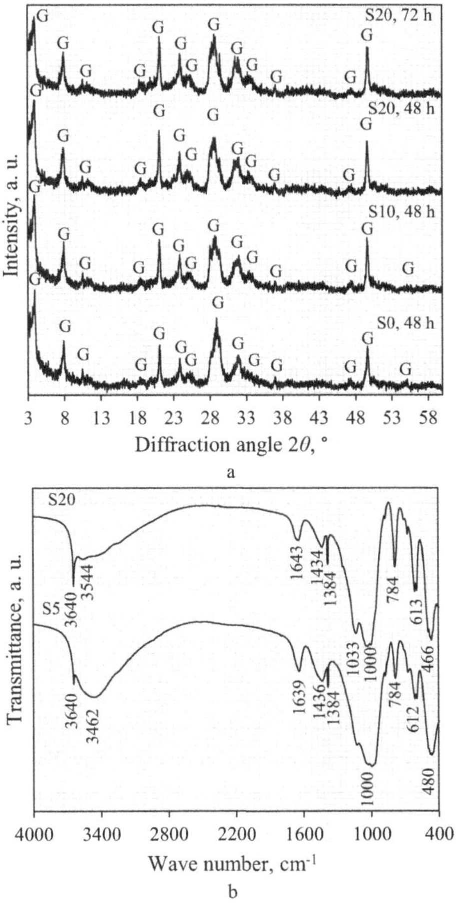

Fig. 3.36 XRD patterns (**a**) and FT-IR spectra (**b**) of the synthesis products with hydrothermal synthesis duration at 200 °C temperature equal to 48 and 72 h. *Indexes* G—gyrolite

Table 3.9 pH values of liquid medium after 48 h of synthesis

Sample	pH value
S0	8.99
S5	8.71
S10	8.52
S15	8.34
S20	7.91

3.5.2 Porosity of Synthetic Gyrolite with Intercalated Nitrate Anions

Assumptions about the shape of the pores of samples S10 and S20 were made based on the nature of the nitrogen adsorption–desorption isotherms which are featured in

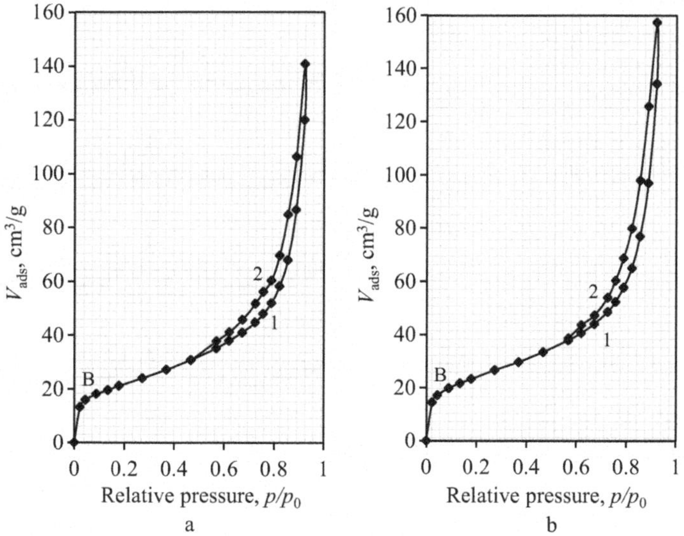

Fig. 3.37 Adsorption-desorption isotherms of samples S10 (**a**) and S20 (**b**) after 48 h of synthesis

Fig. 3.37. As in the case of the pure synthetic sample (Fig. 2.13b), the adsorption isotherms of the synthesis products can be attributed to Type II or Type IV, and they are characterized by the occurrence of the hysteresis loop [128]. Thus, mesoporous pores are predominant in the samples S10 and S20 structure.

It was also established that the calcium nitrate solution exerts an impact on the shape of the hysteresis loops of the synthesis products (Figs. 2.13b and 3.37). The pure sample is characterized by a wide loop which can be assigned to Type H3 category (Fig. 2.13b). This type of hysteresis loop suggests that mesopores are formed between parallel crystal planes (slit-shape pores). Meanwhile, the hysteresis loops of samples S10 and S20 become narrower and completely close at the higher value of p/p_0 (Fig. 3.37). The shape of these hysteresis loops is intermediate between H1 and H3 [128]. Probably, both types of pores are formed in this sample.

Since the adsorption isotherms of samples S10 and S20 are characterized as Type II and/or Type IV and have a point B, the specific surface area was determined by employing the Brunauer–Emmett–Teller (BET) method [128–131]. As expected, the BET equation gives a linear plot in both cases, and straight lines with the correlation coefficient $R^2 > 0.999$ were obtained (Fig. 3.38). The derived parameters of the BET equation are given in Table 3.10.

The calculations revealed that both S10 and S20 samples are characterized by the optimal values of the constant C_{BET}, i.e. between 50 and 150 (Table 3.10). Furthermore, it was determined that, by increasing the concentration of the calcium nitrate solution used for gyrolite synthesis, the value of the specific surface area of the

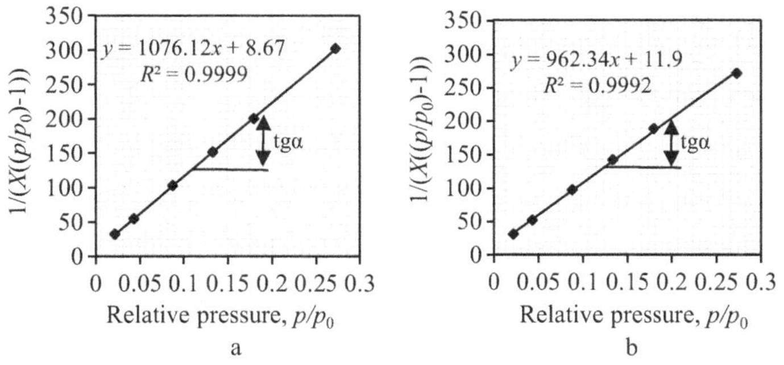

Fig. 3.38 Adsorption-desorption isotherms in the BET plot of samples S10 (**a**) and S20 (**b**) after 48 h of synthesis

Table 3.10 Numerical values of parameters of synthetic gyrolite specific surface area (S_{BET})

Sample	Sample mass m, g	BET equation constants		Capacity of monolayer X_m, g (× 10^4)	S_{BET}, m^2/g	C_{BET}	Reliability coefficient R^2
		Slope S = $tg\alpha$	Intercept I				
SC	0.0323	1128.43	5.79	8.82	95.12	195.82	0.9996
S10	0.0410	1076.12	8.67	9.22	122.61	125.18	0.9999
S20	0.0402	962.34	11.90	10.26	136.52	81.87	0.9992

samples went up from 95 m^2/g (pure sample) to ~136 m^2/g (sample S20). The increment of the surface area can be attributed to the changes in the pore shape or due to the nitrate anions adsorbed by the synthesis products.

To confirm the above-discussed results, calculations of the dominant pore shape by using the corrected Kelvin equation and the scheme devised by Orr et al. by employing the entire N_2 desorption isotherm at 77 K were performed [129–132]. A pure synthetic gyrolite sample is characterized by well-defined slit-shape pores formed between parallel planes because the difference between S_{BET} and ΣA is negligible (4.43%) (Table 2.6). Meanwhile, the slit-shape pore model is not applicable for samples S10 and S20 since the difference between the values of S_{BET} and ΣA is more than 50% (Table 3.11). Lower differences were obtained when using a cylindrical-shape pores model (29–35%); however, the mismatch is still very high. This data aligns fairly well with the classification of the hysteresis loop to the intermediate case between H1 and H3 (Fig. 3.37). Thus, pores of both shapes are present in samples S10 and S20.

The total pore volume and the distribution of the sizes of the pores in samples S10 and S20 synthesized for 48 h were calculated by employing the cylindrical-shape pores model because the difference between the values of S_{BET} and ΣA is lower (Figs. 3.39 and 3.40). It was calculated that the total pore volume of samples S10 and S20 equaled to 220 mm^3/g and 240 mm^3/g, respectively (Fig. 3.39), which is more

Table 3.11 ΣA calculation values of synthetic samples

Sample	S_{BET}, m^2/g	Results for the cylindrical pores model			Results for the slit-shape pores model		
		ΣA, m^2/g	$\|S_{BET}—ΣA\|$, m^2/g	$\|S_{BET}—ΣA\|$, %	ΣA, m^2/g	$\|S_{BET}—ΣA\|$, m^2/g	$\|S_{BET}—ΣA\|$, %
S0	95.12	150.72	55.60	58.45	99.33	4.21	4.43
S10	122.61	78.83	43.78	35.71	47.00	75.61	61.67
S20	136.52	95.84	40.68	29.80	80.96	80.96	59.30

than 2 times lower than the value of the pure synthetic gyrolite sample (Fig. 2.18, curve 2). The lower value of the pore volume can be causally attributed to the intercalation of nitrate anions in the pores of synthetic gyrolite during hydrothermal synthesis.

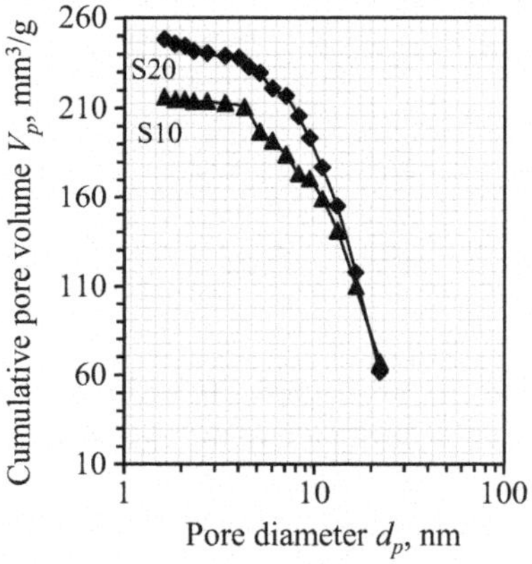

Fig. 3.39 Total pore volume of samples S10 and S20 after 48 h of synthesis

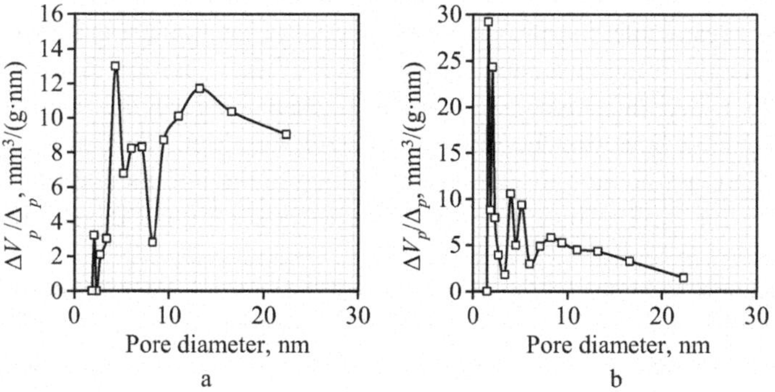

Fig. 3.40 Differential pore volume of samples S10 and S20 after 48 h of synthesis

It was established that the pore diameter of samples S10 and S20 varied in the range of 2–30 nm (Fig. 3.40). The dominant pore diameter of sample S10 is similar to the pure sample (Figs. 2.19b and 3.40a); however, in sample S10, more pores with a higher diameter are present. A different result was obtained in sample S20 because pores with smaller diameters are dominant there (Fig. 3.40b).

3.6 Summary

In conclusion, in terms of the obtained data, it can be emphasized that the calcium nitrate solution used for synthesizing calcium silicate hydrates affects the formation of gyrolite and its properties. It was determined that pure synthetic gyrolite synthesized for 48 h features well-defined slit-shape pores. Meanwhile, gyrolite synthesized by using calcium nitrate solutions is denoted by the cylindrical-shape and slit-shape pores. BET calculations revealed that, by increasing the concentration of the calcium nitrate solution used for gyrolite synthesis, the value of the specific surface area of the samples went up from 95 m^2 /g (pure sample) to ~136 m^2 /g (sample S20). Finally, the total pore volume of samples S10 and S20 was less than half the value of the pure synthetic gyrolite sample.

Although the calcium nitrate solution promotes the crystallization of gyrolite, nitrate anions intercalate into the structure of the synthesis products. Also, because of the intercalation of these anions, the total pore volume of synthetic gyrolite decreased, which can negatively affect the adsorption properties of gyrolite. Thus, it is recommended to use calcium oxide instead of calcium nitrate to synthesize gyrolite with a high pore volume.

References

1. Majdinasab A, Yuan Q (2020) Synthesis of Al-substituted 11Å tobermorite using waste glass cullet: a study on the microstructure. Mater Chem Phys 250:123069
2. Zhu G, Li H, Wang X, Li S, Hou X, Wu W, Tang Q (2016) Synthesis of calcium silicate hydrate in highly alkaline system. J Am Ceram Soc 99:2778–2785
3. Tian H, Stephan D, Lothenbach B, Lehmann C (2021) Influence of foreign ions on calcium silicate hydrate under hydrothermal conditions: a review. Constr Build Mater 301:124071
4. Meller N, Hall C, Phipps JS (2005) A new phase diagram for the $CaOAl_2O_3SiO_2H_2O$ hydroceramic system at 200 °C. Mater Res Bull 40:715–723
5. Meller N, Kyritsis K, Hall C (2009) The mineralogy of the CaO–Al_2O_3–SiO_2–H_2O (CASH) hydroceramic system from 200 to 350 °C. Cem Concr Res 39:45–53
6. Lothenbach B, Jansen D, Yan Y, Schreiner J (2022) Solubility and characterization of synthesized 11 Å Al-tobermorite. Cem Concr Res 159:106871
7. Dambrauskas T, Knabikaite I, Eisinas A, Baltakys K, Palou MT (2020) Influence of Cr^{3+}, Co^{2+} and Cu^{2+} on the formation of calcium silicates hydrates under hydrothermal conditions at 200 °C. J Asian Ceram Soc 8:753–763

8. Matekonis G, Šiaučiūnas R, Vaičiukynienė D (2010) Hydrothermal synthesis and characterization of Na^+ and $[Al^{3+} + N^{a+}]$-substituted tobermorite in $CaO\text{-}SiO_2{\cdot}nH_2O\text{-}H_2O$ system. Medziagotyra 16:242–248
9. Wang Z, Ma S, Zheng S, Wang X (2017) Incorporation of Al and Na in hydrothermally synthesized tobermorite. J Am Ceram Soc 100:792–799
10. NocuÒ-Wczelik W (1999) Effect of Na and Al on the phase composition and morphology of autoclaved calcium silicate hydrates. Cem Concr Res 29:1759–1767
11. Baltakys K, Siauciunas R (2007) Formation of gyrolite in the $CaO\text{-quartz-}Na_2O\text{-}H_2O$ system. Mater Sci Poland 25:1081–1100
12. Blakeman EA, Gard JA, Ramsay CG, Taylor HFW (1974) Studies on the system sodium oxide-calcium oxide-silica-water. J Appl Chem 24:239–245
13. Baltušnikas A, Šiaučiūnas R, Lukošiūtė I, Baltakys K, Eisinas A, Kriūkienė R (2015) Crystal structure refinement of synthetic pure gyrolite. Medziagotyra 21:111–116
14. Merlino S (1988) Gyrolite: its crystal structure and crystal chemistry. Mineral Mag 52:377–387
15. Garbev K, Black L, Beuchle G, Stemmermann P (2002) Inorganic polymers in cement based materials. Wasser Geotechnologie 1:19–30
16. Miyake M, Iwaya M, Suzuki T, Kakehi H, Mitsuda T (1990) Aluminum-substituted gyrolite as cation exchanger. J Am Ceram Soc 73:3524–3527
17. Nocun-Wczelik W (1997) Application of NMR in the studies of some autoclaved calcium silicate hydrates, synthesized with admixtures. Adv Cem Chem Proc Inter Colloq 59–66
18. El-Koroshy SA, Al-Wakeel EI, El-Hemaly SA, Rizk MA (2002) Influence of synthesis conditions on structure, thermal behaviour and cation exchange capacity of 11 Å-tobermorites. Mater Res Bull 45:723–741
19. Baltakys K, Siauciunas R (2006) The influence of $\gamma\text{-}Al_2O_3$ and Na_2O on the formation of gyrolite in the stirring suspension. J Mater Sci 41:4799–4805
20. Baltakys K, Siauciunas R (2007) The influence of $\gamma\text{-}Al_2O_3$ and Na_2O on the formation of calcium silicate hydrates in the $CaO\text{–quartz–}H_2O$ system. Mater Sci Poland 25:185–198
21. Shaw S, Henderson CMB, Clark SM (2002) In-situ synchrotron study of the kinetics, thermodynamics, and reaction mechanisms of the hydrothermal crystallization of gyrolite, $Ca_{16}Si_{24}O_{60}(OH)_8{\cdot}14H_2O$. Am Mineral 87:533–541
22. Różycka A, Kotwica Ł, Małolepszy J (2014) Synthesis of single phase gyrolite in the $CaO\text{-quartz-}Na_2O\text{-}H_2O$ system. Mater Lett 120:166–169
23. Różycka A, Kotwica Ł (2020) Effect of alkali on the synthesis of single phase gyrolite in the system $CaO\text{-quartz-}H_2O$. Constr Build Mater 239:117799
24. Stumm A, Garbev K, Beuchle G, Black L, Stemmermann P, Nüesch R (2005) Incorporation of zinc into calcium silicate hydrates, Part I: formation of C-S-H(I) with C/S=2/3 and its isochemical counterpart gyrolite. Cem Concr Res 35:1665–1675
25. Baltakys K, Siauciunas R (2010) Influence of gypsum additive on the gyrolite formation process. Cem Concr Res 40:376–383
26. Nocuń-Wczelik W (1997) Effect of some inorganic admixtures on the formation and properties of calcium silicate hydrates produced in hydrothermal conditions. Cem Concr Res 27:83–92
27. Siauciunas R, Baltakys K (2010) Calcium silicate hydrates: synthesis, properties and use of compounds of the phyllosilicate group. Technologija, Kaunas
28. Elton NJ, Hooper JJ, Holyer VAD (1998) Al-free gyrolite from the Lizard, Cornwall, England. Mineral Mag 62:271–272
29. Okada Y, Masuda T, Ishida H (1995) Behavior of silicate anions on the formation processes of gyrolite followed by Si-29 NMR. J Ceram Soc Jpn 103:124–127
30. Eisinas A, Dambrauskas T (2023) Cu^{2+} ions adsorption by dense calcium silicate hydrate samples and their catalytic activity. Ceram Int 49:14886–14894
31. Števula L, Petrovič J (1983) Formation of an intermediate C-S-H phase during the hydrothermal synthesis of gyrolite. Cem Concr Res 13:684–688
32. Siauciunas R, Baltakys K (2004) Formation of gyrolite during hydrothermal synthesis in the mixtures of CaO and amorphous SiO_2 or quartz. Cem Concr Res 34:2029–2036

33. Yamaguchi N, Kanda T, Fujimori H, Ioku K, Goto S (2001) Gyrolite fine particles synthesized from glass in the system of CaO-SiO_2 by hydrothermal reaction. J Mater Sci Lett 20:489–490
34. Eisinas A, Dambrauskas T, Eisinaite V, Sipailiene A (2022) Synthesis of antibacterially active wollastonite by using dense calcium silicate hydrate samples. J Therm Anal Calorim 147:14163–14174
35. Elehinafe FB, Olomukoro OG, Ayeni AO, Okedere OB (2022) A short review on land/soil pollution: the pollutants and the treatment techniques. In: Ayeni AO, Oladokun O, Orodu OD (eds) Advanced manufacturing in biological, petroleum, and nanotechnology processing. Green Energy and Technology. Springer Cham
36. Rathi BS, Kumar PS, Vo DN (2021) Critical review on hazardous pollutants in water environment: occurrence, monitoring, fate, removal technologies and risk assessment. Sci Total Environ 797:149134
37. Sicard P, Agathokleous E, Anenberg SC, De Marco A, Paoletti E, Calatayud V (2023) Trends in urban air pollution over the last two decades: a global perspective. Sci Total Environ 858:160064
38. Eurostat SE (2022) Periodical waste statistics. Available via: https://ec.europa.eu/eurostat/statistics-explained/index.php?title=Waste_statistics. Accessed 15 July 2023
39. Luo H, Guo HL, Gao YS, Liu JJ (2018) A study of using construction waste and recyclable materials in modern landscape design, vol 186, p 012060
40. Morfeldt C (1993) Landscaping with waste. Eng Geol 34:135–143
41. Sheth PN, Babu BV (2010) Production of hydrogen energy through biomass (waste wood) gasification. Int J Hydrogen Energy 35:10803–10810
42. Czajczyńska D, Krzyżyńska R, Jouhara H, Spencer N (2017) Use of pyrolytic gas from waste tire as a fuel: a review. Energy 134:1121–1131
43. Fercoq A, Lamouri S, Carbone V, Lelièvre A, Lemieux AA (2013) Combining lean and green in manufacturing: a model of waste management. IFAC Proc Volumes 46:117–122
44. Blackman Jr WC (2016) Basic hazardous waste management. Lewis Publishers
45. Wang LK, Hung Y, Lo HH, Yapijakis C (2004) Handbook of industrial and hazardous wastes treatment. Marcel Dekker
46. Wang J, Shih Y, Wang PY, Yu YH, Su JF, Huang C (2019) Hazardous waste treatment technologies. Water Environ Res 91:1177–1198
47. Directive EC. Directive 2008/98/EC of the European Parliament and of the Council of 19 November 2008 on waste and repealing certain Directives. Document 02008L0098-20180705. 2018 https://eur-lex.europa.eu/legal-content/EN/TXT/?uri=CELEX%3A02008L0098-20180705. Accessed 15 July 2023
48. Gharfalkar M, Court R, Campbell C, Ali Z, Hillier G (2015) Analysis of waste hierarchy in the European waste directive 2008/98/EC. Waste Manage 39:305–313
49. Tan B, He L, Dai Z, Sun R, Jiang S, Lu Z, Liang Y, Ren L, Sun S, Zhang Y, Li C (2022) Review on recent progress of bioremediation strategies in Landfill leachate—a green approach. J Water Process Eng 50:103229
50. Wu G, Wang L, Yang R, Hou W, Zhang S, Guo X, Zhao W (2022) Pollution characteristics and risk assessment of heavy metals in the soil of a construction waste landfill site. Ecol Inform 70:101700
51. Manikanda Bharath K, Muthulakshmi AL, Natesan U (2023) Microplastic contamination around the landfills: distribution, characterization and threats: a review. Curr Opin Environ Sci 31:100422
52. European Commission (2015) Closing the loop—an EU action plan for the circular economy. Document 52015DC0614. https://eur-lex.europa.eu/legal-content/EN/TXT/?uri=CELEX%3A52015DC0614. Accessed 15 July 2023
53. European Commission (2020) A new circular economy action plan for a cleaner and more competitive Europe. Document 52020DC0098. https://eur-lex.europa.eu/legal-content/EN/TXT/?uri=CELEX%3A52020DC0098. Accessed 15 July 2023
54. Hänsch K, Kinkel K (1995) European parliament and council directive 94/62/EC of 20 December 1994 on packaging and packaging waste. J Environ Law 7:323–337

55. European Commission (2018) Impacts of circular economy policies on the labour market. Cambridge Econometrics, Directorate-General for Environment (European Commission), ICF, and Trinomics
56. Ondiviela JA (2020) Beyond smart cities: creating the most attractive cities for talented citizens. Springer Cham
57. El-Gamal SMA, Selim FA (2017) Utilization of some industrial wastes for eco-friendly cement production. Sustain Mater Technol 12:9–17
58. Nasr MS, Shubbar AA, Abed ZAR, Ibrahim MS (2020) Properties of eco-friendly cement mortar contained recycled materials from different sources. J Build Eng 31:101444
59. Habert G, Miller SA, John VM, Provis JL, Favier A, Horvath A, Scrivener KL (2020) Environmental impacts and decarbonization strategies in the cement and concrete industries. Nat Rev Earth Environ 1:559–573
60. Wong S, Ghafar NA, Ngadi N, Razmi FA, Inuwa IM, Mat R, Amin NAS (2020) Effective removal of anionic textile dyes using adsorbent synthesized from coffee waste. Sci Rep 10:2928
61. Bóta A, László K, György Nagy L, Subklew G, Schlimper H, Schwuger MJ (1997) Adsorbents from waste materials, vol 3, pp 81–91
62. Yoldi M, Fuentes-Ordoñez EG, Korili SA, Gil A (2019) Zeolite synthesis from industrial wastes. Microporous Mesoporous Mater 287:183–191
63. Liu W, Liaw B, Chang H, Wang Y, Chen P (2017) From waste to health: synthesis of hydroxyapatite scaffolds from fish scales for lead ion removal. JOM 69:713–718
64. Dreveton A (2012) Manufacture of aluminium fluoride of high density and anhydrous hydrofluoric acid from fluosilicic acid. Procedia Eng 46:255–265
65. O'Driscoll M (2017) China supply shortages hit consumers as new sources emerge. Alum Int Today 30:26–28
66. Krahl T, Kemnitz E (2017) Aluminium fluoride—the strongest solid Lewis acid: structure and reactivity. Catal Sci Technol 7:773–796
67. Dreveton A (2016) Overview of the fluorochemicals industrial sectors. Procedia Eng 138:240–247
68. European Commission (2007) Selected illustrative LVIC-S industry products. Integrated pollution prevention and control reference document on best available techniques for the manufacture of large volume inorganic chemicals—solids and other industry. Available via: https://eippcb.jrc.ec.europa.eu/reference/large-volume-inorganic-chemicals-solids-and-others-industry. Accessed 15 July 2023
69. Samrane K, Al-hjouj M, Doudin F, Kossir A (2011) Fluosilicic acid: recovery system and aluminum fluoride production. Arab Fertil Assoc Publ 11:1–20
70. Dreveton A (2014) Economic aspects of utilizing fluosilicic acid as raw material for the manufacture of hydrofluoric acid and aluminium fluoride. Procedia Eng 83:279–285
71. Saadaoui E, Ghazel N, Ben Romdhane C, Massoudi N (2017) Phosphogypsum: potential uses and problems—a review. Int J Environ Stud 74:558–567
72. Valančius Z, Vaickelionienė R, Vaickelionis G, Makčinskas P (2022) Use of an industrial by-product phosphogypsum in the production of white textured paints. J Clean Prod 380:134888
73. Indian Minerals Yearbook 2017 (2017) Part- III: mineral reviews, 56th edn, Fluorite. Government of India, Ministry of Mines, Indian Bureau of Mines
74. Villalba G, Ayres RU, Schroder H (2007) Accounting for fluorine: production, use, and loss. J Ind Ecol 11:85–101
75. International Fertilizer Association. Phosphate products. Available via: https://www.ifastat.org/supply/Phosphate%20Products/Phosphate%20Rock. Accessed 15 July 2023
76. European Phosphate Fertilizer Alliance. Data and statistics of phosphate rock. Available via: http://aeep.eu/data-and-statistics/. Accessed 15 July 2023
77. Frazier AW, Lehr JR, Dillard EF (1977) Chemical behavior of fluorine in production of wet-process phosphoric acid. Environ Sci Technol 11:1007–1014
78. Gouider M, Feki M, Sayadi S (2009) Separative recovery with lime of phosphate and fluoride from an acidic effluent containing H_3PO_4, HF and/or H_2SiF_6. J Hazard Mater 170:962–968

79. Ramteke LP, Sahayam AC, Ghosh A, Rambabu U, Reddy MRP, Popat KM, Rebary B, Kubavat D, Marathe KV, Ghosh PK (2018) Study of fluoride content in some commercial phosphate fertilizers. J Fluorine Chem 210:149–155
80. Best available techniques for pollution prevention and control in the European fertilizer industry. Booklet No. 4 of 8 Production of phosphoric acid. EFM Association, 2000
81. Speight JG (2017) Chapter three—industrial inorganic chemistry. In: Environmental inorganic chemistry for engineers. Butterworth-Heinemann, pp 111–169
82. Will RK (2016) The benefits of isolating and utilizing fluorine from phosphate operations. Procedia Eng 138:267–272
83. Bayat M, Taeb A, Rastegar S (2002) Investigation of the filtration rate of silica in aluminum fluoride production from silicic acid. Chem Eng Sci 57:2879–2884
84. Coggins A (2019) Aluminium fluoride and aluminium: 2019 trends and outlook to 2029. Fluorine Forum 2019
85. Barbier O, Arreola-Mendoza L, Del Razo LM (2010) Molecular mechanisms of fluoride toxicity. Chem Biol Interact 188:319–333
86. Walna B, Kurzyca I, Bednorz E, Kolendowicz L (2013) Fluoride pollution of atmospheric precipitation and its relationship with air circulation and weather patterns (Wielkopolski National Park, Poland). Environ Monit Assess 185:5497–5514
87. Sahu P (2019) Fluoride pollution in groundwater. In: Sikdar PK (ed) Groundwater development and management: issues and challenges in South Asia. Springer International Publishing, Cham, pp 329–350
88. Tolkou AK, Manousi N, Zachariadis GA, Katsoyiannis IA, Deliyanni EA (2021) Recently developed adsorbing materials for fluoride removal from water and fluoride analytical determination techniques: a review. Sustainability 13:7061
89. Singh J, Singh P, Singh A (2016) Fluoride ions vs removal technologies: a study. Arab J Chem 9:815–824
90. Jadhav SV, Bringas E, Yadav GD, Rathod VK, Ortiz I, Marathe KV (2015) Arsenic and fluoride contaminated groundwaters: a review of current technologies for contaminants removal. J Environ Manage 162:306–325
91. Mohapatra M, Anand S, Mishra BK, Giles DE, Singh P (2009) Review of fluoride removal from drinking water. J Environ Manage 91:67–77
92. Gogoi S, Nath SK, Bordoloi S, Dutta RK (2015) Fluoride removal from groundwater by limestone treatment in presence of phosphoric acid. J Environ Manage 152:132–139
93. Aldaco R, Garea A, Irabien A (2007) Calcium fluoride recovery from fluoride wastewater in a fluidized bed reactor. Water Res 41:810–818
94. Waghmare SS, Arfin T (2015) Fluoride removal from water by various techniques. Int J Innov Sci Eng Technol 2:560–571
95. Jadhav SV, Gadipelly CR, Marathe KV, Rathod VK (2014) Treatment of fluoride concentrates from membrane unit using salt solutions. J Water Process Eng 2:31–36
96. Mandeikytė N, Rumša R, Urbonas L (2002) Binding materials on the basis of technogenical raw materials. Cheminė technologija 2:19–24
97. Vaičiukynienė D, Vaitkevičius V, Kantautas A, Sasnauskas V (2012) Utilization of by-product waste silica in concrete-based materials. Mater Res 15:561–567
98. Vaičiukynienė D, Kantautas A, Vaitkevičius V, Sasnauskas V (2009) Using of modified AlF_3 production waste in cement-based materials. Medziagotyra 15:255–261
99. Borg RP, Briguglio C, Bocullo V, Vaičiukynienė D (2017) Preliminary investigation of geopolymer binder from waste materials. Rev Rom Mater 47:370–378
100. Vaičiukynienė D, Borg RP, Kielė A, Kantautas A (2018) Alkali-activated blends of calcined AlF_3 production waste and clay. Ceram Int 44:12573–12579
101. Krivenko P, Vaičiukynienė D, Kantautas A, Vaitkevičius V, Šerelis E (2019) Effect of AlF_3 production waste on the processes of hydration and hardening of the alkali-activated Portland cement with sodium silicate hydrate. J Therm Anal Calorim 138:879–887
102. Kaminskas R, Kubiliute R (2014) Artificial pozzolana from silica gel waste–clay–limestone composite. Adv Cem Res 26:155–168

103. Kaminskas R, Cesnauskas V, Kubiliute R (2015) Influence of different artificial additives on Portland cement hydration and hardening. Constr Build Mater 95:537–544
104. Kubiliūtė R, Kaminskas R (2013) The pozzolanic activity of calcined clay–silica gel composites. Medziagotyra 19:453–460
105. Vaičiukynienė D, Kantautas A, Vaitkevičius V, Jakevičius L, Rudžionis Ž, Paškevičius M (2015) Effects of ultrasonic treatment on zeolite NaA synthesized from by-product silica. Ultrason Sonochem 27:515–521
106. Palubinskaitė D, Kantautas A (2005) Synthesis of Zeolite using technogenical silicagel. Cheminė technologija 4:35–40
107. Sarkar R, Ghosh S, Das SK (2007) Waste silica from aluminum fluoride industries used for ceramic whitewares. Am Ceram Soc Bull 86:9201
108. Krysztafkiewicz A, Rager B, Maik M (1996) Silica recovery from waste obtained in hydrofluoric acid and aluminum fluoride production from fluosilicic acid. J Hazard Mater 48:31–49
109. Krysztafkiewicz A, Świt Z, Jesionowski T (2005) Evaluation of waste silica precipitated in the process of hydrofluoric acid production from fluosilicic acid. Physicochem Probl Miner Process 39:165–176
110. Dambrauskas T, Baltakys K, Grineviciene A, Rudelis V (2021) The effect of various hydroxide and salt additives on the reduction of fluoride ion mobility in industrial waste. Sustainability 13:1554
111. Yang GY, Shi YC, Liu XD, Mujumdar AS (2007) TG-DTG analysis of chemically bound moisture removal of $AlF_3 \cdot 3H_2O$. Drying Technol 25:675–680
112. Delong X, Yongqin L, Ying J, Longbao Z, Wenkui G (2000) Thermal behavior of aluminum fluoride trihydrate. Thermochim Acta 352–353:47–52
113. Iljina A, Baltakys K, Baltakys M, Siauciunas R (2014) Neutralization and removal of compounds containing fluoride ions from waste silica gel. Rev Rom Mater 44:265–271
114. Duque-Redondo E, Bonnaud PA, Manzano H (2022) A comprehensive review of C-S-H empirical and computational models, their applications, and practical aspects. Cem Concr Res 156:106784
115. Richardson IG (2008) The calcium silicate hydrates. Cem Concr Res 38:137–158
116. Babushkin VI, Matveev GM, Mchedlov-Petrosyan OP (1962) Thermodynamics of silicates. Gosstoiizdat, Moscow
117. Mchedlov-Petrosyan OP (1967) Thermodynamics and thermochemistry of cement. In: 6th International Congress on the chemistry of cement. Moscow, pp 6–16
118. Blanc P, Bourbon X, Lassin A, Gaucher EC (2010) Chemical model for cement-based materials: temperature dependence of thermodynamic functions for nanocrystalline and crystalline C-S-H phases. Cem Concr Res 40:851–866
119. Matschei T, Lothenbach B, Glasser FP (2007) Thermodynamic properties of Portland cement hydrates in the system $CaO–Al_2O_3–SiO_2–CaSO_4–CaCO_3–H_2O$. Cem Concr Res 37:1379–1410
120. Lothenbach B, Winnefeld F (2006) Thermodynamic modelling of the hydration of Portland cement. Cem Concr Res 36:209–226
121. Koeser J, Carvalho TS, Pieles U, Lussi A (2014) Preparation and optimization of calcium fluoride particles for dental applications. J Mater Sci Mater Med 25:1671–1677
122. Chukanov NV (2013) Infrared spectra of mineral species: extended library. Springer Science & Business Media
123. Morsy R, Abuelkhair R, Elnimr T (2017) Synthesis of microcrystalline wollastonite bioceramics and evolution of bioactivity. Silicon 9:489–493
124. Cruz-Ramírez A, Romo-Castañeda J, Hernández-Pérez, María de los Ángeles, Vargas-Ramírez M, Romero-Serrano A, Hallen-López M (2011) An application of infrared analysis to determine the mineralogical phases formation in fluxes for thin slab casting of steel. J Fluorine Chem 132:323–326
125. He Y, Zhao X, Lu L, Struble LJ, Hu S (2011) Effect of C/S ratio on morphology and structure of hydrothermally synthesized calcium silicate hydrate. J Wuhan Univ Technol Mat Sci Edit 26:770–773

126. Murakami J, Vargheese V, Bando KK (2022) Methane activation on Pt^{2+} sites on Pt/TiO_2 with adsorbed nitrate species as evidenced by in situ FTIR spectroscopy. J Phys Chem C 126:11075–11081
127. Goebbert DJ, Garand E, Wende T, Bergmann R, Meijer G, Asmis KR, Neumark DM (2009) Infrared spectroscopy of the microhydrated nitrate ions $NO_3^-(H_2O)_{1-6}$. J Phys Chem A 113:7584–7592
128. Thommes M, Kaneko K, Neimark AV, Olivier JP, Rodriguez-Reinoso F, Rouquerol J, Sing KS (2015) Physisorption of gases, with special reference to the evaluation of surface area and pore size distribution (IUPAC Technical Report). Pure Appl Chem 87:1051–1069
129. Rouquerol J, Rouquerol F, Llewellyn P, Maurin G, Sing KS (2013) Adsorption by powders and porous solids: principles, methodology and applications. Academic Press
130. Lowell S, Shields JE, Thomas MA, Thommes M (2006) Characterization of porous solids and powders: surface area, pore size and density. Springer Science & Business Media
131. Naderi M (2015) Surface area: Brunauer–Emmett–Teller (BET). In: Progress in filtration and separation. Academic Press, pp 585–608
132. Webb PA, Orr C (1997) Analytical methods in fine particle technology. Micromeritics Instrument Corporation

Chapter 4
Application of Synthetic Gyrolite for Adsorption

The fourth chapter of the monograph describes the necessity of adsorption of heavy metal ions and the feasibility of application of synthetic gyrolite for this process. The adsorption offers flexibility in design and operation compared with the alternative water treatment methods. Adsorption depends on the nature and properties of the adsorbate and the adsorptive. Thus, academic knowledge on the synthesis along with the research of low-cost and effective adsorbents is in great demand. Synthetic gyrolite is characterized by a high value of the thickness of the interlayer sheets (2.2 nm); thus, it can absorb a high amount of different ions. For this reason, adsorption experiments in solutions with different chemical compositions were performed. As a result, it was determined that synthetic gyrolite could be used for both single and simultaneous adsorption of metal ions; however, the adsorption rate and capacity depend on the nature of the adsorbate. Finally, the selectivity sequences of synthetic gyrolite according to the adsorbed amount of metal ions were presented by summarizing the experimental results.

4.1 Water Pollution with Heavy Metal Ions

Chapter 3 of this monograph introduced the necessity of reducing and utilizing solid waste. However, the intense growth of the human population numbers and the concurrent progress of technologies has led not only to the generation of an enormous quantity of solid waste, but also elevated the issue of wastewater. According to the literature [1], every year in the world, 390–500 km^3 of wastewater is being generated, which may include various industrial pollutants (drugs, microplastics, heavy metals, petroleum products, etc.) (Fig. 4.1). The purification and reuse of wastewater is a huge challenge [2–4]. The data from the 'first world' suggests that somewhere between one third and two thirds of wastewater is treated, while the share of reused water is less than 10% [1]. Globally, more than 80% of wastewater is released into the

K. Baltakys and T. Dambrauskas, *Synthetic Gyrolite*,
https://doi.org/10.1007/978-3-031-49259-4_4

Fig. 4.1 Wastewater (Photo ID: 1364760615, *Source* iStock by Getty Images)

environment without sufficient treatment. Thus, the purification and reuse of wastewater has been attracting considerable academic attention, which should potentially lead to the development of environment-friendly alternatives within the industry.

One of the most common contaminants in wastewater and groundwater is heavy metals which significantly influence not only the human health but also the environment [5, 6]. Heavy metals are the metals with an atomic number greater than 20 [7, 8]. Some metals are necessary for plants and living organisms (these are microelements), while other metals are toxic and unnecessary for the body. It is worth noting that all metals tend to accumulate in organisms even when their concentration in the water is relatively low; thus, in the long term, their concentration in the organism can reach a toxic limit [1, 7–9]. Heavy metals are long-term pollutants which accumulate in soil and plants because they do not decompose in the natural environment. The primary sources of these pollutants are chemical, electronics, metallurgical, and other industries (Table 4.1).

Unfortunately, every year, more and more heavy metals, including lead, cadmium, chromium, nickel, and mercury, are used in the industry. Hence, their concentration in the environment and water increases. For this reason, scientists and industry specialists are looking for economically attractive ways to purify wastewaters. According to the literature, various treatment processes can be used for water purification: chemical precipitation, foam flotation, electrochemical treatment, filtration, ion exchange, membrane technologies, and evaporation [2, 3, 5, 6]. However, most of these technologies incur essential disadvantages: complex operation, incomplete removal, high-energy requirements, secondary pollution, etc. Among the presently mentioned techniques, adsorption provides a variety of design and operation options; thus, the search for low-cost and effective adsorbents has been in focus recently [13, 14]. Adsorbents include minerals, zeolites, organic compounds, industrial by-products, and other materials with a high surface area. However, when these adsorbents come

Table 4.1 Anthropogenic sources of heavy metal ions [10–12]

Source of pollutant	Emitted heavy metal ions
Processing of metals	All heavy metals: Mo, Co, Ni, V, Cr, W, Ba, Mn, Nb, Ti, Y, Yb, Sc, Ga, etc.
Manufacturing and repairing of electronic devices	Co, Pb, Cu, Mn, Ni, Cd, Cr, Zn, Hg, Sn, Ag
Galvanic	Ni, Zn, Cu, Cd, Cr, Co
Oil-fired power plants	V, Ni, Cr, Mn, Zn, Cu
Coal-fired furnaces	As, Mo, Ni, Ag, Ga, Ba, Sc
Printing industry	Zn, Pb, Sn, Cd, Cu
Chemical industry, production of plastics	Pb, Mn, Cu, Cd, Ni, Cr, Zn
Production of phosphoric fertilizers	Pb, Zn, Cu, Cd, Ni, Cr
Production of leather, light industry	Cr, Cu, Cd, Zn, Ni, etc.
Production of concrete, glass, and thermal insulation materials	Cr, Cu, Cd, Zn, Ni, Mn, Pb
Domestic wastewater	Cr, Cu, Cd, Zn, Ni, Mn, Pb

to the limit of their adsorption capacity, in most cases, they get stored in landfills as they are discarded, which eventually leads to the problem of secondary pollution [14–16]. Thus, the design and exploration of novel adsorbents which could be reused (for example, as additives in ordinary Portland cement or in the manufacturing of alternative cementitious materials) is still necessary.

4.2 Types of Adsorbents

To effectively remove heavy metals from wastewater, it is essential to ensure that the concentration of heavy metals does not go over the highest allowed concentration limits at the end of the process. Thus, the conventional processes and equipment are not sufficient. One of the most effective ways to remove heavy metals is the use of adsorbents.

Adsorbents are usually highly porous materials which can absorb large amounts of chemical elements on their surface [13, 14, 17, 18]. The specific surface area of these materials depends on the volume, size, and number of pores. Essentially, it can vary from 10 to 1400 m^2/g [16]. The adsorption efficiency is determined not only by the specific surface area but also by the similarity of the adsorbent and the sorbed substances, i.e. by the chemical structure of the sorbent surface and the sorbate, the number of functional groups, and their nature (Table 4.2). In order to use adsorbents efficiently and effectively, they must meet the basic requirements:

- Sufficient adsorption capacity and selectivity.
- Large specific surface.

Table 4.2 Adsorption capacity of some inorganic adsorbents for metal ions [13, 16, 19–21]

Adsorbent	Metal ions	Surface area, m^2/g	Adsorption capacity, mg/g
Activated alumina	Cd^{2+}	126	6.16
	Pb^{2+}		9.15
	Cr^{3+}	370	3.12
	Zn^{2+}	–	13.69
Activated carbon	Cr^{3+}	796	9
	Zn^{2+}		4
	Pb^{2+}		0.4
Biochar	Zn^{2+}	48	221
	Pb^{2+}		119
Carbon nanotubes	Ni^{2+}	–	3900
	Zn^{2+}		3650
	As^{3+}		3500
	Co^{2+}		3800
Clay	Cu^{2+}	63	196
	Ni^{2+}		164
Clinoptilolite	Pb^{2+}	40	200.54
Fe(III)-modified zeolite tuff	As^{5+}	–	1.55
Graphene	As^{5+}	151	191
	In^{3+}	177	147
Hydrochar	Cd^{2+}	28	91
LDH	Pb^{2+}	405	591
	CrO_4^-		331
Manganese-oxide coated-zeolite	U	24.87	15.1
Metal oxide	Cr^{3+}	278	92
MOF	Hg^{2+}	296	1080
	Pb^{2+}		510
Mxene	Ba^{2+}	10	180
	Sr^{2+}		225
Ordered carbon	U	187	150
Silica gel	Co^{2+}	–	10
	Ni^{2+}	–	10
Silica	Ru	58	55
Zeolite	Fe^{3+}	15.88	6.41
	Cu^{2+}		0.54
	Mn^{2+}		0.52
	Zn^{2+}		2.21

- Chemically inert.
- Homogeneous.
- Particles of the required size.
- Mechanically resistant.

Compounds that meet these requirements may vary in terms of being either natural or synthetic [13, 14, 17, 18]. Natural organic adsorbents include peat, straw, hay, bark, sawdust, and other similar substances. They can be used to remove heavy metals and organic contaminants. Meanwhile, natural inorganic adsorbents can be clay, diatomite, perlite, zeolite, vermiculite, sand, volcanic ash, activated carbon, opoka, etc. Synthetic sorbents can be various compounds with a high surface area and adsorption capacity for pollutants (organic and inorganic compounds, calcium silicates, calcium aluminates, etc.). The most widely used adsorbents in various industries are activated carbon, zeolites, silica gel, alumina, magnesium oxide, and magnesium silicate.

Activated alumina is an aluminum oxide with a high surface area, porosity, mechanical resistance, and amphoteric properties [14]. Thus, it is widely used as a support of catalysts and as a sorbent. Activated alumina shows good adsorption properties for removing heavy metal and fluoride ions from water [13, 14, 16, 22]. Also, it can be used as an adsorbent in gas purification processes, to produce advanced ceramics or wear-resistant composites, etc.

Zeolites (hydrated aluminosilicates) are another group of adsorbents widely used in the industrial processes (Fig. 4.2). The structure of zeolites includes alkaline and alkaline earth metal cations (normally, Na^+, K^+, Ca^{2+}, Mg^{2+}, etc.) as well as water, which can be easily replaced by other cations and anions [23, 24]. Also, they are denoted by a large surface area and porosity. For these reasons, natural and synthetic zeolites show high cation exchange and molecular network properties [13, 14]. Zeolites offer a high ion exchange capacity for ammonium and heavy metal ions (Table 4.2) and organic matter from aqueous solutions. Zeolites are mainly used in drying processes, water purification, agriculture, metallurgy, mining and quarrying, production of construction materials, energy recovery and storage systems, etc. [13, 14, 23, 24]. Although zeolites are the most widely used adsorbents, they still suffer from some disadvantages: their properties strongly depend on impurities, quarry space, synthesis conditions, limited reuse after adsorption, etc.

Clay minerals (Fig. 4.3) consist essentially of silicon, alumina, magnesia oxides, and water [25, 26]. The crystal lattice of these minerals consists of alternating silica tetrahedral (SiO_4) and aluminum octahedral [$Al(OH)_3$] layers, usually at a ratio of 2:1 or 1:1. The surface of clay minerals has a negative charge; consequently, clay minerals have a low adsorption capacity of anions. Also, the adsorption capacity of heavy metal ions is lower than that of natural zeolites. Still, many studies use clay minerals as adsorbents because they are cheap, widespread in the environment, have a large surface area, and can adsorb water in their interlayer space.

Activated carbon (Fig. 4.4) has a distinctive structure, a well-developed pore structure, a high surface area and a high total pore volume, low density, and high chemical resistance [13, 16, 27–29]. Thus, it is widely used to adsorb various organic

Fig. 4.2 Synthetic clinoptilolite (Photo ID: 1446091359, *Source* iStock by Getty Images)

Fig. 4.3 Kaolin quarry (Photo ID: 1372479433, *Source* iStock by Getty Images)

compounds (dyes, drugs, etc.). Activated carbon has also been employed as a catalyst support, in medicine, in the production of electrode materials, air filters, and gas storage. Therefore, the worldwide demand for activated carbon has been increasing, and, as a result, its price has also been increasing significantly [30].

Another type of adsorbent is *calcium silicate hydrates*. The use of this type of compounds offers more advantages in comparison with other adsorbents: by changing the conditions of synthesis (the nature of the primary materials, the molar ratio of CaO/SiO_2 or water/solid, the temperature and duration of isothermal treatment, the intensity of stirring, the rate of heating and cooling, etc.), the mineral composition and the parameters of the structure (the surface area, the porosity, the size of crystallites, the crystal shape, the composition, etc.) can be controlled [31–36]. Calcium silicate

Fig. 4.4 Activated carbon granules (Photo ID: 1272838375, *Source* iStock by Getty Images)

hydrates can adsorb both organic/inorganic anions and cations, and they demonstrate a low affinity for these compounds. According to the scientific literature, calcium silicate hydrates prepared under hydrothermal synthesis function as adsorbents in the process of the removal of heavy metal ions. The research of calcium silicate hydrates as adsorbents started over four decades ago, and any improvements are still in great demand.

The first adsorption properties of calcium silicate hydrates were found more than three decades ago. Komarneni et al. [37–40] found that the synthetic and natural minerals of the tobermorite group (Fig. 4.5) have particularly good adsorption properties for heavy metal ions. These properties can be improved by intercalating Al^{3+} and/or Na^{+} ions into the crystal structure of tobermorite [40–43]. Shrivastava and Komarneni [44, 45] determined that ion exchange reactions proceeding during adsorption experiments by synthetic tobermorite are analogous to those occurring in zeolites and clay minerals. These scientists also concluded that synthetic 1.13 nm tobermorite with intercalated Al^{3+} and Na^{+} ions can be obtained by using various zeolites or $NaAlO_2$ as a source of aluminum and sodium. Natural tobermorite has a lower adsorption capacity for alkali metal ions comparing to synthetic tobermorite with the conversion of Si^{4+} to [Al^{3+}/Na^{+}] or [Fe^{3+}/Na^{+}] [37, 38, 42].

According to the literature, pure synthetic tobermorite does not adsorb at all, or only adsorbs a low quantity of cesium ions. However, the adsorption capacity of synthetic tobermorite for cesium ions can be improved by replacing silicon ions with aluminum ions [43, 45]. The authors also found that tobermorite synthesized at higher temperatures and pressures has higher selectivity for cesium ions than tobermorite obtained during the conventional hydrothermal synthesis. This property is highly beneficial in the cement industry, nuclear and hazardous waste disposal, or catalysis. Further, investigation showed that the selectivity of synthetic tobermorite with intercalated [Al^{3+}/Na^{+}] ions for cesium ions depends not only on the amount of aluminum ions, but also on the other ions which are present in the solution [45]. The selectivity for radioactive cesium ions (from nuclear waste) increases in the following order: $Mg^{2+} > Li^{+} > Ca^{2+} > Na^{+} > Ba^{2+} > K^{+}$. The source of Cs^{+} was radioactive cesium from nuclear waste. Later on, Coleman [47] proved that synthetic

Fig. 4.5 SEM images of the synthetic tobermorite samples [46]. Reprinted from *Materials*, 2017, Vol. 10, Zhao et al., Characterization and Hexavalent Chromium Adsorption Characteristics of Aluminum- and Sucrose-Incorporated Tobermorite, No. 597. Open access under a CC BY 4.0 license, https://creativecommons.org/licenses/by/4.0/

1.13 nm tobermorite with intercalated aluminum ions can absorb 8 times more cesium ions in comparison to tobermorite with aluminum and sodium ions in its crystalline structure.

O. P. Shrivastava and R. Shrivastava determined that the selectivity sequence of synthetic tobermorite for Sr^{2+} ions in solutions containing Na^+, K^+, Ca^{2+}, Mg^{2+} or Ba^{2+} ions is the same as for cesium ions, i.e. $Mg^{2+} > Li^+ > Ca^{2+} > Na^+ > Ba^{2+} > K^+$[48]. These authors also determined that the release of Sr^{2+} from the mixture of Portland cement and synthetic tobermorite decreases by increasing the amount of intercalated aluminum ions from 10 to 40% into the structure of synthetic tobermorite.

Further, investigation showed that metal ions can replace calcium ions; for example, all Ca^{2+} ions which are present in the structure of synthetic tobermorite and synthetic xonotlite can be stoichiometrically replaced by Co^{2+} or Ni^{2+} ions [49]. It is worth highlighting that partial replacement does not cause changes in crystallinity, but the complete replacement of ions destroys the structure of the latter compounds. Also, experimental results showed that the replacement of Ca^{2+} ions by Co^{2+} ions depends on both the concentration of cobalt in the initial solution and on the duration of the reaction. Other scientists [44] determined that magnesium can only partially replace calcium ions in the structure of synthetic tobermorite (3.5 wt%) and synthetic xonotlite (3.71 wt%). Finally, it has been found that synthetic tobermorite adsorbs metal ions in the following sequence: $Pb^{2+} < Cd^{2+} < Ag^{2+} < Mn^{2+} < Cu^{2+} < Zn^{2+} < Co^{2+} < Ni^{2+} < Fe^{2+}$ [50].

Over time, the ion exchange and sorption properties of synthetic tobermorite and some other synthetic calcium silicate hydrates have been studied in detail. Also, foreign ions were incorporated into the structure of synthetic calcium silicate hydrates, which significantly augmented their capacity to adsorb heavy metal ions from aqueous solutions [51–56]. Unfortunately, the sorption properties of other calcium silicate hydrates have not been investigated completely yet. For example, synthetic dibasic calcium silicate hydrate—α-C_2SH—was only applied in the adsorption process as recently as in the 2010s. For this reason, the application of calcium silicate hydrates for adsorption is of great interest [57, 58].

4.3 Synthetic Gyrolite Application for Single Metal Ions Adsorption Process

4.3.1 Literature Review on Synthetic Gyrolite Application for Adsorption Process

Calcium silicate hydrates with a similar crystalline structure to gyrolite can be employed as adsorbents in the treatment of wastewater, some biotechnological and pharmaceutical processes, etc. Supposedly, gyrolite can absorb more chemical elements because the interlayer sheets with a thickness of about 2.2 nm (which ranks among the highest values of all the groups of calcium silicate hydrates) in gyrolite are available for the intercalation of new ions by controlling the charge of the calcium or silicon ions [59–61].

The currently available, yet limited scientific data suggests that synthetic gyrolite can be used as an adsorbent for heavy metal (Mn^{2+}, Fe^{2+}, Co^{2+}, Ni^{2+}, Pb^{2+}, Cs^{+}, Hg^{2+}) ions from aqueous solutions or wastewater. The crystal structure of gyrolite contains a higher number of calcium ions and water molecules which can participate in the ion exchange process in comparison with the well-known adsorbents: tobermorite (Fig. 4.5) and xonotlite (Fig. 4.6) [31, 59–63]. According to the literature [39, 40, 50–58], the adsorption of heavy metal ions is dependent on the nature of calcium silicate hydrates and the adsorbate. For example, the adsorption capacity of transition metal ions (Mn^{2+}, Fe^{2+}, Co^{2+}, Ni^{2+}, Cu^{2+}, Zn^{2+}, Pb^{2+}, Cs^{+}, Cd^{2+}, Hg^{2+}) by synthetic xonotlite, synthetic gyrolite, and synthetic β-wollastonite increased in the following order: β-wollastonite < gyrolite < xonotlite [50, 64–66]. Furthermore, the uptake of Fe^{2+}, Co^{2+}, and Ni^{2+} ions by synthetic xonotlite and synthetic wollastonite was higher than the other cations, while synthetic gyrolite showed lower selectivity for these ions.

Fig. 4.6 Mineral xonotlite as found in nature (Photo ID: 134143015, Photographer: Ekaterina Kriminskaya, *Source* Dreamstime.com)

The adsorption of metal ions and the release of calcium ions proceed in a non-stoichiometric pattern; thus the amorphization of calcium silicate hydrates and the formation of calcium hydroxide or carbonates takes place. Amorphization depends on the adsorption time, concentration, and on the nature of the adsorbate. For example, partial amorphization of calcium silicate hydrates proceeds during the adsorption of Zn^{2+}, Cu^{2+}, Mn^{2+}, Cd^{2+}, and Pb^{2+} ions, while decomposing is observed in solutions containing Fe^{2+}, Co^{2+}, and Ni^{2+} ions. It is worth noting that, during adsorption experiments, not only adsorption and ion exchange reactions occur, but also new products may form due to the increment of the pH value of the liquid medium. When synthetic xonotlite or synthetic wollastonite (Fig. 4.7) adsorb Fe^{2+}, Pb^{2+}, Cd^{2+}, or Hg^{2+} ions, goethite [FeO(OH)], hydrocerussite [$Pb_3(CO_3)_2(OH)_2$], otavite [$CdCO_3$], or HgCO·2HgO can precipitate, whereas gerhardite [$Cu_2(OH)_3NO_3$] or zincate [$CaZnO_2{\cdot}xH_2O$] precipitate during Cu^{2+} or Zn^{2+} adsorption by synthetic wollastonite or synthetic gyrolite [50, 64–66].

El-Korashy [66] determined that synthetic gyrolite can be used as a cation exchanger to remove sodium and potassium ions from aqueous solutions via the $Ca^{2+} \leftrightarrow 2M^+$ exchange process. The amount of the adsorbed ions increases by boosting the concentration of cations in the liquid medium or the time assigned for the reaction to proceed. The experimental results showed that 100 g of synthetic gyrolite can adsorb 40.96 mmol of K^+ and 39.97 mmol of Na^+ when the length of the experiments was set at 24 h. We should specify that the adsorption capacity of

Fig. 4.7 Mineral wollastonite as found in nature (Photo ID: 215939469, Photographer: Ruslan Minakryn, *Source* Dreamstime.com)

synthetic gyrolite for K^+ and Na^+ is larger than the uptake of these ions by synthetic xonotlite. Due to the relative activity of the ions and their radius, the quantity of K^+ ions adsorbed by gyrolite is always higher than that of Na^+. On the other hand, Na^+ ions intercalated into the crystal lattice of synthetic gyrolite seem to positively influence the adsorption capacity of synthetic gyrolite for various metal ions. For example, in the case of the adsorption of Cu^{2+} ions, the ion exchange capacity of the Na^+-substituted synthetic gyrolite is almost two times higher than that for (Al, Na)-substituted synthetic tobermorite [50, 67].

Synthetic gyrolite with intercalated Na^+ and Al^{3+} ions can be used to the adsorption for K^+ and Cs^+ ions [67]. The ion exchange reaction of synthetic gyrolite with intercalated Al^{3+} ions proceeds faster with Cs^+ ions compared to K^+ ions. It is worth noting that the adsorbent did not lose its crystallinity during experiments, and the main interplanar distance (d ~2.2 nm) did not change. The opposite results were obtained when using synthetic tobermorite with intercalated Al^{3+} ions for the adsorption of cesium ions [40–42, 45, 48]. The selectivity and adsorption capacity of synthetic tobermorite for Cs^+ ions increases by increasing the amount of Al^{3+} ions which replace Si^{4+} ions in the crystal lattice. Al-substituted synthetic gyrolite may be used to detect heavy metal ions and get them removed from wastewater, and this type of gyrolite is well known for its selectivity to DNA [68].

However, there is scarce information in the scientific literature about the single and simultaneous adsorption of metal ions, the influence of the pH value of the liquid medium, the kinetic parameters of adsorption, the stability of the adsorbent, and, finally, about the utilization of gyrolite with intercalated metal ions.

4.3.2 Synthetic Gyrolite Application for Zn^{2+} Ions Adsorption Process

According to the literature, zinc is one of the essential and indispensable functional metals in the human body which can bind ~10% of the human genome encodes; however, it is also assigned to the second class of toxicity defined by low mutagenic and carcinogenic properties [69, 70]. By consuming food and drinking water featuring a high concentration of zinc, major hazard to human health is being faced: corneal ulcers may form, and the esophagus may get negatively affected. Zinc enters the environment, food, and water due to natural processes and anthropogenic activity (metal processing; manufacturing of electrical, radio, and electronic equipment; chemical industry; production of plastics; manufacturing of fertilizers; domestic wastewater) (Table 4.1) [10–12]. Adsorption is one of the most efficient and appealing techniques for the elimination of zinc from aqueous solutions [13, 16, 21, 71–74]; however, so far, zinc adsorption by synthetic gyrolite has not been thoroughly investigated. For these reasons, it is important to determine the influence of the concentration of zinc ions and the pH value of the liquid medium on the adsorption process as well as the impact of kinetic parameters (k, R^2, q_e) on adsorption reactions, which gives information about the mechanism of adsorption [75–77]. To describe the kinetics of adsorption, the pseudo-first- and pseudo-second-order empirical kinetic equations were used. The *pseudo-first-order adsorption kinetics* model was selected which is analogous to the Lagergren model and which can be represented by the equation provided below:

$$\frac{dX_t}{dt} = k_1\left(\sum X_e - X_\tau\right), \tag{4.1}$$

where ΣX_e and X_t are the adsorption capacity at equilibrium and at time τ_e, respectively (mg g^{-1}), k_1 is the rate constant of pseudo-first-order adsorption (min^{-1}). After the integration and after applying boundary conditions $\tau = 0$ to $\tau = \tau_e$ and $X_\tau = 0$ to $X_\tau = \Sigma X_e$, the integrated form becomes

$$\log\left(\sum X_e - X_\tau\right) = \sum X_e - \frac{k_1}{2.303}\tau. \tag{4.2}$$

The *pseudo-second-order adsorption kinetic* model was selected; it is essentially identical to the models presented in [75–77] and can be written as

$$\frac{dX_\tau}{d\tau} = k_2\left(\sum X_e - X_\tau\right)^2, \tag{4.3}$$

where k_2 is the rate constant of the pseudo-second-order adsorption (g mg^{-1} min^{-1}). For the boundary conditions $\tau = 0$ to $\tau = \tau_e$ and $X_\tau = 0$ to $X_\tau = \Sigma X_e$, the integrated form of the equation becomes as follows (we use the integrated rate law for the pseudo-second-order reaction):

$$\frac{1}{\sum X_e - X_\tau} = \frac{1}{\sum X_e} + k_2\tau. \tag{4.4}$$

Adsorption experiments in acidic solution

Adsorption experiments were implemented at a temperature of 25 °C in a thermostatic absorber *Grant SUB14* by stirring 1 g of synthetic gyrolite in 100 cm^3 of zinc nitrate solutions containing 250 or 300 g Zn^{2+}/dm^3 with the adsorption times varying in between 0.5 and 60 min. The concentrations of heavy metal ions were measured by using the spectrometer *Perkin-Elmer Analyst 4000* (*Perkin Elmer*, Waltham, MA, USA). The pH value was measured with a *Hanna* instrument (*Hi 9321*, microprocessor pH meter, *Hanna Instruments*, Woonsocket, RI, USA).

It is known that adsorption depends on many factors; thus, at the beginning of the experiments, the concentration of zinc ions in the solution equaled to 250 g/dm^3, while the pH value of the liquid medium was equal to 5.8. As a result, it was determined that approximately 12 mg Zn^{2+}/g (i.e. 50%) was adsorbed by synthetic

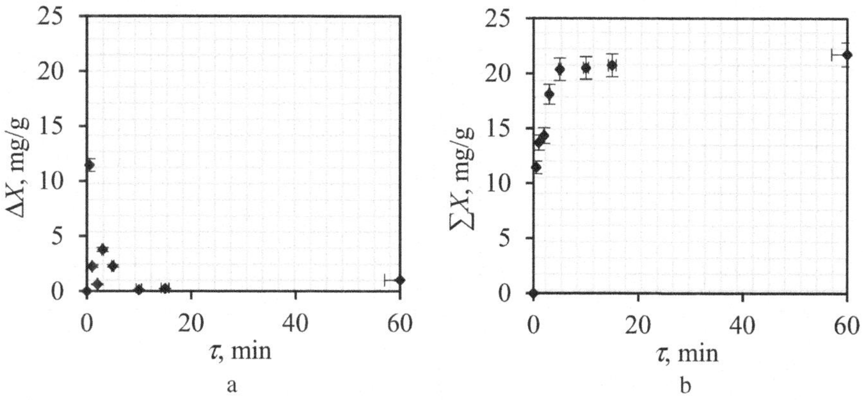

Fig. 4.8 Differential (**a**) and integral (**b**) kinetic curves of Zn^{2+} adsorption on synthetic gyrolite with the primary concentration of zinc equaling to 250 mg/dm^3

gyrolite within 30 s of the adsorption experiment (Fig. 4.8). By prolonging the duration of adsorption, the total number of the adsorbed ions went upward, and 21.75 mg of Zn^{2+} (i.e. 87%) was adsorbed by 1 g of synthetic gyrolite after 30 min of the experiment.

By employing the pseudo-first-order kinetic rate equation in its linear form (Eq. 4.2), the equilibrium adsorption capacity ($\Sigma X_{e(exp)}$) and the first-order constant k_1 (1/min) were calculated from the slope, and the intercept plot of $\log(\Sigma X_{e(exp)} - X_\tau)$ versus τ was determined. It was determined that the pseudo-first-order model did not conform to the experimental data as the value of the correlation coefficient (R^2) was as low as 0.754, although $\Sigma X_{e(cal)}$ was equal to the experimental $\Sigma X_{e(exp)}$ value (Table 4.3). The comparison of $\Sigma X_{e(cal)}$ and $\Sigma X_{e(exp)}$ values showed that these values are equal only at the end of the adsorption, i.e. after 15 min. Yet, when the pseudo-second-order kinetic equation (Eq. 4.4) was used, the equilibrium adsorption capacity ($\Sigma X_{e(cal)}$) and the second-order constants k_2 [g/(mg min)] were calculated from the slope and the intercept of the plot τ/X_τ versus τ. It was calculated that the correlation coefficient equaled to 0.998, while the difference between $\Sigma X_{e(cal)}$ and $\Sigma X_{e(exp)}$ was insignificant in all experimental points (Table 4.3 and Fig. 4.9). The adsorption rate (k_2) of zinc on synthetic gyrolite, when the starting concentration of zinc is 250 mg/dm^3, equals to 0.077 g/(mg min).

Table 4.3 Kinetic parameters of pseudo-first- and pseudo-second-order kinetic models of Zn^{2+} ions adsorption on synthetic gyrolite with the initial concentration of zinc equal to 250 mg/dm^3

$\Sigma X_{e(exp)}$	Pseudo-first-order			Pseudo-second-order		
	k_1, 1/min	$\Sigma X_{e(cal)}$, mg/g	R^2	k_2, g/(mg min)	$\Sigma X_{e(cal)}$, mg/g	R^2
21.754	0.186	21.754	0.754	0.077	21.452	0.998

Fig. 4.9 Calculated and experimental differential kinetic curves of Zn^{2+} ions adsorption on synthetic gyrolite with the primary concentration of zinc equaling to 250 mg/dm^3

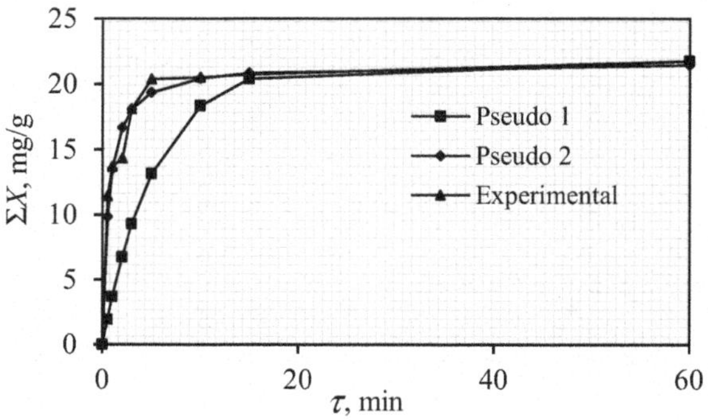

$\Sigma X_{e(exp)}$ represents the equilibrium adsorption capacity, mg/g, obtained from data of experiments; $\Sigma X_{e(cal)}$ is the equilibrium adsorption capacity, mg/g, derived by employing Eqs. (4.2) and (4.4) of kinetic models.

In order to increase the intercalation of ions into the structure of synthetic gyrolite and accelerate adsorption, the initial concentration of zinc ions in the liquid medium was increased to 300 mg/dm^3. We established that the increment of the concentration of zinc ions has a positive effect on the adsorption because more than 68% Zn^{2+} ions initially present in the liquid medium (20.43 mg Zn^{2+}/g) were intercalated into the synthetic gyrolite structure (Fig. 4.10).

By extending the length of the adsorption process to 5 min, the amount of the adsorbed zinc ions increased to 22.82 mg/g, while, at the final point of the experiment (60 min), this value was as low as 23.47 mg/g. The calculations by using pseudo-first- and pseudo-second-order kinetic equations showed that the adsorption process fit the pseudo-second-order kinetic equation ($R^2 = 1$, $\Sigma X_{e(cal)} \approx \Sigma X_{e(exp)}$) (Table 4.4 and Fig. 4.11). The adsorption rate (k_2) of zinc on synthetic gyrolite when the primary concentration of zinc equaled to 300 mg/dm^3 was determined to be 0.273 g/(mg min),

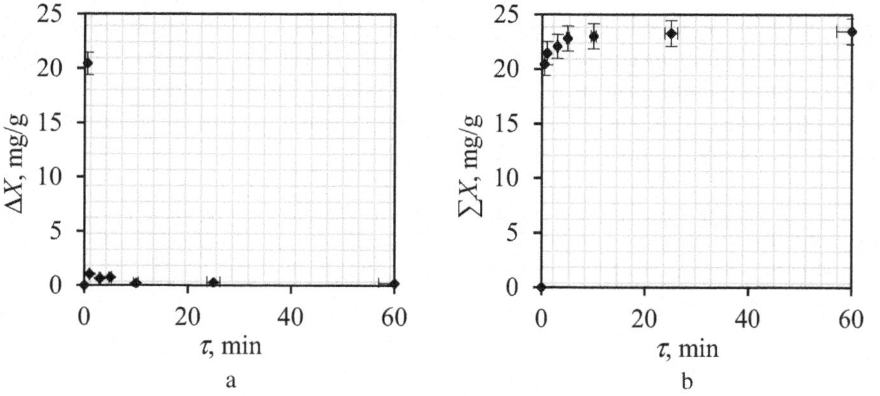

Fig. 4.10 Differential (**a**) and integral (**b**) kinetic curves of Zn^{2+} ions adsorption on synthetic gyrolite with the primary concentration of zinc equaling to 300 mg/dm^3

which is more than 3.5 times higher in comparison with the value measured during adsorption in a solution of a lesser concentration. Thus, it can be stated that the increment of the concentration of zinc ions insignificantly affects the total number of the adsorbed Zn^{2+} ions on synthetic gyrolite, but strongly affects the adsorption rate (k_2).

Also, a significant increment of the pH value from 5.6 (the initial solution) to 6.7 (after 5 min of adsorption) in the liquid medium was observed (Fig. 4.12). During the adsorption of zinc ions on synthetic gyrolite, both Ca^{2+} cations and OH^- anions are probably released; thus, the pH value of the liquid medium increases.

Table 4.4 Kinetic parameters of the pseudo-first- and pseudo-second-order kinetic models of Zn^{2+} ions adsorption on synthetic gyrolite with initial concentration of zinc equal to 300 mg/dm^3

$\Sigma X_{e(exp)}$	Pseudo-first-order			Pseudo-second-order		
	k_1, 1/min	$\Sigma X_{e(cal)}$, mg/g	R^2	k_2, g/(mg min)	$\Sigma X_{e(cal)}$, mg/g	R^2
23.47	0.06	23.597	0.552	0.273	23.478	1.000

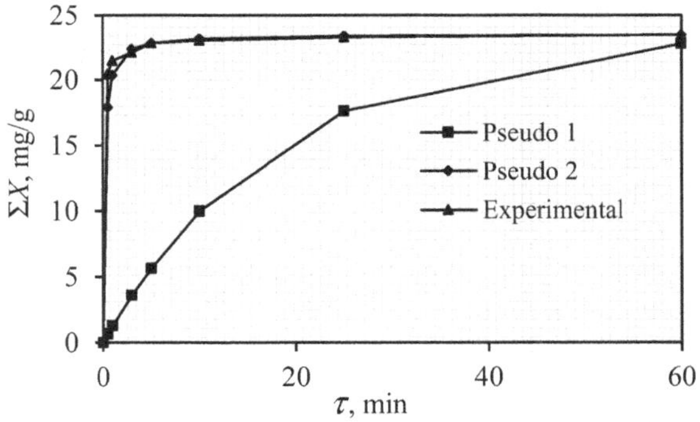

Fig. 4.11 Calculated and experimental differential kinetic curves of Zn^{2+} ions adsorption on synthetic gyrolite with initial concentration of zinc equal to 300 mg/dm^3

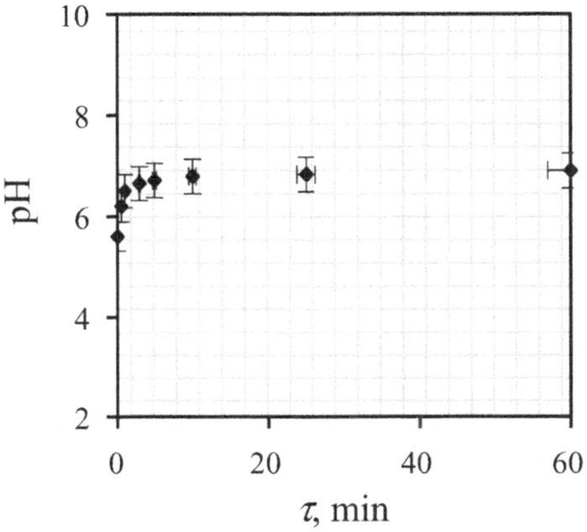

Fig. 4.12 Variation of pH value of liquid medium during Zn^{2+} ions adsorption on synthetic gyrolite

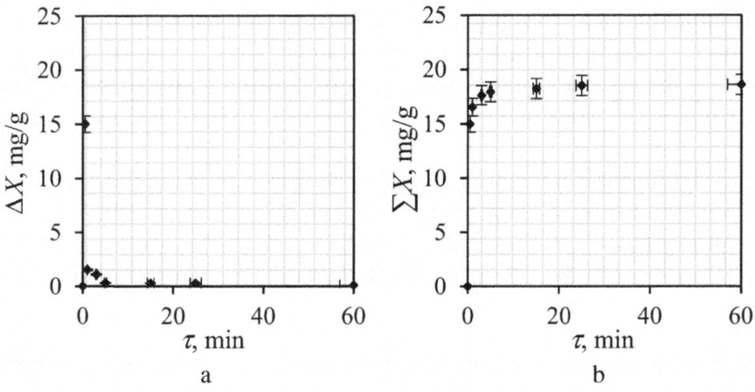

Fig. 4.13 Differential (**a**) and integral (**b**) kinetic curves of the variation of the concentration of Ca^{2+} ions

The results of the atomic absorption spectroscopy were in line with the previously obtained data. It was established that, in the course of the process of adsorption of Zn^{2+} ions on synthetic gyrolite, calcium ions are released from the crystal lattice of the adsorbent into the liquid medium (Fig. 4.13). As expected, most Ca^{2+} ions (17.95 mg Ca^{2+}/g) are released within the earliest part of the adsorption process (5 min). Meanwhile, a further extension of the adsorption duration does not affect the stability of the adsorbent because the number of the released Ca^{2+} ions ultimately reached only the level of 18.61 mg Ca^{2+}/g (Fig. 4.13).

Calculations suggest that, during the adsorption experiments, zinc ions intercalated into the synthetic gyrolite structure only by the substitution reaction: gyrolite-$Ca^0 + Zn^{2+} \leftrightarrow$ gyrolite-$Zn^0 + Ca^{2+}$. These results agree with the data of kinetic calculations because the pseudo-second-order equation describes chemical adsorption.

With the objective of determining the stability of the intercalated zinc ions, a desorption experiment was performed as follows: the adsorbent after adsorption was dried at 50 °C temperature and poured into decarbonized water (25 °C) to hold for 2 h. It was measured that, after the experiment, the concentration of zinc ions in the liquid medium did not surpass ~0.008 mg/dm^3; thus, it can be stated that the adsorption of zinc ions on synthetic gyrolite is irreversible in water.

One of the most important characteristics of adsorbents is the adsorption selectivity and capacity; however, adsorbents must remain stable during adsorption. Therefore, synthetic gyrolite after adsorption was characterized with XRD and STA methods. The data of XRD analysis demonstrated that synthetic gyrolite is not stable during adsorption as, after the experiments, synthetic gyrolite recrystallized into gyrolite gel and C–S–H(I) (Fig. 4.14a).

The XRD results were approved by the data of DSC: the first endothermic effect at ~146 °C (Fig. 4.14b, curve 1) moved to a lower temperature range (60–120 °C) after the adsorption (Fig. 4.14b, curve 2). In a higher temperature range, two exothermic

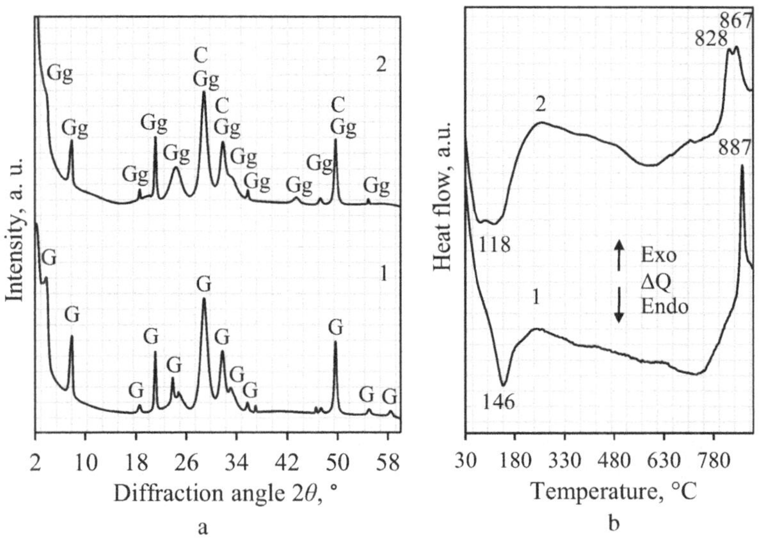

Fig. 4.14 X-ray diffraction patterns (**a**) and DSC curves (**b**) of synthetic gyrolite (curve 1) and synthetic gyrolite after adsorption (curve 2). *Indexes* G—gyrolite; Gg—gyrolite gel; C—C–S–H(I)

effects (at 828 and 867 °C) characteristic of the recrystallization of synthetic gyrolite gel and C–S–H(I) to wollastonite were observed. It is worth specifying that these compounds recrystallize into wollastonite at lower temperatures in comparison with the sample of pure synthetic gyrolite (Fig. 4.14b).

This data agrees with the results obtained by Stumm et al. [78]. These scholars suggested that zinc incorporation into synthetic gyrolite proceeds until reaching Zn/(Zn + Ca) = 1/6, which corresponds to approximately 6 wt%. An increase of the content of zinc led to a gradual decrease of the basal reflection (001) not only of gyrolite but also that of the nanocrystalline phases.

Adsorption experiments in alkaline solution

Our experiments established that, in an acidic solution, not all zinc ions were adsorbed on synthetic gyrolite, and that synthetic gyrolite regroups into related compounds. Thus, the adsorption experiments in an alkaline solution were carried out to boost the stability of the adsorbent and promote adsorption reactions. With the objective to maintain the alkaline value of the solution (pH ~9.0) and to eliminate the possibility of $Zn(OH)_2$ precipitation, aqueous solution (c = 10%) of NH_4OH was employed.

We established that the pH value of the liquid medium strongly affects the adsorption process because, in the alkaline solution (pH ~9.0), intercalation of zinc ions proceeds faster and more efficiently than in an acidic solution (pH ~5.6). The AAS results showed that, when the initial concertation of zinc ions is equal to 300 mg/dm^3, more than 95% of Zn^{2+} ions that had been present in the liquid medium (28.50 mg Zn^{2+}/g) were adsorbed on synthetic gyrolite after 60 s of the experiment (Fig. 4.15a).

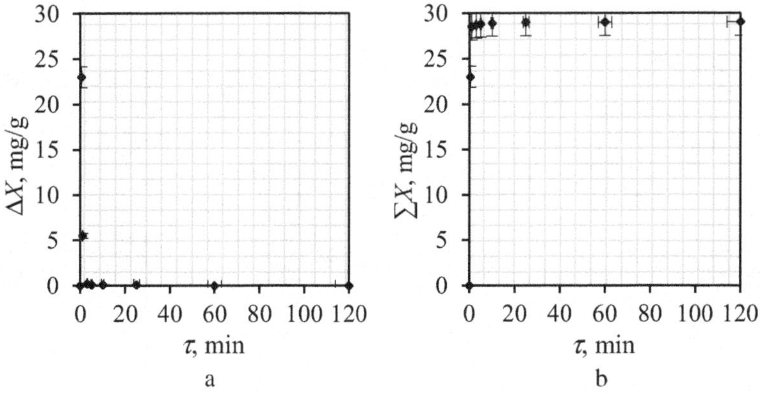

Fig. 4.15 Differential (**a**) and integral (**b**) kinetic curves of the adsorption of Zn^{2+} ions on synthetic gyrolite in alkaline solution

However, by extending the length of adsorption to 120 min, the total number of the adsorbed ions grew negligibly, i.e. it only reached 29.12 mg Zn^{2+}/g (Fig. 4.15b).

After the adsorption, the adsorbent was dried at 50 °C and later poured into decarbonized water for the leaching experiment. It was established that, after 120 min of leaching, the concentration of zinc ions in the solution did not get higher than 0.1 mg/dm^3. Therefore, it can be stated that the adsorption process in the alkaline solution is irreversible, even though the total number of the released zinc ions was 10 times greater than in the course of the experiments in acidic solutions. Probably, more zinc is chemosorbed during the adsorption in an acidic solution than during the adsorption in an alkaline solution.

With the objective to establish the kinetic parameters of adsorption, kinetic models were applied for the adsorption process of Zn^{2+} ions on synthetic gyrolite. It was found out that the pseudo-first-order model did not conform to the pattern of adsorption in alkaline solutions because R^2 was as low as 0.511, and the $\Sigma X_{e(cal)}$ value disagreed with the $\Sigma X_{e(exp)}$ value (Table 4.5). Meanwhile, by using a pseudo-second-order model, the agreement between $\Sigma X_{e(exp)}$ and $\Sigma X_{e(cal)}$ values was detected (Table 4.5). Also, the R^2 value of linear dependence was equal to 1.000 (Fig. 4.16); hence, the process of adsorption in the alkaline solution is chemisorption. It was calculated that the value of the adsorption rate constant is 0.401 g/(mg min), which significantly exceeds the value obtained in an acidic solution [0.273 g/(mg min)] (Table 4.4).

The analysis of calcium ions in the liquid medium showed that calcium ions are released from the structure of the adsorbent in an alkaline solution, although this process is more intensive in acidic solutions. It was measured that most Ca^{2+} ions are released within 5 min because their concentration in the liquid medium $\left(\sum X_{Ca^{2+}}\right)$ equals to 13.92 mg Ca^{2+}/g, and, after two h, $\sum X_{Ca^{2+}}$ reached as low as 14.53 mg Ca^{2+}/g (Table 4.6). It is worth indicating that the amount of the released calcium is ~4 mg/g lower than in an acidic solution (18.61 mg Ca^{2+}/g) (Fig. 4.12b).

Table 4.5 Kinetic parameters of the pseudo-first- and pseudo-second-order kinetic models of the adsorption of Zn^{2+} ions on synthetic gyrolite when the primary concentration of zinc equals to 300 mg/dm^3

$\Sigma X_{e(exp)}$	Pseudo-first-order			Pseudo-second-order		
	k_1, 1/min	$\Sigma X_{e(cal)}$, mg/g	R^2	k_2, g/(mg min)	$\Sigma X_{e(cal)}$, mg/g	R^2
29.120	0.097	26.61	0.511	0.401	29.12	1.000

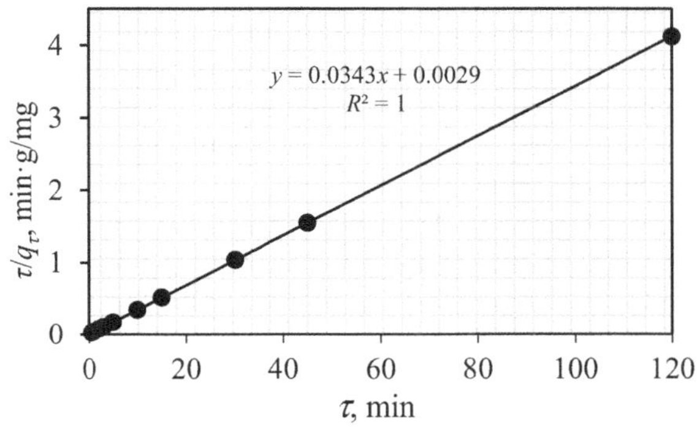

Fig. 4.16 Pseudo-second-order kinetic curve of the adsorption of Zn^{2+} ions on synthetic gyrolite in alkaline solution

Table 4.6 Amount of desorbed Ca^{2+} ions in alkaline solution

Duration, min	$\Delta X(Ca^{2+})$, mg/g	$\Sigma X(Ca^{2+})$, mg/g
0	0	0
0.5	10.82	10.82
1	1.20	12.02
3	1.00	13.02
5	0.90	13.92
15	0.30	14.22
25	0.20	14.42
60	0.10	14.52
120	0.01	14.53

As calculated, the substitution reaction was typical of synthetic gyrolite in the alkaline solution due to the fact that 82% of Zn^{2+} ions were involved in the substitution/chemical reaction (gyrolite-Ca^0 + Zn^{2+} $\leftrightarrow$ gyrolite-Zn^0 + Ca^{2+}), whereas the remaining share of these ions were adsorbed on synthetic gyrolite according to the addition interaction. The pH value in the alkaline solution during the adsorption process did not significantly change: pH was equal to 9.0 $\pm$ 0.1. It is thus worth emphasizing that, during adsorption in an acidic solution, zinc ions intercalated into the synthetic gyrolite structure only by the substitution reaction.

With the objective to determine the adsorbent's stability, the products after adsorption were explored by the XRD and DSC analysis (Fig. 4.17). It was established that, during adsorption, synthetic gyrolite remained stable because, in the XRD pattern, all the diffraction peaks characteristic of gyrolite were identified (Fig. 4.17a). However, due to the intercalation of zinc ions, the intensity of reflection with *d*-0.113 nm nearly doubled. Also, new small diffraction peaks in the 2θ range of 10–15° and 40–45°, which can be related to the changes in the interplanar spacing of synthetic gyrolite or the formation of new compounds, were observed. Unfortunately, these peaks could not be identified by using the PDF-4 database (2021). In the DSC curve of synthetic gyrolite with intercalated zinc ions, identical thermal effects (the endothermic effect at 138 °C characteristic of dehydration and the exothermic effect at 855 °C characteristic of wollastonite formation) were determined (Fig. 4.17b, curve 2) as in pure synthetic gyrolite (Fig. 4.17b, curve 1). Thus, synthetic gyrolite substituted with Zn^{2+} ions recrystallizes into wollastonite at lower temperatures (855 °C) than pure synthetic gyrolite (887 °C).

On the grounds of the obtained data, we can suggest that the adsorption process of zinc ions depends on the pH value of the liquid medium and on the primary concentration of zinc ions. It was determined that, regardless of the adsorption conditions, the pseudo-second-order kinetic model describes zinc adsorption on synthetic gyrolite, which shows that chemisorption takes place. The results of kinetic calculations were confirmed by leaching experiments (the adsorption reactions are irreversible). It was calculated that the adsorption rate (k_2) increased by boosting the concentration of zinc ions and the pH value of the liquid medium. We determined that, after 60 s of adsorption, more than 95% of Zn^{2+} ions (28.50 mg Zn^{2+}/g) intercalated into the synthetic gyrolite structure in the alkaline solution (pH ~9.0), whereas, in the acidic solution (pH ~5.6), a lower share, specifically, 71%, was intercalated (21.45 mg Zn^{2+}/

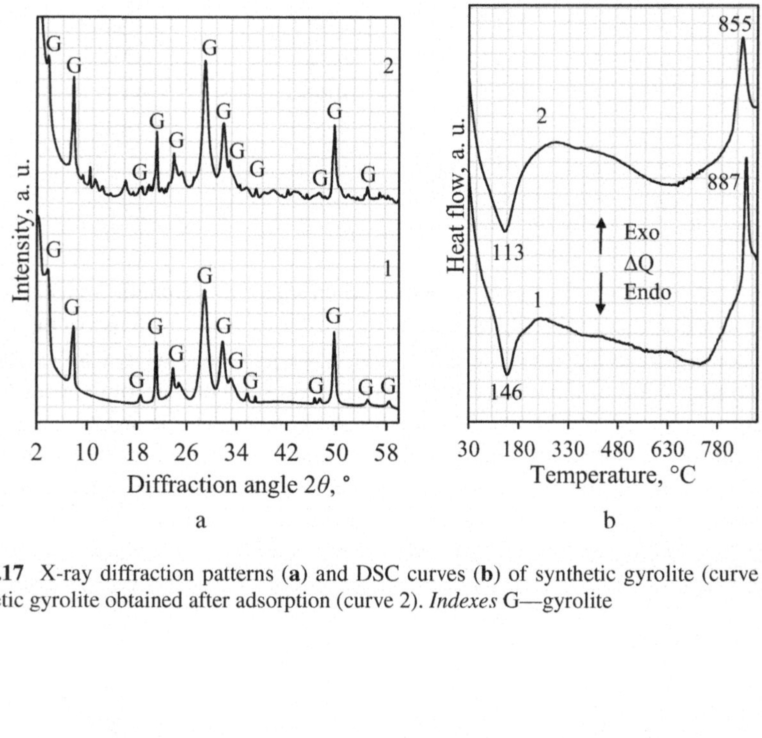

Fig. 4.17 X-ray diffraction patterns (**a**) and DSC curves (**b**) of synthetic gyrolite (curve 1) and synthetic gyrolite obtained after adsorption (curve 2). *Indexes* G—gyrolite

g). At the end of adsorption (60 min), the relative weight of the adsorbed ions in an alkaline solution (>29 mg Zn^{2+}/g) was higher than in an acidic solution (23.47 mg Zn^{2+}/g).

In the alkaline solution, 82% of Zn^{2+} ions intercalated into synthetic gyrolite by the substitution reaction (gyrolite-Ca^0 + Zn^{2+} $\leftrightarrow$ gyrolite-Zn^0 + Ca^{2+}), whereas the remaining share of these ions were present in synthetic gyrolite according to the addition interaction. In this context, in the acidic solution, the adsorption reaction took place differently as almost all zinc ions only participated in the substitution reaction.

Influence of crystallinity on adsorption in alkaline solution

It is known that the adsorption capacity and rate depends not only on the parameters of the liquid medium, but also on the structure of the adsorbent. For this reason, the adsorption experiments were performed by using synthetic gyrolite with different crystallite sizes and porosity. A detailed description of the mineralogical composition and porosity is reported in Sect. 2.4 *Effect of hydrothermal synthesis duration at 200 °C temperature on the properties of gyrolite*, while summarized information is presented in Table 4.7.

It was determined that the properties of synthetic gyrolite strongly affect the process of adsorption. For example, it was obtained that, after 30 s of adsorption, gyrolite which was synthesized for 32 h adsorbed more than 88% (26.48 mg Zn^{2+}/g) of zinc ions that had been present in the liquid medium (Fig. 4.18). Meanwhile, the uptake of these ions (after 30 s of adsorption) into gyrolite decreased by increasing the duration of the synthesis:

- Gyrolite synthesized for 48 h adsorbed 77% (23.00 mg Zn^{2+}/g) of zinc ions.
- Gyrolite synthesized for 72 h adsorbed 71% (21.36 mg Zn^{2+}/g) of zinc ions.
- Gyrolite synthesized for 120 h adsorbed 62% (18.51 mg Zn^{2+}/g) of zinc ions.
- Gyrolite synthesized for 168 h adsorbed 56% (16.67 mg Zn^{2+}/g) of zinc ions.

By extending the length of adsorption to 45 min, the amount of the adsorbed ions went up in all samples (Fig. 4.19); however, the value of the maximum amount of adsorbed Zn^{2+} ions decreased by increasing the duration of hydrothermal synthesis:

Table 4.7 Parameters of gyrolite obtained after different durations of isothermal treatment

Duration of synthesis, h	Peak (*d*-spacing 2.273 nm) intensity of gyrolite, cps	Crystallite size	Mass loss within temperature range of 100–175 °C, %	S_{BET}, m^2/g
32	23.2	43.84	5.4	81.17
48	35.2	64.9	5.0	95.12
72	45.9	77.34	4.7	44.52
120	112.6	83.2	4.1	39.35
168	130.6	95.38	3.8	47.42

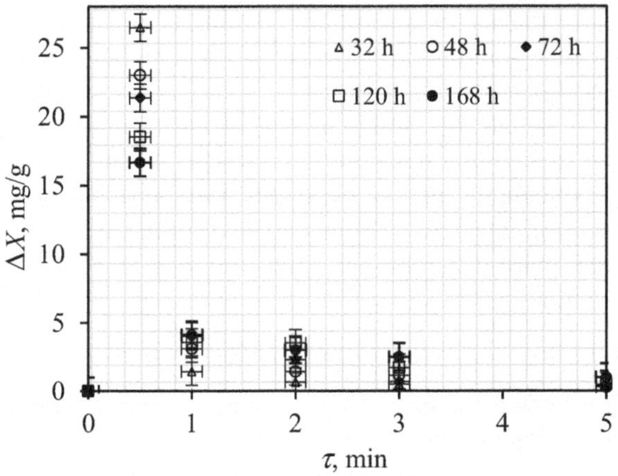

Fig. 4.18 Differential kinetic curves during Zn^{2+} ion adsorption on gyrolite samples obtained after different durations of hydrothermal synthesis

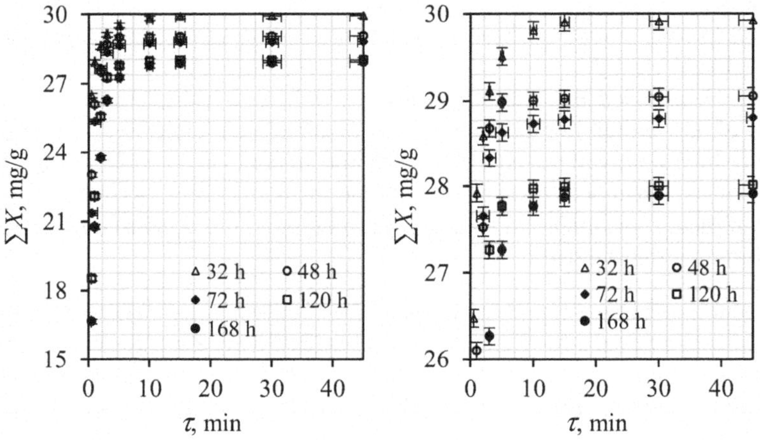

Fig. 4.19 Integral kinetic curves during Zn^{2+} ion adsorption on gyrolite samples obtained after different durations of hydrothermal synthesis

- Gyrolite synthesized for 32 h adsorbed 99.7% (29.92 mg Zn^{2+}/g) of zinc ions.
- Gyrolite synthesized for 48 h adsorbed 96.8% (29.05 mg Zn^{2+}/g) of zinc ions.
- Gyrolite synthesized for 72 h adsorbed 96.0% (28.80 mg Zn^{2+}/g) of zinc ions.
- Gyrolite synthesized for 120 h adsorbed 93.4% (28.01 mg Zn^{2+}/g) of zinc ions.
- Gyrolite synthesized for 168 h adsorbed 93.0% (27.91 mg Zn^{2+}/g) of zinc ions.

Thus, gyrolite which had been synthesized for 32 h adsorbed almost all zinc ions that were present in the liquid medium (99.7%), while long-term synthesized (168 h) gyrolite adsorbed only 93% (Fig. 4.19). It was determined that, in all cases, the process of adsorption is irreversible as, after the immersion of synthetic gyrolite with substituted zinc ions in distilled water, the concentration of zinc ions in the liquid medium was below the value of 0.1 mg/dm^3.

We established that the release of calcium ions from the crystal lattice of the adsorbent into the solution is dependent on the amount of the adsorbed zinc ions and the properties of synthetic gyrolite samples (Figs. 4.20 and 4.21). As in the previous cases, most Ca^{2+} ions were released during the initial minutes, and their amount modestly increased when the adsorption process was extended (Figs. 4.20 and 4.21). The most significant amount of Ca^{2+} ions (15.09 mg Ca^{2+}/g) was released into the solution from the gyrolite sample synthesized for 32 h (Fig. 4.21). This sample adsorbed the highest amount of zinc ions (29.92 mg/g). Meanwhile, the lowest amount of Ca^{2+} ions (13.21 mg Ca^{2+}/g) was released from the gyrolite sample synthesized for 168 h.

We identified that, in all cases, the substitution reaction of Zn^{2+} ions (gyrolite-Ca^0 + Zn^{2+} $\leftrightarrow$ gyrolite-Zn^0 + Ca^{2+}) is dominant. Calculations show that, when applying gyrolite synthesized for 32 h, 82% of zinc ions entered into the structure of

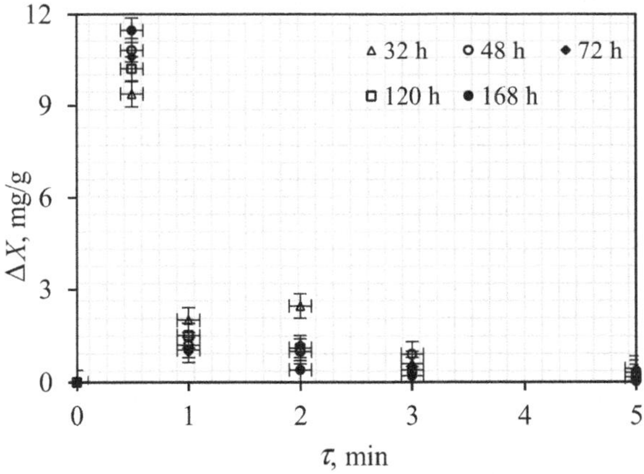

Fig. 4.20 Differential kinetic curves of Ca^{2+} ion description process from gyrolite samples obtained after different durations of hydrothermal synthesis

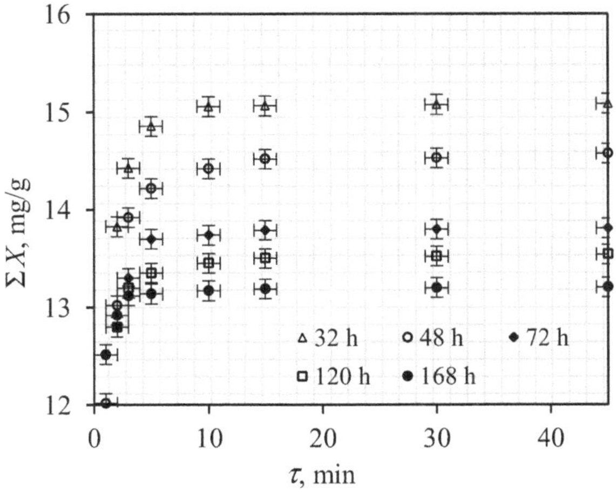

Fig. 4.21 Integral kinetic curves of Ca^{2+} ion desorption process from gyrolite samples obtained after different durations of hydrothermal synthesis

the adsorbent by a substitution reaction, while, when employing gyrolite synthesized for 168 h, only 77% of zinc ions entered by a substitution reaction. The remaining share of the ions was adsorbed on synthetic gyrolite by an addition reaction (~20%). Also, it was calculated that from 3.87% (in gyrolite synthesized for 168 h) to 5.42% (in gyrolite synthesized for 32 h) of Ca^{2+} ions that had been present in the synthetic gyrolite structure were replaced by zinc ions.

It can be stated that the adsorption process mainly depends on the crystallite size of synthetic gyrolite and the content of water in the interlayers (by increasing the amount of water in the interlayer, the interplanar spacing increases). Therefore, by increasing the crystallite size and decreasing the quantity of water in the interlayers, the synthetic gyrolite capacity of adsorption decreased, yet the stability was improved.

Kinetic models were employed to establish the influence of gyrolite synthesis on the kinetic parameters of adsorption. It was observed that the pseudo-first-order model did not conform well to the experimental data because the values of R^2 were low (from 0.781 to 0.8), and $\Sigma X_{e(cal)}$ values showed poor agreement with the $\Sigma X_{e(exp)}$ values (Table 4.8 and Fig. 4.22a).

By using a pseudo-second-order model, the linear dependence in plot t/X_t versus t was obtained ($R^2 > 0.999$) (Table 4.8 and Fig. 4.22b). Also, we determined that the $\Sigma X_{e(exp)}$ and $\Sigma X_{e(cal)}$ values overlapped to a great extent. The calculations revealed that the adsorption rate constant (k_2) decreased from 0.446 to 0.142 by extending the length of the gyrolite synthesis from 32 to 168 h. The established values of the adsorption rate constant confirmed the experimental data: the synthesis conditions of gyrolite affect its adsorption capacity for zinc ions as the adsorption process takes place under more challenging conditions when employing synthetic gyrolite featuring a more orderly structure.

To determine the stability of synthetic gyrolite during adsorption, the adsorbent obtained after the experiments was characterized by XRD analysis (Fig. 4.23). It was established that the mineralogical composition did not alter while adsorption was taking place because, in the XRD patterns, diffraction peaks characteristic only of gyrolite were identified.

Although zinc ions do not impact the mineralogical composition of the adsorbent, they strongly influence its crystallinity. We established that the crystallite size of

Table 4.8 Kinetic parameters of the pseudo-first- and pseudo-second-order kinetic models of the adsorption of Zn^{2+} ions on gyrolite samples obtained after different durations of hydrothermal synthesis

Duration of synthesis, h	$\Sigma X_{e(exp)}$	Pseudo-first-order			Pseudo-second-order			
		$\Sigma X_{e(cal)}$	$\Delta\Sigma X_e$	R^2	$\Sigma X_{e(cal)}$	$\Delta\Sigma X$	R^2	k_2
32	29.92	2.72	27.20	0.78	29.94	0.02	1.00	0.45
48	29.05	2.15	26.90	0.66	29.15	0.10	1.00	0.44
72	28.80	2.59	26.21	0.72	28.90	0.10	1.00	0.35
120	28.01	3.58	24.43	0.74	28.17	0.16	1.00	0.23
168	27.91	5.68	22.23	0.80	28.07	0.16	1.00	0.14

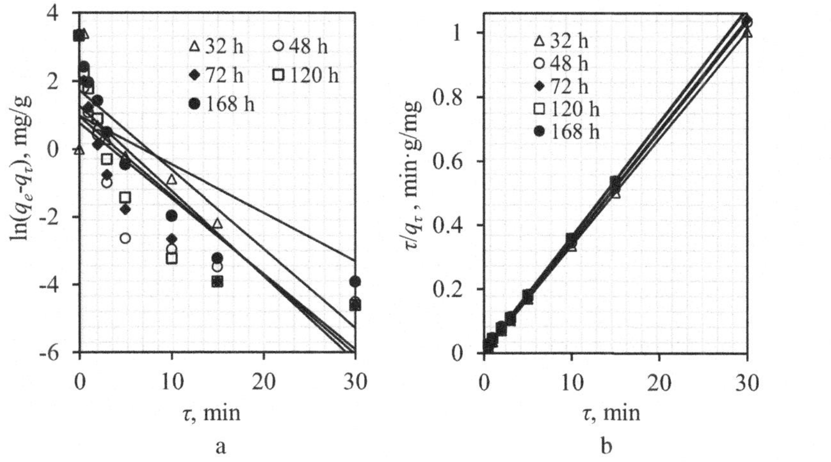

Fig. 4.22 Pseudo-first-order (**a**) and pseudo-second-order (**b**) kinetic curves of Zn^{2+} ions adsorption on gyrolite samples obtained after different durations of hydrothermal synthesis

Fig. 4.23 X-ray diffraction patterns of gyrolite synthesized in the $CaO–SiO_2–H_2O$ system after the adsorption process. Duration of hydrothermal synthesis at 200 °C, in h: 1 at 32 h; 2 at 48 h; 3 at 72 h; 4 at 120 h; 5 at 168 h

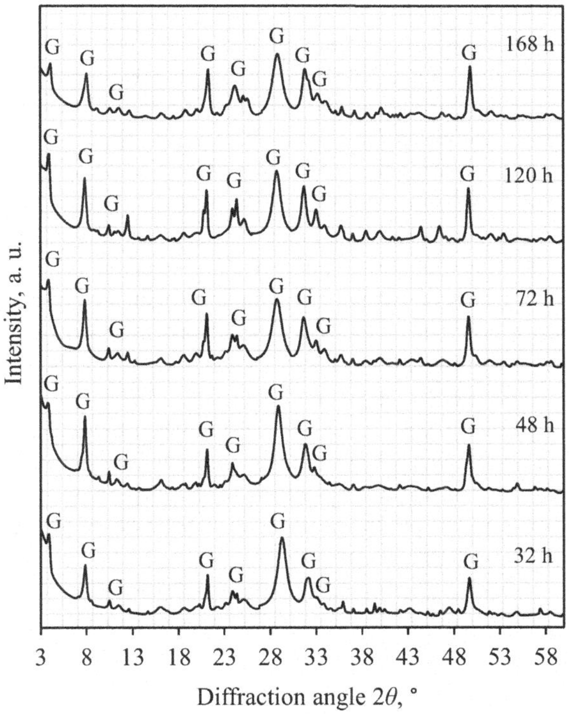

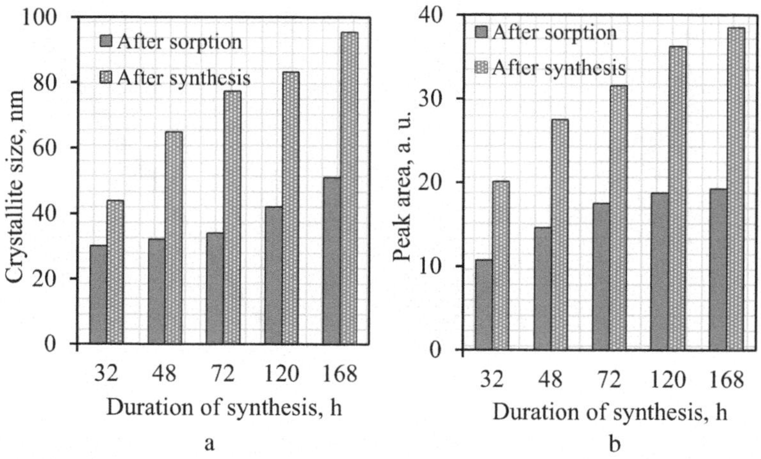

Fig. 4.24 Dependence of the crystallite size (**a**) and the diffraction peak area (*d*—2.2 nm) (**b**) of gyrolite on the duration of hydrothermal synthesis before and after the adsorption process

gyrolite synthesized for 32 h decreased from 44 nm (after synthesis) to 30 nm (after adsorption) (Fig. 4.24a). Meanwhile, the decrement is more intensive in more ordered synthetic gyrolite samples because the crystallite size of gyrolite synthesized for 48–168 h decreased by approximately two times. The same tendency was observed when comparing the area of the diffraction peak with *d*—2.2 nm (Fig. 4.24b). These results conform fairly well to the data of calcium ions release from the crystal lattice of synthetic gyrolite and zinc ions adsorption on synthetic gyrolite.

To find out the impact of Zn^{2+} ions on the shape of pores and the specific surface area of synthetic gyrolite, nitrogen adsorption–desorption experiments were performed. The isotherms of the adsorbent before the adsorption and after the adsorption are presented in Figs. 4.25 and 4.26, while the full description of pure synthetic gyrolite porosity is given in Sect. 2.2.2 *Effect of hydrothermal synthesis duration on the porosity of gyrolite*. It was established that the shape of all adsorption isotherms can be assigned to Type II (the absence of a plateau at higher p/p_0 values) or Type IV (the presence of a hysteresis loop) [79]. Thus, as in pure synthetic gyrolite, mesoporous pores predominate in the structure of all synthetic gyrolite samples with intercalated zinc ions.

Notably, zinc ions influence the shape of synthetic gyrolite pores because hysteresis loops of the samples synthesized for 32 and 48 h changed from Type H3 (mesoporous pores formed between parallel crystal planes (slit-shape pores) or macroporous pores with a pore network) to Type H1 which is present in mesoporous materials with cylindrical pores or in materials consisting of spherical particles (Figs. 4.25 and 4.26). Meanwhile, the changes of the hysteresis loops in the samples with higher crystallinity (72–168 h) are negligible.

Since the adsorption isotherms of the synthetic gyrolite samples with intercalated zinc ions are assigned to Type IV and have a point B, the specific surface area can

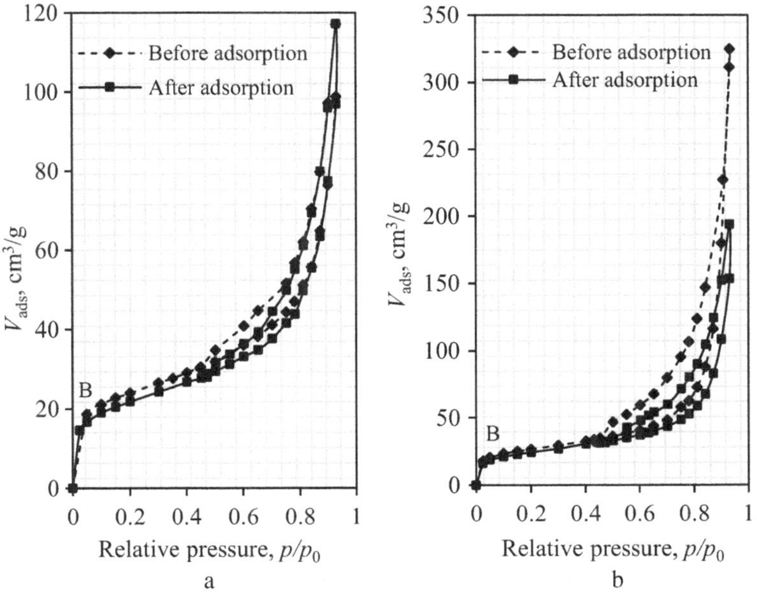

Fig. 4.25 Adsorption–desorption isotherms of gyrolite obtained at a temperature of 200 °C after 32 h (**a**) and 48 h (**b**)

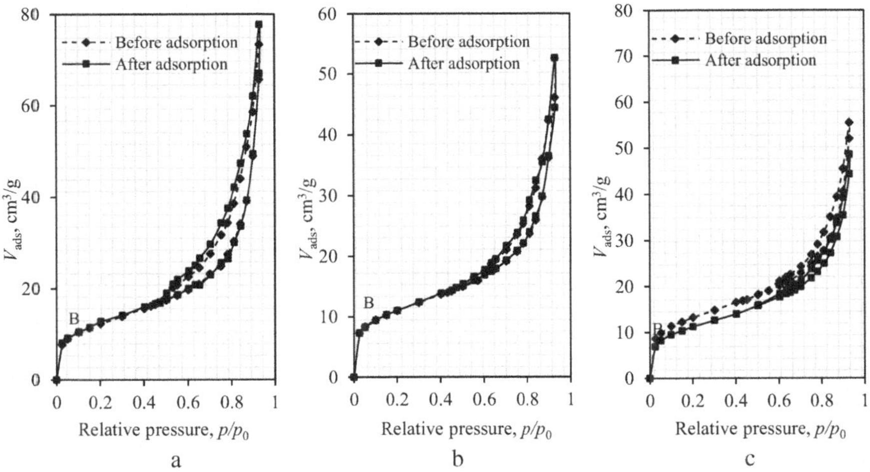

Fig. 4.26 Adsorption-desorption isotherms of gyrolite obtained at a temperature of 200 °C after 72 h (**a**), 120 h (**b**), and 168 h (**c**)

be calculated by employing the Brunauer–Emmett–Teller (BET) method (Eq. 2.2) [79].

The BET equation of synthetic gyrolite samples with intercalated Zn^{2+} ions gives a linear plot within a range of relative pressures $0.05 \leq p/p_0 \leq 0.35$ ($R^2 > 0.99$) (Fig. 4.27 and Table 4.9). It is worth highlighting that, due to gas condensation in the narrow pores or because of the strong interaction between the adsorbent and the adsorptive, for the BET calculations of the pure synthetic gyrolite samples, lower relative pressure intervals ($p/p_0 = 0.05$–0.2) were used.

The remarkable changes in the S_{BET} values were observed in the samples obtained after 32, 48, and 168 h of synthesis, while in those obtained after 72 and 120 h of synthesis, changes are negligible (Table 4.9). Calculations determined that the specific surface area of the gyrolite sample synthesized for 32 h decreased by 7.6%, that of the sample synthesized for 48 h decreased by 11.8%, and that of the sample synthesized for 168 h decreased by 17.1% (Tables 4.7 and 4.9). Meanwhile, the value of the constant C_{BET} increased from 1.4 to 16.4 times, which shows the increased interaction between the adsorbent and the adsorptive. Probably, intercalated zinc ions acted as active centers; thus, the presently discussed interaction increased.

To determine the effect of zinc ions on the micropore volume of synthetic gyrolite, the Halsey equation in a low p/p_0 range of 0.02–0.4 (Eq. 2.3) was used. In the t-plot graphs, straight lines with $R^2 > 0.99$ were obtained for all samples, which shows that the Halsey equation correctly described the experimental data (Fig. 4.28 and Table 4.10). The obtained results are listed in Table 4.10. It was established that, after the adsorption experiments, the sample synthesized for 32 h had some micropores as the difference between the S_{BET} and S_t values was ~8 m^2/g (11%). Furthermore, it was calculated that the micropore volume of this sample was equal to 4.028×10^{-3} cm^3/g, i.e. it was two times lower than for the sample before adsorption. The calculations showed that other synthetic gyrolite samples with intercalated zinc ions do not have micropores since the difference between the S_{BET} and S_t values is insignificant, i.e. it is less than 5%.

To find out the average diameter of pores and the total pore volume of synthetic gyrolite with intercalated zinc ions, the corrected Kelvin equation and the scheme suggested by Orr et al. by referring to the entire N_2 desorption isotherm at 77 K were used (Eqs. 2.6–2.17) [80–83]. Since the hysteresis loops of the samples were assigned to Type H3, desorption isotherms were calculated by employing the models of slit-shape pores between parallel plates and cylindrical pores (Tables 4.11–4.14). The primary data for ΣA calculations of gyrolite (synthesized for 32 and 168 h) with intercalated zinc ions is given in Tables 4.11 and 4.13, while the results of these calculations are listed in Tables 4.12 and 4.14. The calculations were run until $\Delta t \Sigma_A$ exceeded ΔV_L. Analogous calculations were performed for other adsorbent samples, and the generalized outline of results is presented in Table 4.15.

The calculations of the desorption isotherms confirmed that zinc ions affect the shape of synthetic gyrolite pores (Table 4.15, Figs. 4.25 and 4.26). Furthermore, it was determined that, when using the slit-shape pores between parallel plates model, high errors between the measured S_{BET} and calculated ΣA (30–50% in all cases) were observed (Table 4.15). Meanwhile, the employment of the cylindrical-shape

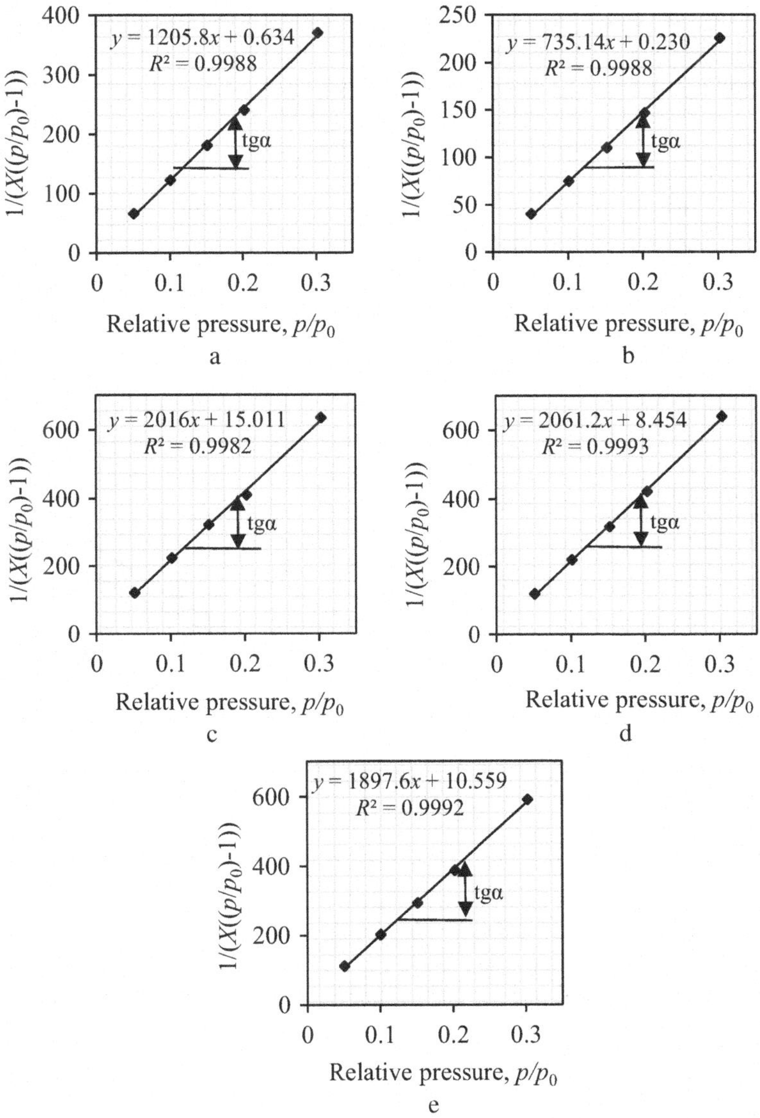

Fig. 4.27 Adsorption–desorption isotherms in the BET plot of gyrolite synthesized for 32 h (**a**), 48 h (**b**), 72 h (**c**), 120 h (**d**), and 168 h (**e**) and substituted with zinc ions

pores model yields a difference between S_{BET} and $\sum A$ which is lower than 20% in all the investigated cases, and thus, it was thus was proven that the employed model is valid. Furthermore, it was obtained that the shape of the pores of lower crystallinity synthetic gyrolite changed. In contrast, the shape of the pores of higher crystallinity synthetic gyrolite remained unchanged:

Table 4.9 Specific surface area (S_{BET}) parameters of gyrolite synthesized in the CaO–SiO_2–H_2O system after the process of adsorption of Zn^{2+} ions

Duration of synthesis, h	Sample mass, g	BET equation constants		Capacity of monolayer X_m, g	S_{BET}, m^2/g	C_{BET}	Reliability coefficient R^2
		Slope $S = tg\alpha$	Intercept, I				
32	0.039	1205.8	0.634	0.0008	75.03	1903.19	0.9988
48	0.057	735.1	0.230	0.0014	83.93	3204.22	0.9988
72	0.039	2016.0	15.011	0.0005	44.45	135.30	0.9982
120	0.044	2061.2	8.454	0.0005	38.57	244.82	0.9993
168	0.047	1897.6	10.559	0.0005	39.33	180.71	0.9992

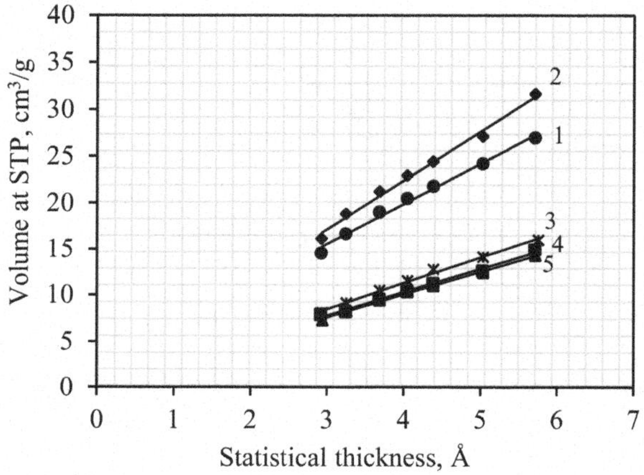

Fig. 4.28 t-plot of gyrolite obtained after adsorption when the duration of hydrothermal treatment, h: 1—32; 2—48; 3—72; 4—120; 5—168

Table 4.10 Data of micropores calculations of synthetic samples

Duration of synthesis, h	S_{BET}, m^2/g	Slope	Intercept	R^2	S_t, m^2g	S_{BET}—S_t, m^2g	S_{BET}—S_t, %	Volume of micropores, ($\times 10^3$) mm^3/g
32	75.03	43.19	2.604	0.990	66.82	8.21	10.95	4.028
48	83.93	52.38	1.445	0.991	81.03	2.90	3.46	2.235
72	44.45	28.08	0.030	0.997	43.44	1.01	2.28	0.047
120	38.57	24.47	0.252	0.997	37.85	0.72	1.86	0.390
168	39.33	25.22	0.197	0.991	39.02	0.31	0.78	0.305

- The shape of the pores of gyrolite, synthesized for 32 h, changed from undefined to cylindrical-shape pores.
- The shape of the pores of gyrolite, synthesized for 48 h, changed from slit-shape pores to cylindrical-shape pores.

Table 4.11 Primary data of ΣA calculations and pore size distribution of synthetic gyrolite (32 h) with intercalated Zn^{2+} ions

No.	Relative pressure, p/p_0	N_2 volume adsorbed on surface V_{ad}, ncm^3/g	Kelvin radius of the pores, r_K, Å	Thickness of N_2 layer t, Å	True pore radius, Å
1	0.93	117.21	135.14	14.64	149.79
2	0.90	96.00	93.21	12.94	106.15
3	0.87	79.89	70.29	11.78	82.07
4	0.84	69.40	55.95	10.91	66.86
5	0.81	61.09	46.24	10.24	56.48
6	0.78	55.13	39.13	9.69	48.82
7	0.75	49.88	33.83	9.23	43.06
8	0.70	44.46	27.25	8.59	35.83
9	0.65	39.26	22.55	8.06	30.61
10	0.61	36.44	19.01	7.62	26.62
11	0.56	33.68	16.24	7.23	23.46
12	0.51	31.60	13.98	6.87	20.86
13	0.50	27.92	13.58	6.81	20.39
14	0.46	27.27	12.13	6.56	18.69

- The shape of the pores of gyrolite, synthesized for 72, 120, or 168 h, remained unchanged.

It was observed that, after adsorption, the total pore volume of gyrolite synthesized for 32 h increased from 170 mm^3/g (Fig. 2.18, curve 1) to 183 mm^3/g (Table 4.15). The increment of the pore volume can be related to the high amount of released Ca^{2+} ions from the structure of the adsorbent (Fig. 4.21). Significant decrement of the total pore volume (from 509 to 294 mm^3/g) in the gyrolite sample synthesized for 48 h was observed (Fig. 2.18, curve 2, and Table 4.15). Such a change can be explained by transforming slit-shape pores into cylindrical-shape pores. Further, analysis showed that zinc ions do not impact the total pore volume of high crystallinity synthetic gyrolite (Fig. 2.18, curves 2–3, and Table 4.15).

To generalize on the obtained data, it is evident that the adsorption process of zinc ions depends on the crystallinity of synthetic gyrolite. We established that the total number of the adsorbed ions decreased by increasing the crystallite size of the adsorbent and by reducing the quantity of water in the interlayers. In all cases, the substitution reactions of Zn^{2+} ions are dominant and can be described by the pseudo-second-order kinetic model which shows that chemisorption takes place. The results of the kinetic calculations were confirmed by leaching experiments (the adsorption reactions are irreversible). It was determined that zinc ions prominently affect the surface area and the porosity of low crystallinity samples.

Table 4.12 Primary data of ΣA calculations and pore size distribution of synthetic gyrolite (32 h) with intercalated Zn^{2+} ions

Average Kelvin radius of pores r_K, Å	Average true pore radius r_p, Å	Change Δt, Å	Change ΔV_{ad}, ncm^3/g	Change of evaporated liquid adsorbate volume ΔV_L, × 10^3 cm^3/g	$\Delta t \Sigma A$, × 10^3 cm^3/g	True volume of pores, V_p, × 10^3 cm^3/g	Surface of pore walls A, m^2/g	Total surface ΣA, m^2/g
Calculations using the model of parallel plates pores								
114.18	141.76	1.71	21.21	32.66	0.00	40.55	5.72	0.17
81.75	106.47	1.16	16.11	24.80	0.02	32.25	6.06	6.23
63.12	85.81	0.86	10.49	16.16	0.54	20.51	4.78	11.01
51.09	72.25	0.67	8.30	12.79	0.74	15.99	4.43	15.43
42.69	62.62	0.55	5.97	9.19	0.86	10.97	3.50	18.94
36.48	55.40	0.46	5.25	8.08	0.87	9.63	3.48	22.41
30.54	48.35	0.64	5.42	8.35	1.44	8.67	3.59	26.00
24.90	41.54	0.53	5.20	8.01	1.37	8.80	4.24	30.24
20.78	36.45	0.45	2.82	4.34	1.35	2.88	1.58	31.82
17.62	32.46	0.39	2.77	4.26	1.24	3.28	2.02	33.84
15.11	29.21	0.35	2.08	3.20	1.19	1.60	1.10	34.93
13.78	27.47	0.07	3.67	5.65	0.23	10.34	7.53	42.46
12.86	26.22	0.25	0.65	1.01	1.07			
Calculations using the cylindrical pores model								
114.18	127.97	1.71	21.21	32.66	0.00	41.03	6.41	0.17
81.75	94.11	1.16	16.11	24.80	0.02	32.84	6.98	7.15
63.12	74.46	0.86	10.49	16.16	0.62	21.63	5.81	12.96
51.09	61.67	0.67	8.30	12.79	0.87	17.36	5.63	18.59
42.69	52.65	0.55	5.97	9.19	1.03	12.41	4.71	23.30
36.48	45.94	0.46	5.25	8.08	1.07	11.12	4.84	28.14
30.54	39.44	0.64	5.42	8.35	1.81	10.92	5.54	33.68
24.90	33.22	0.53	5.20	8.01	1.77	11.10	6.69	40.37
20.78	28.62	0.45	2.82	4.34	1.80	4.81	3.36	43.73
17.62	25.04	0.39	2.77	4.26	1.71	5.16	4.12	47.85
15.11	22.16	0.35	2.08	3.20	1.68	3.28	2.96	50.81
13.78	20.63	0.07	3.67	5.65	0.34	11.91	11.54	62.35
12.86	19.54	0.25	0.65	1.01	1.57			

Table 4.13 Primary data of ΣA calculations and pore size distribution of synthetic gyrolite (168 h) with intercalated Zn^{2+} ions

No.	Relative pressure, p/p_0	N_2 volume adsorbed on surface V_{ad}, ncm^3/g	Kelvin radius of the pores r_K, Å	Thickness of N_2 layer t, Å	True pore radius, Å
1	0.93	48.59	134.74	14.63	149.37
2	0.90	39.23	92.96	12.93	105.88
3	0.87	33.77	70.11	11.77	81.88
4	0.84	30.20	55.94	10.91	66.85
5	0.81	27.59	46.28	10.24	56.52
6	0.78	25.65	39.19	9.69	48.88
7	0.75	23.75	33.86	9.23	43.09
8	0.70	22.38	27.30	8.59	35.89
9	0.68	20.35	24.30	8.26	32.56
10	0.66	19.82	22.59	8.07	30.65
11	0.64	19.23	21.04	7.88	28.91
12	0.61	18.37	19.02	7.62	26.64
13	0.55	16.85	16.00	7.19	23.19
14	0.51	15.93	14.00	6.88	20.87
15	0.41	13.68	10.57	6.26	16.83
16	0.30	12.38	8.04	5.72	13.75
17	0.25	11.44	6.98	5.45	12.43
18	0.19	11.10	5.83	5.13	10.96

Whereas the synthetic gyrolite sample showed good adsorption capacity for Zn^{2+} ions, the adsorption experiments of iron ions were performed in the next step of the research. For the experiments, gyrolite synthesized for 48 h was chosen.

4.3.3 *Synthetic Gyrolite Application for Fe^{3+} Ions Adsorption Process*

According to the literature, iron ions are a necessary element for live organisms and plants [84]. This element plays a fundamental role in the process of biological redox and is important for other biological processes [85, 86]. However, the expansion of industry leads to toxic concentrations of iron in the environment, especially in water [10–12]. Adsorption is one of the most effective ways of controlling the iron concentration in wastewater. Thus, in the next step of this research, synthetic gyrolite was used as an adsorbent for cleaning water contaminated with iron ions.

We established that synthetic gyrolite is denoted by good adsorption capacity for iron ions when the initial value of concentration of these ions is 250 mg/dm^3, and

Table 4.14 Primary data of ΣA calculations and pore size distribution of synthetic gyrolite (168 h) with intercalated Zn^{2+} ions

Average Kelvin radius of pores r_K, Å	Average true pore radius r_p, Å	Change Δt, Å	Change ΔV_{ad}, ncm^3/g	Change of evaporated liquid adsorbate volume ΔV_L, × 10^3 cm^3/g	$\Delta t \Sigma A$, × 10^3 cm^3/g	True volume of pores, V_p, × 10^3 cm^3/g	Surface of pore walls A, m^2/g	Total surface ΣA, m^2/g
Calculations using the model of parallel plates pores								
113.85	141.40	1.70	9.36	14.41	0.00	17.90	2.53	0.17
81.53	106.23	1.16	5.46	8.41	0.02	10.90	2.05	2.22
63.02	85.70	0.85	3.57	5.50	0.19	6.96	1.62	3.85
51.11	72.26	0.67	2.61	4.02	0.26	4.96	1.37	5.22
42.73	62.67	0.55	1.94	2.99	0.29	3.54	1.13	6.35
36.52	55.45	0.46	1.90	2.92	0.29	3.55	1.28	7.63
30.58	48.40	0.64	1.38	2.12	0.49	1.81	0.75	8.38
25.80	42.65	0.33	2.02	3.12	0.27	4.25	1.99	10.37
23.44	39.77	0.20	0.53	0.82	0.21	0.69	0.34	10.71
21.81	37.75	0.19	0.59	0.91	0.20	0.88	0.47	11.18
20.03	35.52	0.26	0.86	1.32	0.29	1.31	0.74	11.92
17.51	32.32	0.43	1.52	2.34	0.51	2.43	1.51	13.42
15.00	29.06	0.31	0.93	1.43	0.42	1.14	0.78	14.21
12.28	25.42	0.61	2.25	3.47	0.87	3.56	2.80	17.01
9.30	21.28	0.55	1.30	2.00	0.93	0.33	0.31	17.32
7.51	18.68	0.26	0.94	1.44	0.45	1.32	1.41	18.73
6.40	16.99	0.32	0.34	0.53	0.60			
Calculations using the cylindrical pores model								
113.85	127.63	1.70	9.36	14.41	0.00	18.11	2.84	0.17
81.53	93.88	1.16	5.46	8.41	0.02	11.12	2.37	2.54
63.02	74.36	0.85	3.57	5.50	0.22	7.35	1.98	4.52
51.11	61.69	0.67	2.61	4.02	0.30	5.42	1.76	6.28
42.73	52.70	0.55	1.94	2.99	0.35	4.02	1.52	7.80
36.52	45.98	0.46	1.90	2.92	0.36	4.06	1.77	9.57
30.58	39.49	0.64	1.38	2.12	0.61	2.51	1.27	10.84
25.80	34.23	0.33	2.02	3.12	0.35	4.86	2.84	13.68
23.44	31.61	0.20	0.53	0.82	0.27	0.99	0.63	14.31
21.81	29.78	0.19	0.59	0.91	0.27	1.20	0.81	15.11
20.03	27.78	0.26	0.86	1.32	0.39	1.78	1.28	16.40
17.51	24.91	0.43	1.52	2.34	0.70	3.31	2.66	19.06
15.00	22.03	0.31	0.93	1.43	0.60	1.79	1.63	20.68

(continued)

Table 4.14 (continued)

Average Kelvin radius of pores r_K, Å	Average true pore radius r_p, Å	Change Δt, Å	Change ΔV_{ad}, ncm^3/g	Change of evaporated liquid adsorbate volume ΔV_L, × 10^3 cm^3/g	$\Delta t \Sigma A$, × 10^3 cm^3/g	True volume of pores, V_p, × 10^3 cm^3/g	Surface of pore walls A, m^2/g	Total surface ΣA, m^2/g
12.28	18.85	0.61	2.25	3.47	1.27	5.17	5.49	26.17
9.30	15.29	0.55	1.30	2.00	1.43	1.54	2.02	28.19
7.51	13.09	0.26	0.94	1.44	0.74	2.13	3.25	31.45
6.40	11.70	0.32	0.34	0.53	1.00			

the pH value of the liquid medium equals to 2. The results of atomic adsorption analysis showed that more than 98% of Fe^{3+} ions (24.49 mg Fe^{3+}/g) were adsorbed on synthetic gyrolite within 30 s (Fig. 4.29). When increasing the length of adsorption to 60 min, 1 g of synthetic gyrolite adsorbed 24.97 mg of Fe^{2+} ions (99.9%).

It was obtained that, as in the previous case (adsorption of Zn^{2+} ions), in the course of adsorption of Fe^{3+} ions, Ca^{2+} ions were released from the crystal lattice of synthetic gyrolite (Fig. 4.30). At the beginning of the adsorption, 18.02 mg of Ca^{2+} from 1 g of the adsorbent were released (Fig. 4.30). By extending the length of the experiment to 60 min, the share of the released calcium ions ultimately reached as high as 20.96 mg/g.

It was determined that, together with calcium ions, OH^- anions are released; thus, the pH value of the liquid medium increased from 2 (before the experiment) to 6.39 (after 60 min of the adsorption) (Fig. 4.31).

We also established that, during adsorption of Fe^{3+} ions on synthetic gyrolite, more than 77% of Fe^{3+} ions participated in a substitution reaction, i.e. trivalent iron ions replaced divalent calcium ions. The remaining share of Fe^{3+} ions was adsorbed by synthetic gyrolite as described with the addition reaction. Therefore, it can be stated that the adsorption of Fe^{3+} ions on synthetic gyrolite is a fairly complex process. The leaching experiment was performed to confirm the nature of the adsorption process: synthetic gyrolite obtained after adsorption was dried and later immersed in distilled water (25 °C) for 60 min. The analysis of the liquid medium indicated that the concentration of Fe^{3+} ions in the liquid medium was lower than the detection limit, which means that the adsorption process of Fe^{3+} on synthetic gyrolite is irreversible.

The adsorption process was described by the pseudo-first-order and pseudo-second-order models to determine the kinetic parameters of adsorption. The calculations demonstrated that the pseudo-first-order kinetic equation did not fit the iron ions adsorption mechanism because $\Sigma X_{e(cal)}$ disagreed with the $\Sigma X_{e(exp)}$ value, and the value of R^2 was very low (0.48) (Fig. 4.32a and Table 4.16).

Different results were received when employing the pseudo-second-order kinetic equation: the calculated value of the adsorbed Fe^{3+} ions ($\Sigma X_{e(cal)}$) agreed with the

Table 4.15 Data of ΣA calculations of synthetic samples

Duration of synthesis, h	S_{BET}, m^2/g	Results of using the cylindrical-shape pores model			Total pore volume, mm^3/g	Results of using the slit-shape pores model		
		ΣA, m^2/g	$\lvert S_{BET}—\Sigma A\rvert$, m^2/g	$\lvert S_{BET}—\Sigma A\rvert$, %		ΣA, m^2/g	$\lvert S_{BET}—\Sigma A\rvert$, m^2/g	$\lvert S_{BET}—\Sigma A\rvert$, %
32	75.03	62.35	12.68	16.90	183.57	42.46	32.57	43.41
48	83.93	76.27	7.66	9.13	294.54	48.18	35.75	42.60
72	44.45	48.04	3.59	8.08	128.48	31.75	12.70	28.57
120	38.57	30.05	8.52	22.09	81.28	17.84	20.73	53.75
168	39.33	31.45	7.88	20.04	75.39	18.73	20.60	52.38

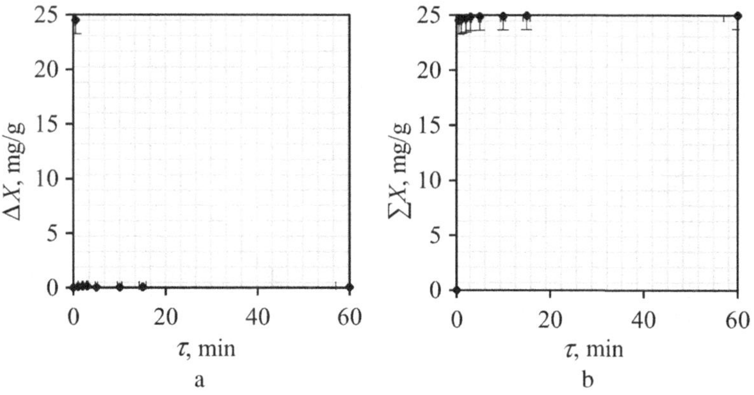

Fig. 4.29 Differential (**a**) and integral (**b**) kinetic curves of Fe^{3+} ions adsorption on synthetic gyrolite

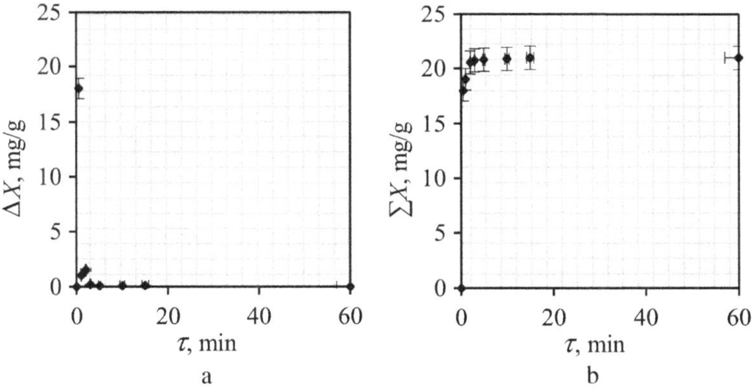

Fig. 4.30 Differential (**a**) and integral (**b**) kinetic curves of Ca^{2+} ions concentration variation in acidic solution

Fig. 4.31 Variation of pH value of liquid medium

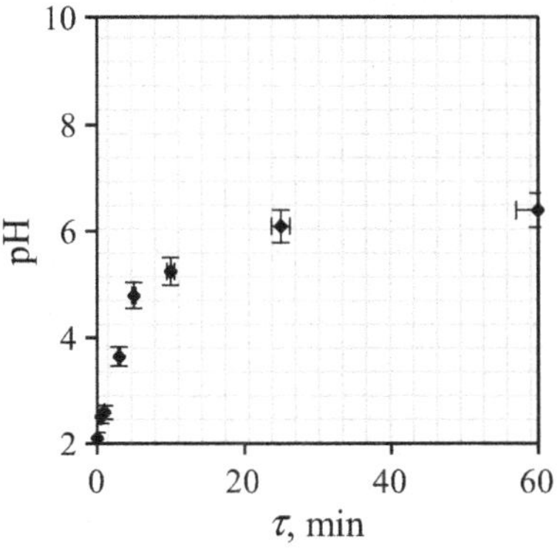

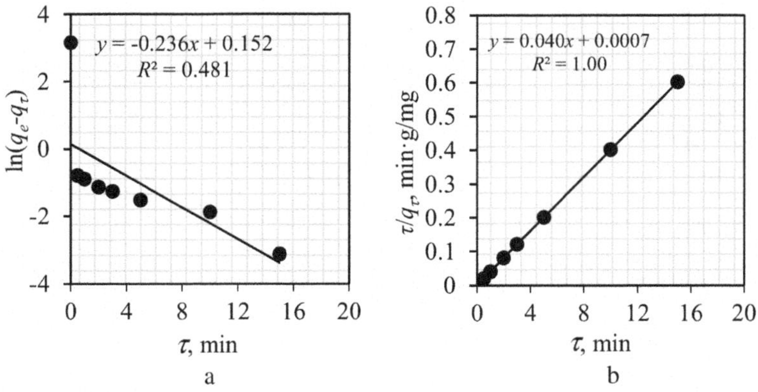

Fig. 4.32 Pseudo-first-order (**a**) and pseudo-second-order (**b**) kinetic plots in acidic solutions

Table 4.16 Kinetic parameters of pseudo-first- and pseudo-second-order kinetic models of Fe^{3+} ions adsorption on synthetic gyrolite

$\Sigma X_{e(exp)}$	Pseudo-first-order			Pseudo-second-order		
	k_1, 1/min	$\Sigma X_{e(cal)}$, mg/g	R^2	k_2, g/(mg min)	$\Sigma X_{e(cal)}$, mg/g	R^2
24.97	–	1.02	0.481	2.30	24.94	1.00

experimental values ($\Sigma X_{e(exp)}$) (Table 4.16 and Fig. 4.33), and the value of the correlation coefficient of the plot t/q_t versus t was equal to 1.000 (Fig. 4.32b). Calculations showed that the Fe^{3+} ions adsorption rate (k_2) on synthetic gyrolite is equal to 2.30 g/mg min (Table 4.16), which is significantly higher than those of zinc adsorption [0.08–0.40 g/(mg min)] (Tables 4.3–4.5 and 4.8). Thus, Fe^{3+} ions adsorption reactions proceed more rapidly than the reactions involving zinc ions.

The stability of synthetic gyrolite can be identified during adsorption if the product of adsorption is dried and characterized by XRD and DSC analysis (Fig. 4.34). The results of the XRD analysis demonstrated that synthetic gyrolite is stable during

Fig. 4.33 Calculated and experimental differential kinetic curves of Fe^{3+} ions adsorption on synthetic gyrolite

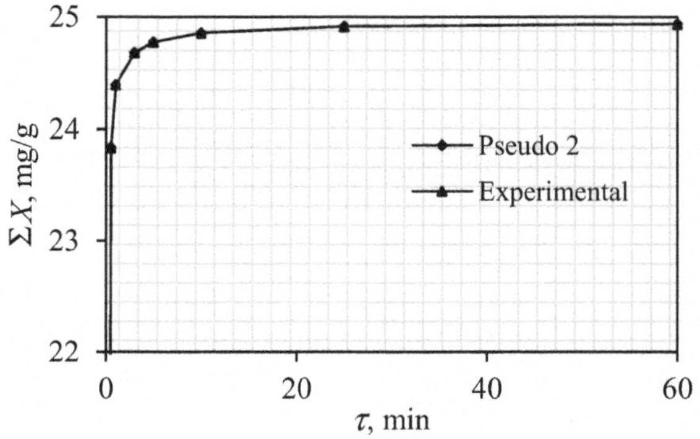

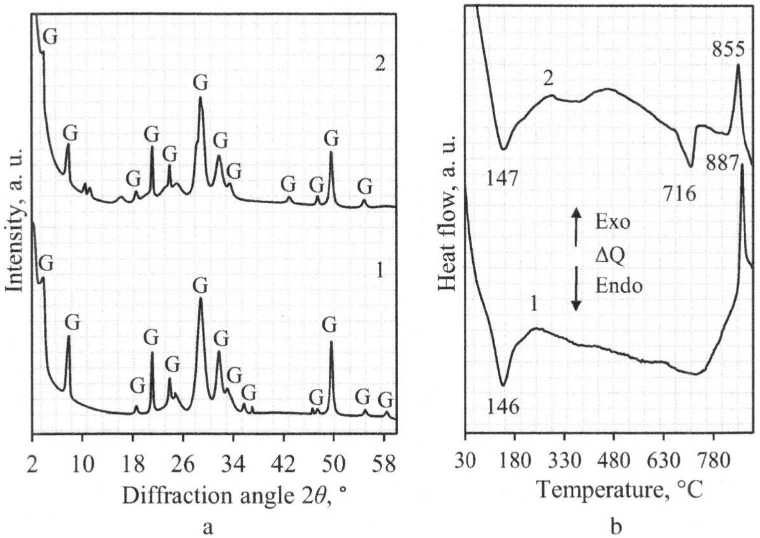

Fig. 4.34 X-ray diffraction patterns (**a**) and DSC curves (**b**) of synthetic gyrolite (curve 1) and synthetic gyrolite after adsorption (curve 2). *Indexes* G—gyrolite

adsorption because all the diffraction peaks of gyrolite were identified, although their intensity slightly decreased (Fig. 4.34a). The decrement can be related to the calcium ions desorption from the structure of the adsorbent (Fig. 4.30). It is to be emphasized that no other compounds were detected after adsorption in the XRD pattern.

The results of XRD were corroborated by DSC analysis. In the DSC curve, identical thermal effects were identified in both samples: an endothermic effect at ~146 °C characteristic of water removal from the structure of synthetic gyrolite and an exothermic effect at ~870 °C characteristic of the formation of wollastonite (Fig. 4.34b). In addition, it was detected that, when adsorption was taking place, carbonization of the adsorbent also proceeded as a small endothermic effect at ~716 °C characteristic of the decomposition of calcium carbonate was observed (Fig. 4.34b, curve 2).

The results of the zinc and iron ion adsorption experiments (Figs. 4.15, 4.19, and 4.29) indicate that the adsorption capacity of gyrolite for heavy metal ions depends on their nature. Thus, in the next step, synthetic gyrolite was employed to adsorb various heavy metal ions.

4.3.4 *Synthetic Gyrolite Application for Ni^{2+}, Co^{2+}, and Mn^{2+} Ions Adsorption Process*

The rapid development of industries leads to the increment of the amount of wastewater with toxic elements [10–12]. As the regulations covering heavy metal contamination are explicitly stringent, their removal has evolved into a severe environmental and industrial problem. However, most adsorbents have good adsorption capacity only to one or very few heavy metal ions. For this reason, the synthesis and application of adsorbents that can adsorb various pollutants is in great demand. One of such adsorbents can be lower basicity calcium silicate hydrates, especially synthetic gyrolite. Experiments in solutions containing 250 mg/dm^{3+} of Ni^{2+}, Cu^{2+}, Co^{2+}, or Mn^{2+} ions were performed to investigate the adsorption capacity of synthetic gyrolite (synthesized for 48 h at 200 °C).

Adsorption of Ni^{2+} ions. The outcomes of the adsorption experiments indicated that the adsorption of heavy metal ions on synthetic gyrolite depends on the chemical nature of the adsorptive. As determined, the uptake of nickel ions takes place intensively during the initial part of the process as, after 30 s, the amount of the ions adsorbed by synthetic gyrolite exceeded 10 mg Ni^{2+}/g (Fig. 4.35). The adsorbed amount of Ni^{2+} ions slightly increased by prolonging the length of the adsorption experiment, and, after 60 min, it equaled to 20.84 mg Ni^{2+}/g (Fig. 4.35b).

Further, research of the liquid medium indicated that, in the course of adsorption of Ni^{2+} ions, Ca^{2+} and OH^- ions are released from the crystal lattice of the adsorbent (Figs. 4.36 and 4.37). As in the case of zinc and iron ions adsorption, most calcium ions (16.7 mg Ca^{2+}/g) are released within 30 s of adsorption (Fig. 4.36a). Later, the concentration of Ca^{2+} ions in the liquid medium slightly increases with the increase of the adsorption duration. At the end of the process (60 min), the amount of the released calcium ions ultimately reached 21.31 mg/g (Fig. 4.36b).

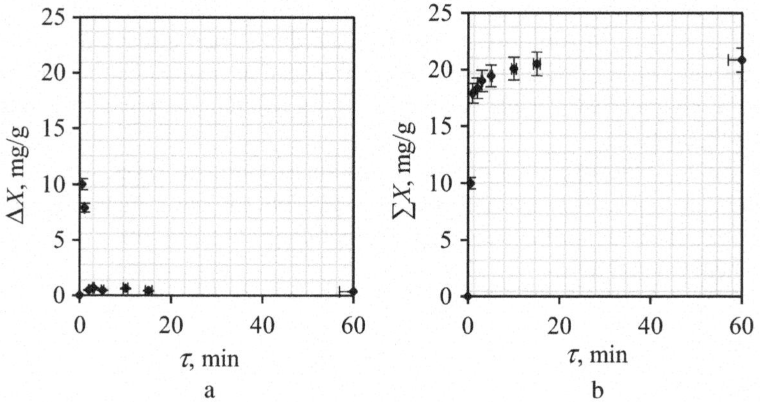

Fig. 4.35 Differential (**a**) and integral (**b**) kinetic curves of Ni^{2+} ions adsorption on synthetic gyrolite

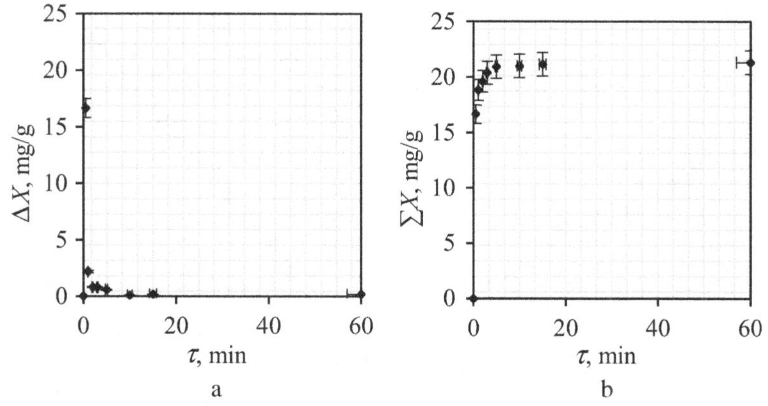

Fig. 4.36 Differential (**a**) and integral (**b**) kinetic curves of Ca^{2+} ions concentration variation

Fig. 4.37 Variation of pH value of the liquid medium

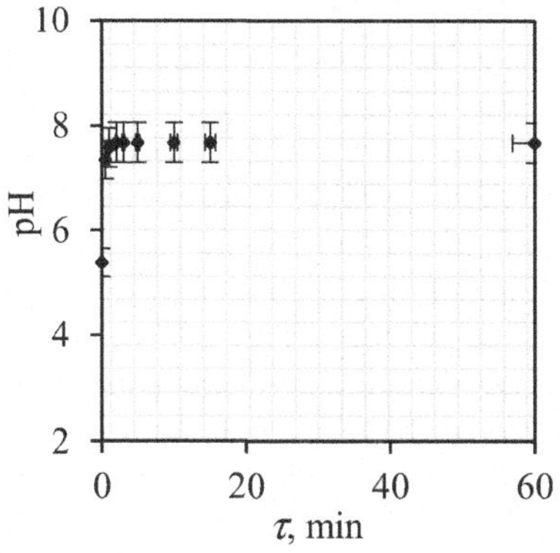

We established that the pH value of the liquid medium rapidly grew, starting with 5.39 (before adsorption), and ultimately reaching 7.58 (after 1 min of adsorption) (Fig. 4.37). After 3 min of adsorption, the equilibrium value of pH got established. These changes are in good agreement with the amount of Ca^{2+} ions released from the structure of synthetic gyrolite in the liquid medium.

It was calculated that all Ni^{2+} ions took part only in the substitution reaction: gyrolite-Ca^0 + Ni^{2+} $\leftrightarrow$ gyrolite-Me^0 + Ca^{2+}, while the results of the leaching test (immersion of the adsorbent with intercalated Ni^{2+} ions in distilled water) showed that this reaction is irreversible.

The calculations of the adsorption kinetic models showed that the pseudo-first-order model did not show a good fit for the nickel ions adsorption mechanism as significant difference between the values of $\Sigma X_{e(cal)}$ and $\Sigma X_{e(exp)}$ was observed, and the value of R^2 was quite low as it only equaled to 0.62 (Fig. 4.38a and Table 4.17). Meanwhile, by employing the pseudo-second-order model, the calculated $\Sigma X_{e(cal)}$ value agreed with the experiment and $\Sigma X_{e(exp)}$ value, and the value of R^2 was equal

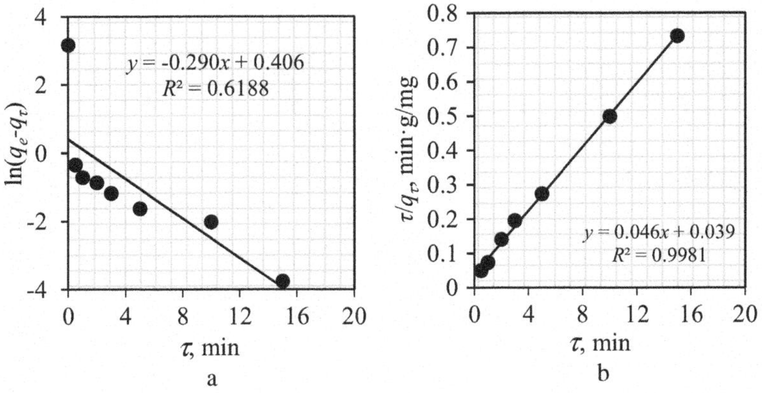

Fig. 4.38 Pseudo-first-order (**a**) and pseudo-second-order (**b**) kinetic plots in acidic solutions

to 1 (Fig. 4.38b and Table 4.17). We determined that the value of the adsorption rate (k_2) of Ni^{2+} ions on synthetic gyrolite is low and equal to as little as 0.05 g/(mg min) (Table 4.17).

Adsorption of Co^{2+} ions. We found out that the adsorption of cobalt ions on synthetic gyrolite proceeds heavily because, 30 s after the start of adsorption, only 3 mg of Co^{2+} ions were adsorbed by 1 g of adsorbent (Fig. 4.39). By extending the length of the experiment to 1 min, the amount of the adsorbed ions more than tripled (9.8 mg Co^{2+}/g) (Fig. 4.39b). The adsorption equilibrium was achieved after 15 min of the experiment when synthetic gyrolite adsorbed the maximum amount of Co^{2+} ions, i.e. 14.86 mg Co^{2+}/g.

The calculations of the adsorption kinetic models showed that linear dependency of both pseudo-first- and pseudo-second-order models was obtained in the plot $\log(\Sigma X_{e(exp)} - X_t)$ versus t and t/X_t versus t, respectively (Fig. 4.40). It was calculated that the difference between the values of $\Sigma X_{e(cal)}$ and $\Sigma X_{e(exp)}$ obtained at the final experimental point is similar in both cases (Table 4.18 and Fig. 4.41). However, when using the pseudo-first-order model, a significant difference between the values of $\Sigma X_{e(cal)}$ and $\Sigma X_{e(exp)}$ was observed at the beginning of the calculations; thus, this model did not show a good fit for the cobalt ions adsorption mechanism. Meanwhile, when using the pseudo-second-order model, $\Sigma X_{e(cal)}$ agreed with $\Sigma X_{e(exp)}$ across the entire range of calculations (Fig. 4.41). The calculated value of the adsorption rate (k_2) of Co^{2+} ions on synthetic gyrolite is low and equal to 0.05 g/(mg min) (Table 4.18).

Table 4.17 Kinetic parameters of pseudo-first- and pseudo-second-order kinetic models of Ni^{2+} ions adsorption on synthetic gyrolite

$\Sigma X_{e(exp)}$	Pseudo-first-order			Pseudo-second-order		
	k_1, 1/min	$\Sigma X_{e(cal)}$, mg/g	R^2	k_2, g/(mg min)	$\Sigma X_{e(cal)}$, mg/g	R^2
20.84	–	11.99	0.62	0.05	21.60	1.00

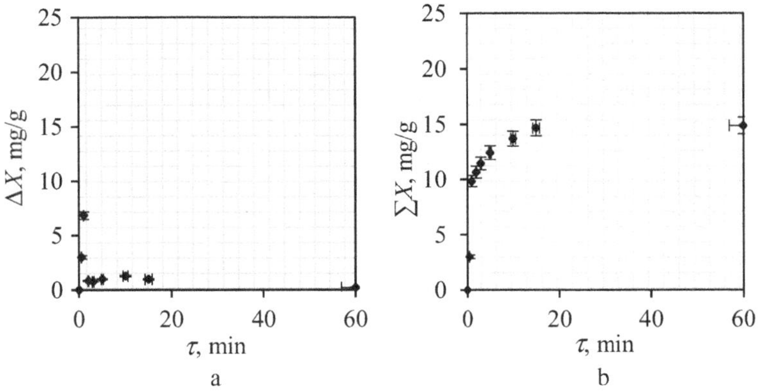

Fig. 4.39 Differential (**a**) and integral (**b**) kinetic curves of Co^{2+} ions adsorption on synthetic gyrolite

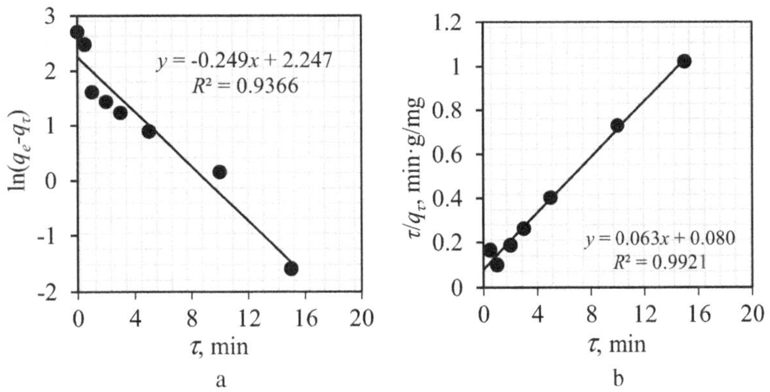

Fig. 4.40 Pseudo-first-order (**a**) and pseudo-second-order (**b**) kinetic plots in acidic solutions

Table 4.18 Kinetic parameters of the pseudo-first- and pseudo-second-order kinetic models of Co^{2+} ions adsorption on synthetic gyrolite

$\Sigma X_{e(exp)}$	Pseudo-first-order			Pseudo-second-order		
	k_1, 1/min	$\Sigma X_{e(cal)}$, mg/g	R^2	k_2, g/(mg min)	$\Sigma X_{e(cal)}$, mg/g	R^2
14.86	–	14.51	0.94	0.05	14.55	0.99

Adsorption of Mn^{2+} ions. We established that, at the beginning of the adsorption of manganese ions on synthetic gyrolite, the process was fairly intensive because, after 30 s of the experiment, more than 17 mg Mn^{2+} ions was adsorbed by 1 g of synthetic gyrolite (Fig. 4.42). By extending the length of the experiment, the amount of the adsorbed ions went up, and, at the end of the experiment, synthetic gyrolite adsorbed 86.7% of Mn^{2+} ions featuring in the initial liquid medium (Fig. 4.42b).

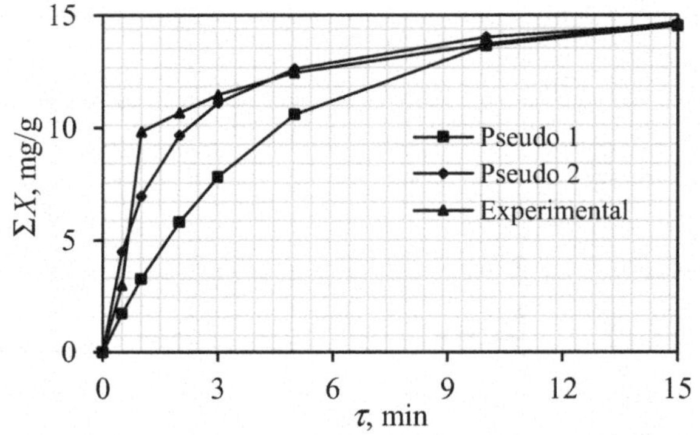

Fig. 4.41 Calculated and experimental differential kinetic curves of Co^{2+} ions adsorption on synthetic gyrolite

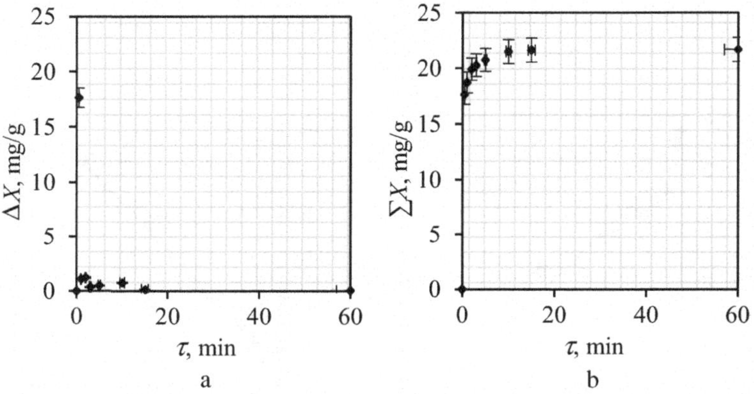

Fig. 4.42 Differential (**a**) and integral (**b**) kinetic curves of Mn^{2+} ions adsorption on synthetic gyrolite

We determined that the pseudo-first-order model did not have a good fit for the kinetics and mechanism of manganese ions adsorption ($R^2 < 0.9$, $\Sigma X_{e(cal)} \neq \Sigma X_{e(exp)}$) (Figs. 4.43, 4.44 and Table 4.19). The calculations employing the pseudo-second-order model showed a good fit between the values of $\Sigma X_{e(cal)}$ and $\Sigma X_{e(exp)}$ (Fig. 4.44 and Table 4.19). We thus established the value of the adsorption rate (k_2) of Mn^{2+} ions on synthetic gyrolite which was actually quite low and equal to as little as 0.24 g/(mg min) (Table 4.19).

Thus, the adsorption of Co^{2+}, Ni^{2+}, Zn^{2+}, Mn^{2+}, and Fe^{3+} from aqueous solutions by synthetic gyrolite is a complex process. As already determined, heavy metal ions were mostly involved in the substitution reaction: gyrolite-$Ca^0 + Me^{x+} \leftrightarrow$ gyrolite-$Me^0 + Ca^{2+}$. The amount of the adsorbed metal ions and the adsorption rate depends on both the nature of the adsorbate and the parameters of the adsorptive (Table 4.20).

It was determined that, in acidic solutions when the primary concentration of metal ions in the liquid medium equals to 250 mg/dm^3, the selectivity sequence (in terms of the adsorbed amount of metal ions) of adsorption by synthetic gyrolite can

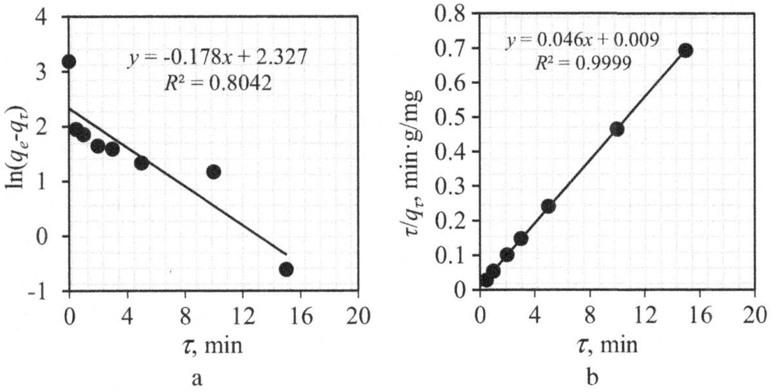

Fig. 4.43 Pseudo-first-order (**a**) and pseudo-second-order (**b**) kinetic plots in acidic solutions

Fig. 4.44 Calculated and experimental differential kinetic curves of Mn^{2+} ions adsorption on synthetic gyrolite

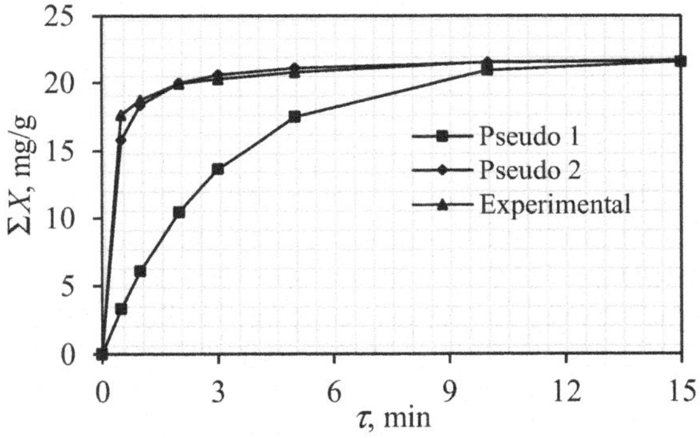

Table 4.19 Kinetic parameters of the pseudo-first- and pseudo-second-order kinetic models of Mn^{2+} ions adsorption on synthetic gyrolite

$\Sigma X_{e(exp)}$	Pseudo-first-order			Pseudo-second-order		
	k_1, 1/min	$\Sigma X_{e(cal)}$, mg/g	R^2	k_2, g/(mg min)	$\Sigma X_{e(cal)}$, mg/g	R^2
21.67	–	21.52	0.80	0.24	21.59	1.00

be represented in the following order:

$$Co^{2+} < Ni^{2+} < Mn^{2+} < Zn^{2+} < Fe^{3+}.$$

Meanwhile, the sequence of the adsorption rate k_2 differs:

$$Co^{2+} = Ni^{2+} < Zn^{2+} < Mn^{2+} < Fe^{3+}.$$

Table 4.20 Amount of adsorbed and leached metal ions

Duration of hydrothermal synthesis, h	Me^{x+} ions	Initial concentration of Me^{x+} ions in liquid medium, mg/dm^3	pH value of in liquid medium	Amount of adsorbed Me^{x+}, mg/g	k_2, g/(mg min)	Amount of leached Ca^{2+}, mg/g
32	Zn^{2+}	300	~9	29.92	0.446	15.09
48				29.05	0.435	14.58
72				28.8	0.352	13.81
120				28.01	0.225	13.54
168				27.91	0.142	13.21
48			5.6	23.47	0.273	18.61
		250	5.8	21.75	0.077	14.5
	Fe^{3+}		2.00	24.97	2.3	20.96
	Ni^{2+}		5.39	20.84	0.05	21.31
	Co^{2+}		4.46	14.86	0.05	30.09
	Mn^{2+}		5.82	21.68	0.24	30.25

Thus, synthetic gyrolite is most suitable for the adsorption of iron and zinc ions. To increase the adsorption capacity, the concentration of the primary metal ions or the pH value of the liquid medium needs to be increased. As already determined, in an alkaline solution, the adsorption rate of Zn^{2+} ions on synthetic gyrolite increased by almost 4 times, i.e. from 0.077 g/(mg min) (in acidic solutions) to 0.273 g/(mg min) (in an alkaline solution).

Furthermore, it was established that gyrolite synthesized for 32 h adsorbed the highest amount of Zn^{2+} ions; however, it also releases the highest value of Ca^{2+} ions from the crystal lattice. The prolongation of the duration of hydrothermal synthesis leads to the decrement of the amount of adsorbed Zn^{2+} ions and the adsorption rate. Thus, synthetic gyrolite is a promising adsorbent for the elimination of heavy metal ions from wastewater; however, optimal adsorption conditions should be evaluated individually.

4.4 Synthetic Gyrolite Application for Simultaneous Metal Ions Adsorption Process

4.4.1 Simultaneous Adsorption of Co^{2+}, Ni^{2+}, Cu^{2+}, Zn^{2+}, Mn^{2+}, and Fe^{3+} Ions by Synthetic Gyrolite

It is known that the adsorption process depends not only on the nature of the adsorbate and the properties of the adsorptive, but also on the other ions which are present in the liquid medium. While taking this into consideration, the adsorption experiments in solutions with different chemical compositions were performed.

To evaluate the adsorption of multiple adsorbates which can be present in wastewater on synthetic gyrolite, adsorption experiments were conducted in a solution containing Co^{2+}, Ni^{2+}, Cu^{2+}, Zn^{2+}, Mn^{2+}, and Fe^{3+} ions. The concentration of each metal ion in the solution equaled to 250 mg/dm^3, while the temperature of adsorption was 25 °C.

The simultaneous adsorption process was determined to significantly differ from the adsorption of single metal ions. At the beginning of the process, synthetic gyrolite adsorbed more than 92% (231 mg Me^{x+}/g) of the metal ions initially contained in the liquid medium. By extending the adsorption time, the amount of the adsorbed ions slightly improved, and, at the final point of the process (60 min), ~96% (144 mg Me^{x+}/g) of the metal ions were adsorbed on synthetic gyrolite. We established that the adsorbed amount of individual metal ions depends on their nature, and one gram of synthetic gyrolite adsorbed 24.73 mg of Zn^{2+}, 24.51 mg of Co^{2+}, 23.71 mg of Ni^{2+}, 23.54 mg of Fe^{3+}, and 23.00 mg of Cu^{2+} ions (Fig. 4.45). In this case, the selectivity sequence of synthetic gyrolite in terms of the adsorbed amount of metal ions can be represented by the following order:

$$Cu^{2+} < Fe^{3+} < Ni^{2+} < Co^{2+} < Mn^{2+} < Zn^{2+}.$$

It is worth noting that synthetic gyrolite had the best adsorption capacity for iron ions and the lowest for cobalt and nickel during the adsorption of a single metal ions. Such a difference can be related to the difference in the radius and electronegativity of metal ions which compete with each other.

As expected, during the intercalation of metal ions into the structure of the adsorbent, a high amount of calcium ions is released from its crystal lattice into the liquid medium (Fig. 4.46). The most intensive release is observed at the beginning of the process because more than 44 mg of calcium ions was released from one gram of the adsorbent (Fig. 4.46). By prolonging the length of adsorption, the amount of the released calcium ions into the liquid medium grew with the increasing duration of the adsorption, and, after 60 min, it was as high as 48.03 mg Ca^{2+}/g (Fig. 4.46).

Due to the released calcium ions and hydroxy anions into the liquid medium, the pH value of the liquid medium increased from 2.28 (the initial mixture) to 3.53 after

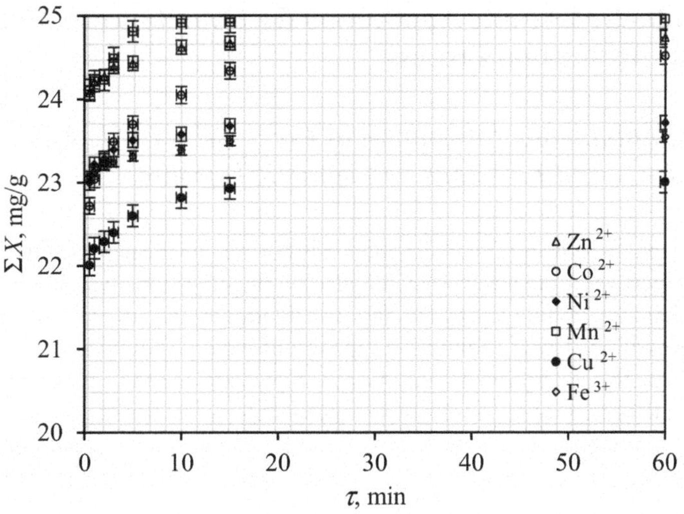

Fig. 4.45 Kinetic curves of metal ions adsorption on synthetic gyrolite

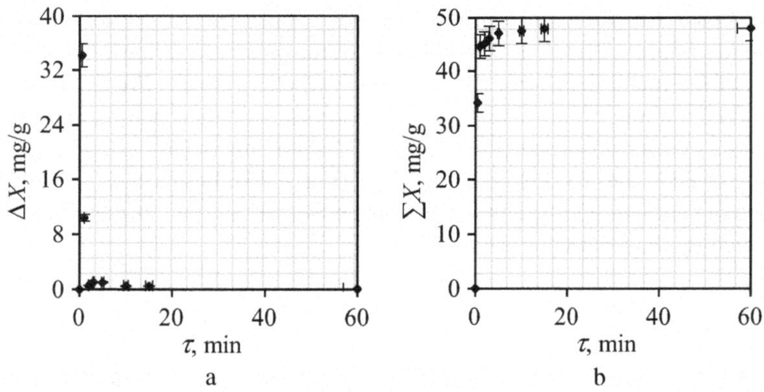

Fig. 4.46 Differential (**a**) and integral (**b**) kinetic curves of Ca^{2+} ions concentration variation

1 min of adsorption (Fig. 4.47). At the final point of the adsorption process, the pH value became more than two times higher; to be exact, it reached 5.12.

As already determined, the intercalation of heavy metal ions into the synthetic gyrolite structure involves two chemical reactions: substitution and addition. It was calculated that ~37% of Ca^{2+} ions took part in the substitution reaction: gyrolite-Ca^0 + Me^{x+} $\leftrightarrow$ gyrolite-Me^0 + Ca^{2+}, while the remaining part of calcium ions was combined by synthetic gyrolite according to the addition reaction. On top of that, the results of the leaching experiment in distilled water demonstrated that adsorption is irreversible because the concentration of metal ions in the liquid medium did not

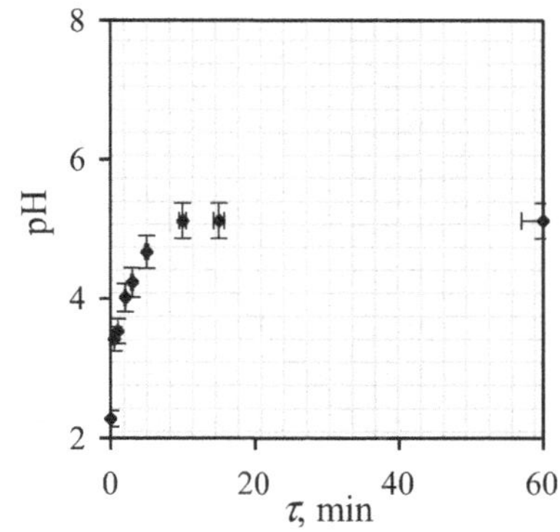

Fig. 4.47 Variation of pH value of the liquid medium

Table 4.21 Kinetic parameters of the pseudo-first-order kinetic model of simultaneous metal ions adsorption on synthetic gyrolite

Me^{x+}	R^2	$\Sigma X_{e(exp)}$, mg/g	$\Sigma X_{e(cal)}$, mg/g
Zn^{2+}	0.54	24.73	1.47
Ni^{2+}	0.62	23.71	1.50
Mn^{2+}	0.69	24.95	1.87
Co^{2+}	0.62	24.51	3.23
Cu^{2+}	0.62	23.00	2.09
Fe^{3+}	0.48	23.54	1.16

exceed the detection limit of instrumental analysis. Thus, synthetic gyrolite acts as a chemisorbent for all metal ions.

To determine the kinetic parameters of the adsorption process, pseudo-first- and pseudo-second-order kinetic models were applied. The calculations revealed that the pseudo-first-order kinetic model did not conform to the experimental data because R^2 values were low (0.48–0.69), and the $\Sigma X_{e(cal)}$ values did not align with the $\Sigma X_{e(exp)}$ values (Table 4.21).

Further, calculations showed that the pseudo-second-order model may serve for the description of the mechanism of the simultaneous adsorption of Me^{x+} ions on synthetic gyrolite as, in all cases, the value of R^2 was more than 0.99 (Fig. 4.48). This data was in good agreement with the results of experiments and confirmed that heavy metal ions are chemisorbed by synthetic gyrolite.

The linear equations obtained during the calculations by the pseudo-second-order kinetic model were employed to determine the equilibrium adsorption capacity $\Sigma X_{e(cal)}$ and the constants of the adsorption rate k_2 [g/(mg min)] (Fig. 4.48).

We found out that, by using a pseudo-second-order kinetic model, the calculated values $\Sigma X_{e(cal)}$ agree with the experimental values $\Sigma X_{e(exp)}$ (Table 4.22). The calculations showed that the value of the adsorption rate k_2 is the highest in the case of Fe^{3+} ions [1.65 g/(mg min)], whereas it is the lowest in the case of Co^{2+} ions [0.45 g/(mg min)] (Table 4.22). We should note that the adsorption rate of all the metal ions is higher during simultaneous adsorption (Table 4.22) than during single metal ions

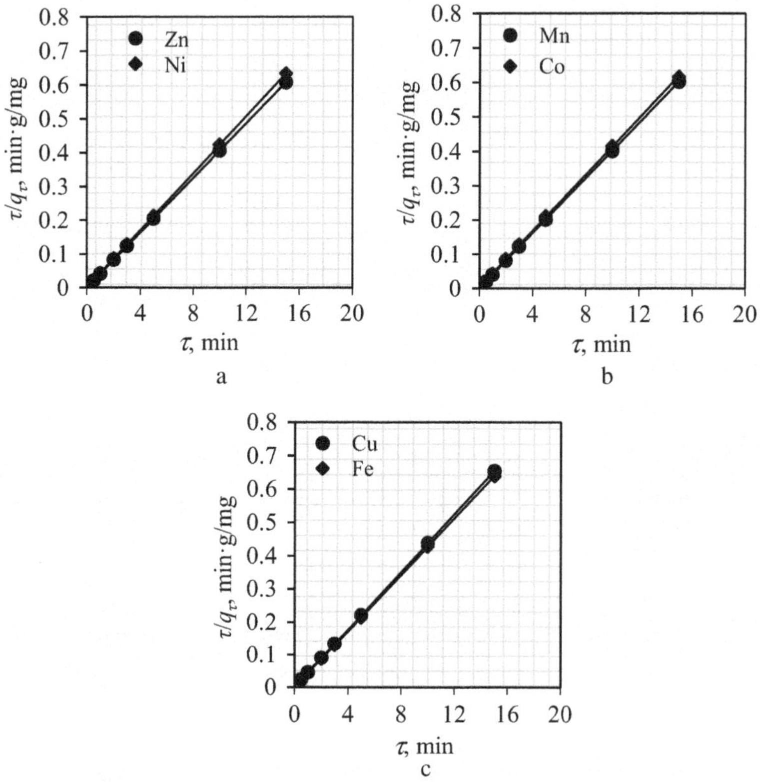

Fig. 4.48 Pseudo-second-order kinetic plots

adsorption (Table 4.20), except for the case of iron ions where a different tendency was observed.

XRD and DSC analyses were performed to evaluate the stability of the adsorbent during adsorption. The results of XRD analysis demonstrated that synthetic gyrolite maintains stability during adsorption as all the diffraction peaks typically defining this

Table 4.22 Kinetic parameters of the pseudo-second-order kinetic model of simultaneous metal ions adsorption on synthetic gyrolite

Me^{x+}	R^2	$\Sigma X_{e(exp)}$, mg/g	$\Sigma X_{e(cal)}$, mg/g	k_2, g/(mg min)
Zn^{2+}	1.00	24.73	24.69	1.26
Ni^{2+}	1.00	23.71	23.70	1.37
Mn^{2+}	1.00	24.95	25.00	1.00
Co^{2+}	0.99	24.51	24.39	0.45
Cu^{2+}	1.00	23.00	22.99	0.79
Fe^{3+}	1.00	23.54	23.47	1.65

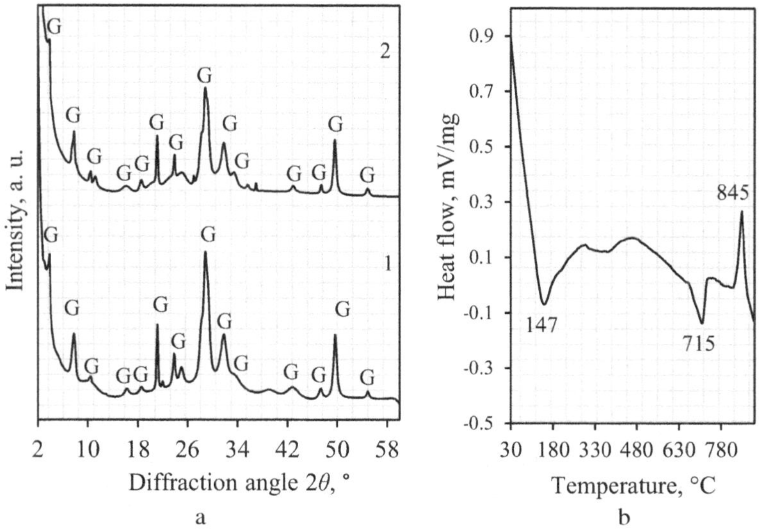

Fig. 4.49 X-ray diffraction patterns (**a**) and DSC curve (**b**) of synthetic gyrolite (curve 1) and synthetic gyrolite after adsorption (curve 2). *Indexes* G—gyrolite

compound were identified, although their intensity slightly decreased (Fig. 4.49a). It should be noted that no evidence has been obtained regarding the possible formation of other compounds.

The XRD results were reiterated after conducting DSC analysis. In the DSC curve, all the thermal effects characteristic of synthetic gyrolite were observed: the endothermic effect at 147 °C is typical of the removal of the adsorbed and structural water, whereas the exothermic effect at 853 °C characterizes the formation of wollastonite (Fig. 4.49b). We should note that, in the course of the adsorption or drying process, the carbonization of synthetic gyrolite proceeds because the intensive exothermic effect at 715 °C characteristic of calcium carbonate decomposition was identified.

It was determined that synthetic gyrolite simultaneously adsorbed almost all cations (Zn^{2+}, Cu^{2+}, Co^{2+}, Ni^{2+}, Mn^{2+}, and Fe^{3+}) in a liquid medium. Research proved that adsorption reactions are irreversible as all heavy metal ions were chemosorbed, which was confirmed by the calculations of kinetic models. The results of instrumental analysis showed that synthetic gyrolite remains stable during adsorption, although part of the calcium ions was released from its crystal structure into the liquid medium. Thus, synthetic gyrolite is suitable for simultaneous adsorption of heavy metal ions. In the next part, synthetic gyrolite was used for the adsorption process in the modeling system in which the chemical composition was chosen according to wastewater formed at JSC *Kauno vandenys*.

4.4.2 *Simultaneous Adsorption of Cd^{2+}, Cu^{2+}, and Zn^{2+} Ions by Synthetic Gyrolite*

Since synthetic gyrolite acts as a suitable adsorbent during simultaneous adsorption of heavy metal ions, synthetic gyrolite was employed as an adsorbent serving to remove Zn^{2+}, Cu^{2+}, and Cd^{2+} ions from acidic and alkaline liquid medium. The nature of metal ions, their concentration, and pH values were chosen based on the wastewater formed at JSC *Kauno vandenys*. Our adsorption experiments used two solutions with different values of pH (Table 4.23). In the case of an alkaline solution, an aqueous solution of NH_4OH [c = 10%] was employed with the objective of preventing the precipitation of metal hydroxides so that the alkalinity of the solution would be maintained.

It was determined that simultaneous adsorption of Zn^{2+}, Cu^{2+}, and Cd^{2+} ions on synthetic gyrolite depends on both the adsorption process duration and the pH value of the liquid medium. The intercalations of all heavy metal ions in an acidic solution took place slower because, at the beginning of the adsorption (30 s), 1 g of synthetic gyrolite adsorbed 10.2 mg Zn^{2+}, 2.2 mg Cu^{2+}, and 0.19 Cd^{2+} ions in the acidic solution (Fig. 4.50); meanwhile, in the alkaline solution, synthetic gyrolite adsorbed all Cu^{2+} and Cd^{2+} ions and more than 99% of Zn^{2+} ions that had been present in the liquid medium (Fig. 4.51). By extending the length of the adsorption to 1 min, all the zinc ions were adsorbed on synthetic gyrolite in the alkaline solution (Fig. 4.51a, b), i.e. the adsorption process was finished. It was determined that the adsorption equilibrium of cadmium and copper ions was reached after 3 min of the experiment in the acidic solution. In contrast, the amount of the adsorbed zinc ions slightly increased during the whole experiment which took 120 min. It was measured that, at the end of the process, ~13.6 mg (95.6%) of Zn^{2+} ions were adsorbed on synthetic gyrolite (Fig. 4.50a, b). Thus, synthetic gyrolite showed superior adsorption properties in alkaline solutions in comparison with acidic solutions. Calcium silicate hydrates are known for their alkaline properties, which probably leads to a higher stability and a higher adsorption capacity of synthetic gyrolite in alkaline solutions.

To confirm that all metal ions were adsorbed on synthetic gyrolite, after completing the adsorption, the adsorbent was rinsed with distilled water, dried at 50 °C, and dissolved in HCl acid. The concentration of metal ions in the liquid medium was measured by performing atomic absorption spectroscopy. The results of dissolution confirmed the previous results as the concentration of heavy metal ions found

Table 4.23 Parameters of primary liquid medium

Sample No.	Concentration of metal ions, mg/dm^3			Amount of NH_4OH, ml	pH value
	Zn^{2+}	Cu^{2+}	Cd^{2+}		
Solution 1	142.23	25.8	2.17	0	5.86
Solution 2	142.23	25.8	2.17	25	10.61

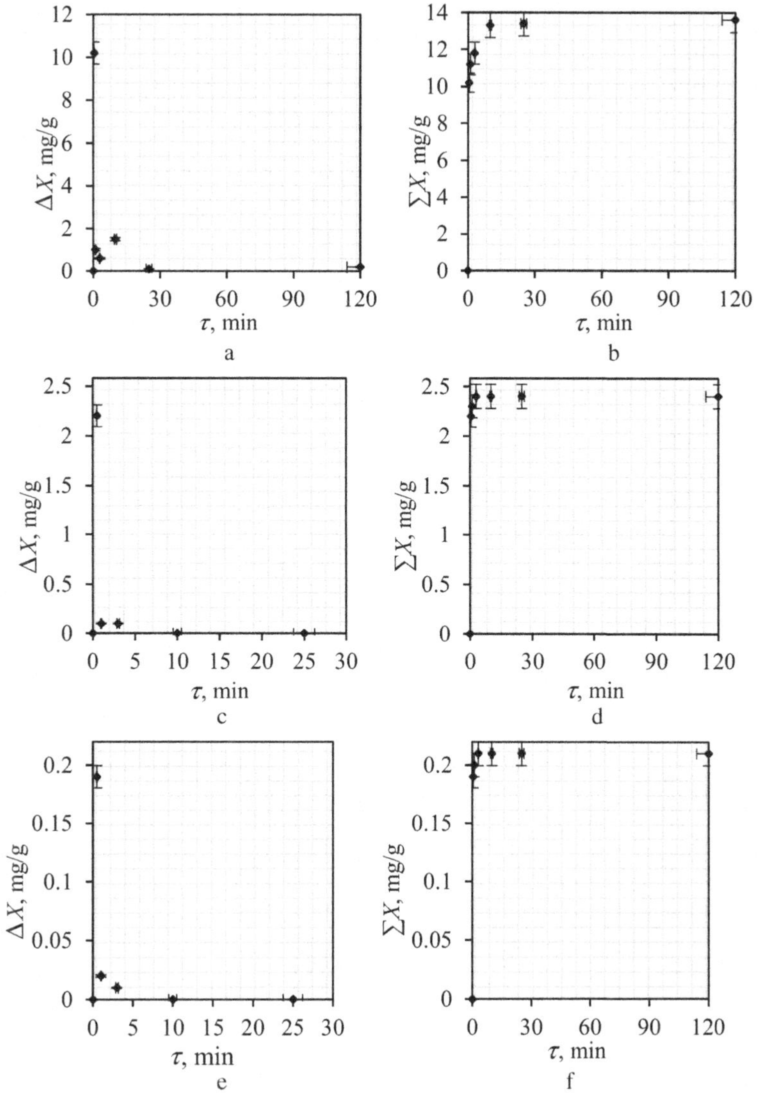

Fig. 4.50 Differential (**a**, **c**, **e**) and integral (**b**, **d**, **f**) kinetic curves of Zn^{2+} (**a**, **b**), Cu^{2+} (**c**, **d**), and Cd^{2+} (**e**, **f**) ions adsorption on synthetic gyrolite in acidic solution

in synthetic gyrolite (Table 4.24) corresponds to the results obtained in the liquid medium after adsorption (Figs. 4.50 and 4.51).

It was determined that, during the experiments, calcium ions were released from the crystal lattice of synthetic gyrolite into the liquid medium (Fig. 4.52). In both cases, the highest amount of calcium ions was released during the first minute of the experiment and increased slightly by extending the duration. However, at the end

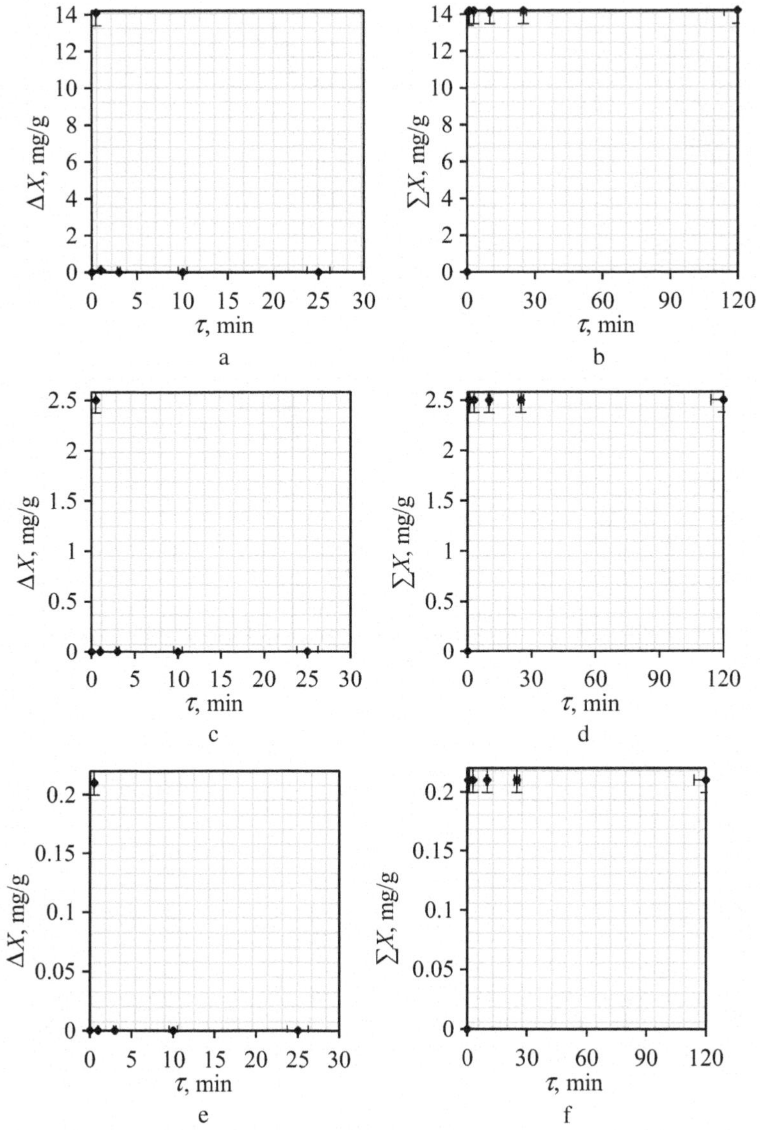

Fig. 4.51 Differential (**a**) and integral (**b**) kinetic curves of Zn^{2+} (**a**, **b**), Cu^{2+} (**c**, **d**), and Cd^{2+} (**e**, **f**) ions adsorption on synthetic gyrolite in alkaline solution

of the experiment, the amount of the released calcium ions in acidic solutions was equal to ~16.2 mg/g, while, in the alkaline solution, the value was more than 8 times lower, i.e. only 1.9 mg/g (Fig. 4.52). These results confirmed that synthetic gyrolite is more stable in an alkaline solution than in an acidic solution.

Table 4.24 Amount of adsorbed metal ions on synthetic gyrolite as determined during the dissolution experiment

Sample No.	Metal ions	Amount of adsorbed metal ions, mg/g
Solution 1	Zn^{2+}	13.8
	Cu^{2+}	2.4
	Cd^{2+}	0.21
Solution 2	Zn^{2+}	14.2
	Cu^{2+}	2.5
	Cd^{2+}	0.22

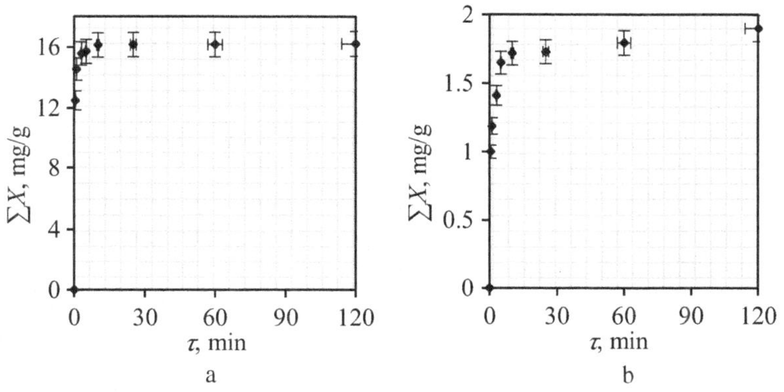

Fig. 4.52 Integral kinetic curves of the variation of Ca^{2+} ions concentration in acidic (**a**) and alkaline (**b**) solutions

The values yielded by the pH measurements conform fairly well to the changes of the released calcium ions into the liquid medium (Figs. 4.52 and 4.53). It was determined that, in an acidic solution, the pH value increased from 5.86 (before the experiment) to ~8 (after 120 min of adsorption) (Fig. 4.53a). The liquid medium pH value went up because, together with Ca^{2+} cations, OH^- anions are also released. Meanwhile, the variation of the pH value in alkaline solutions is insignificant (Fig. 4.53b).

To determine the reversibility of the process, the adsorbent after the adsorption was dried and afterward poured into distilled water for the leaching experiment. We established that, after 120 min of the experiment, the concentration of metal ions in the liquid medium never went above the value of the detection limit; thus, the irreversibility of adsorption was demonstrated.

We calculated that, in an acidic solution, heavy metal ions were adsorbed on synthetic gyrolite according to the substitution reaction (gyrolite-Ca^0 + Me^{2+} $\leftrightarrow$ gyrolite-Me^0 + Ca^{2+}) (Fig. 4.54a). Completely different results were obtained in the case of adsorption in an alkaline solution because only ~20% of heavy metal

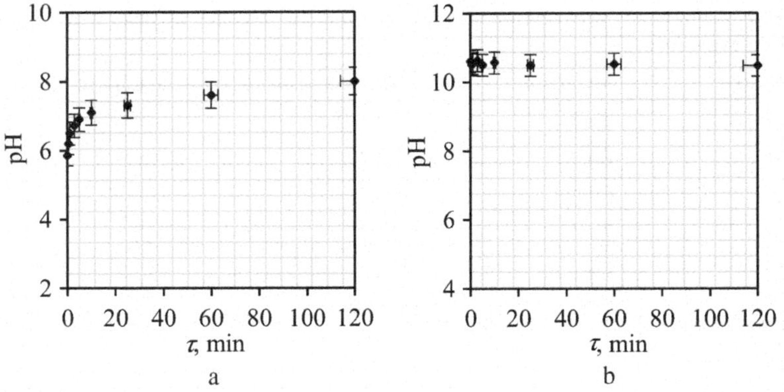

Fig. 4.53 Variation of pH value of the liquid medium in Solution 1 (**a**) and Solution 2 (**b**)

ions participated in the substitution reaction. The remaining share of metal ions was adsorbed according to the addition reaction (Fig. 4.54b).

Since the adsorption equilibrium was reached within 0.5–3 min (Figs. 4.50 and 4.51), the pseudo-first- and pseudo-second-order kinetic models cannot be employed for the description of the adsorption process.

To determine the stability of synthetic gyrolite during adsorption, the product of sorption was explored by employing various methods of instrumental analysis. The results of XRD analysis showed that, in both cases, synthetic gyrolite was stable, and no other phases were formed (Fig. 4.55). In addition, we should highlight that the intensity of the diffraction peaks characteristic of synthetic gyrolite slightly diminished in the acidic solution (due to the released calcium ions) (Fig. 4.55, curve 2), while, in the alkaline solution, it remained almost the same (Fig. 4.55, curve 3).

The results of SEM and STA analysis confirmed that the crystal structure of synthetic gyrolite remains stable in the course of the adsorption process under all

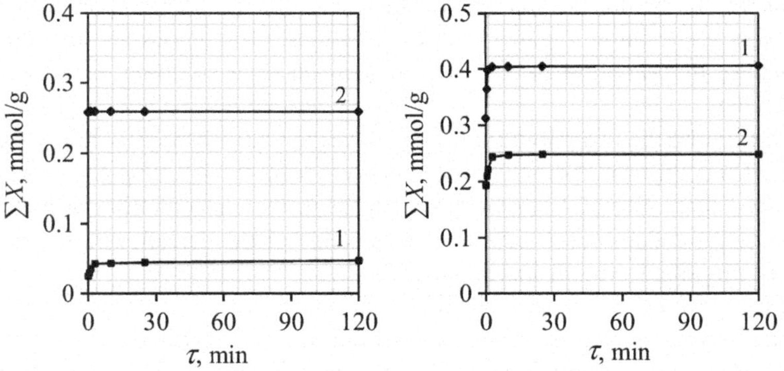

Fig. 4.54 Integral kinetic curves of released Ca^{2+} ions (curve 1) and adsorbed metal ions (curve 2) during adsorption experiments in acidic (**a**) and alkaline (**b**) solutions

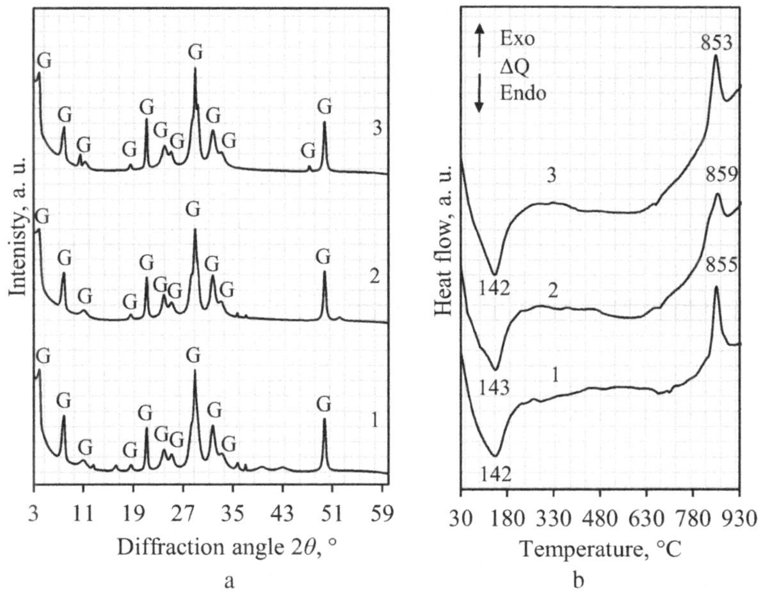

Fig. 4.55 X-ray diffraction patterns (**a**) and DSC curves (**b**) of synthetic gyrolite (curve 1) and synthetic gyrolite after adsorption in acidic (curve 2) and alkaline solutions (curve 3). *Indexes* G—gyrolite

experimental conditions. We thus established that all samples are characterized by two thermal effects typical to gyrolite (Fig. 4.55b). Due to the released calcium ions from the structure of synthetic gyrolite in the acidic solution, the exothermic effect at ~850 °C becomes broader (Fig. 4.55b, curve 2). In the SEM micrograph of synthetic gyrolite after adsorption in the alkaline solution, particles typical of synthetic gyrolite were observed (Fig. 4.56a). Meanwhile, the results of EDS analysis confirmed that heavy metal ions intercalate into the synthetic gyrolite structure because Zn and Cu were identified in the spectra (Fig. 4.56b). It needs to be emphasized that, due to the sensitivity of EDS analysis, the peak characteristic for cadmium ions cannot be distinguished.

When generalizing on the obtained results, we may state that the adsorption capacity of heavy metal ions on synthetic gyrolite is dependent on the duration of adsorption and the pH value of the liquid medium. We thus determined that adsorption in an alkaline solution proceeds intensively because all metal ions were adsorbed within 1 min of the experiment. Meanwhile, in an acidic solution, the adsorbed amount of zinc ions was lower and equal to 95.6%. Also, it was determined that the mechanism of adsorption depends on the pH value of the liquid medium because, in an acidic solution, heavy metal ions were adsorbed on synthetic gyrolite according to the substitution reaction. In contrast, in an alkaline solution, only ~20% of heavy metal ions participated in the substitution reaction. The remaining share of metal ions was adsorbed according to the addition reaction.

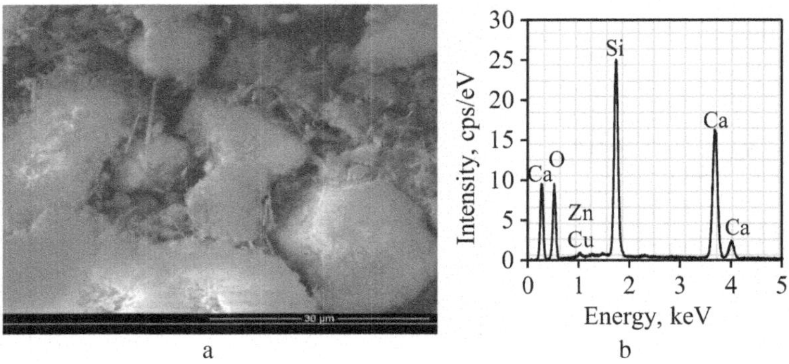

Fig. 4.56 Scanning electron micrograph (**a**) and curve of EDS analysis (**b**) of synthetic gyrolite after adsorption in alkaline solution

Thus, transition metal ions in wastewater can be adsorbed simultaneously on synthetic gyrolite in both acidic and alkaline solutions. However, synthetic gyrolite is a solid material; thus, it can adsorb metal ions only from a liquid medium, but it is impossible to use synthetic gyrolite to purify soils directly. Therefore, the extraction/adsorption of metal ions from EDTA solution by synthetic gyrolite was performed in the following step of the research.

4.4.3 Influence of EDTA on Adsorption of Cd^{2+}, Cu^{2+}, and Zn^{2+} Ions by Synthetic Gyrolite

It is known that the process of removing heavy metal ions from an aqueous medium is a more straightforward process than the extraction of heavy metal ions from soils (Fig. 4.57) or sludge because the traditional methods (chemical precipitation, foam floatation, electrochemical treatment, filtration, ion exchange, membrane technologies, and adsorption) are inefficient [87–91]. One of the most suitable ways to remove metal ions from sludge or soils is the extraction from solid to liquid. In a further step, the obtained liquid medium can be treated by the 'conventional' methods. Chelating complexing agents or strong inorganic acids can be used for the extraction of metal ions [87, 88, 90]. Using inorganic acids is more challenging because they are environmentally harmful, and the obtained products are not biodegradable. Meanwhile, extraction based on chelating agents is characterized by the high extraction capacity as well as the possibility for regeneration and reuse. Also, most chelating compounds are biodegradable, which leads to lower pollution levels of the environment.

According to the literature, all complexing agents that form chelating compounds with metal ions can be used to purify sludge or soils [90, 92–96]. Complexing agents can be used directly for the extraction of metal ions or for improving the

Fig. 4.57 Soil polluted by copper mining factory (Photo ID: 154017367, Photographer: Dvulikaia, *Source* Dreamstime.com)

efficiency of phytoextraction. In order to use complexing agents in the purification process, they are required to react with metal ions and form compounds which are not only stable but also soluble in water. The main complexing compounds are carboxylic acids, especially amino polycarboxylic acids such as ethylenediaminetetraacetic acid (EDTA), ethylenediamine-N,N′-disuccinic acid (EDDS), diethylenetriaminepentaacetic acid (DTPA), nitrilotriacetic acid (NTA), ethylene glycol bis(2-aminoethyl)tetraacetic acid (EGTA), etc.

The efficiency of chelating compounds depends on their ability to form stable compounds with metal ions. The stability constants of some metal chelates are presented in Table 4.25 [97, 98]. It is evident that EDTA and DTPA form the most stable compounds. Although DTPA is characterized by a higher efficiency in terms of the removal of metal ions, its applicability is limited due to its negative effect on the environment [99]. Since EDTA forms stable compounds and is biodegradable [97, 98, 100, 101], it is one of the most extensively investigated chelating compounds for the purification of soils and sludge.

According to the literature [98, 102], EDTA with metal ions forms a chelate with an octahedral ring (Figs. 4.58 and 4.59).

The efficiency of extraction by chelating agents depends on the properties of the chelating agent (specifically, the possibility to form stable compounds, stability at various pH levels and temperatures, etc.) and the composition of soil or sludge

Table 4.25 Stability constants (logK) of metal chelates (2:1) [97, 98]

Chelating agent	logK									
	Ca(II)	Cd(II)	Co(II)	Cr(III)	Cu(II)	Fe(III)	Mg(II)	Mn(II)	Ni(II)	Zn(II)
EDTA	11.0	16.5	16.3	23.4	18.8	25.1	8.7	14.0	18.6	16.5
NTA	6.4	9.5	10.4	>10	13.0	15.9	5.5	7.4	11.5	10.7
DTPA	10.7	19.3	18.4	–	21.5	28.6	9.3	15.6	19.4	18.7
DHEG	–	6.3	6.1	–	8.6	–	1.2	3.9	7.7	6.5
EDDP	1.0	5.6	7.3	–	15.1	13.1	1.6	3.4	9.3	7.6

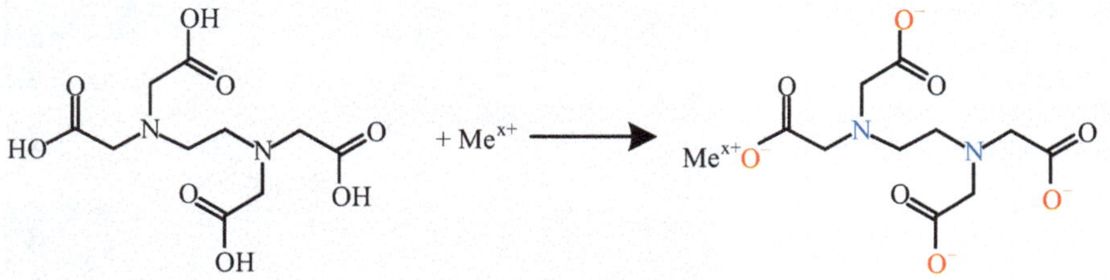

Fig. 4.58 Scheme of EDTA-Me complex formation

Fig. 4.59 EDTA-Me complex

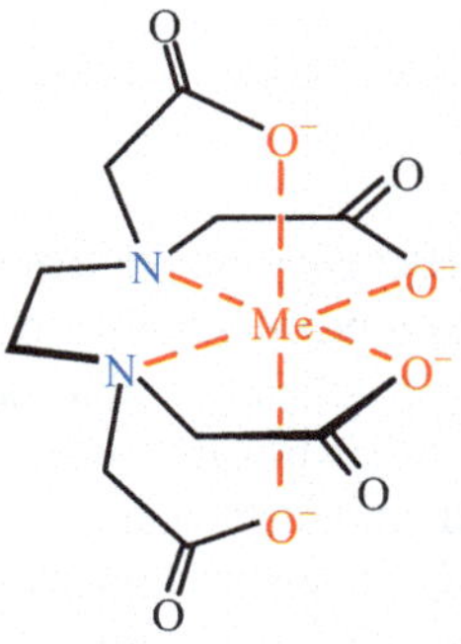

[90, 92–96]. As an example, the sequence of the extraction efficiency by EDTA for different metal ions from the typical urban sewage sludge is Pb > Cd > Zn > Ni ≈ Cu > Cr, while, from the common industrial sewage sludge, it is Cd > Ni > Pb > Cr > Zn > Cu [103, 104]. Also, some chelating agents can dissolve organic compounds containing metal ions, and, as a result, the efficiency of extraction increases. A wide variety of parameters can be changed to improve the efficiency of the extraction of metal ions from soils or sludge: the duration and temperature of extraction, the pH value of the liquid medium, the intensity of mixing, the ratio of solid/water, etc. However, the optimal conditions of extraction are always chosen according to the experimental results [90, 92–96].

The biodegradability of chelates is highly important in the pursuit of avoiding the pollution of water and soil by organic compounds. However, the excessively long or too short duration of decomposition may cause pollution of the environment. Biodegradation of chelates during the purification process leads to the decrement of the extraction efficiency and leaching of heavy metals into surface water and groundwater. Meanwhile, excessively long biodegradation may cause the pollution of water and soil by chelating compounds. Although most chelating compounds are not toxic to mammals (at low concentrations), nevertheless, drinking water should not contain xenobiotic compounds. According to the literature, the biodegradation of EDTA is slow because only 13–46% of EDTA was found to decompose in the topsoil after 15 weeks [98, 100, 101]. Meanwhile, in the deeper layers of soil, this value significantly decreases.

Since the biodegradation process of EDTA is slow, this compound can be reused for the purification of sludge, which leads to the decrement of the purification cost. However, the regeneration of chelating compounds is a challenging task [105, 106]. There are several ways to separate metal ions from chelating compounds: electrolysis, precipitation, adsorption, etc. [15, 107–109]. The regeneration of EDTA by electrolysis is an inefficient and time-consuming process. Meanwhile, the sedimentation process requires additional chemical reagents, such as various hydroxides ($Ca(OH)_2$, NaOH, etc.) and sulfides (Na_2S, H_2S, FeS, etc.). To regenerate chelating agents through adsorption, it is necessary to find adsorbents that could destroy chemical bonds between chelate agents and metal ions, and adsorb metal ions. Also, the adsorbent should not negatively affect the environment and the properties of the chelate agent.

With the objective to investigate the possibility of using synthetic gyrolite as a regenerating agent for EDTA chelates, the adsorption experiment in solutions with various pH values (3.04 or 10.86) containing 600 mg/l of EDTA, 142.23 mg/l of Zn^{2+} ions, 25.8 mg/l of Cu^{2+} ions, and 2.17 mg/l of Cd^{2+} ions was performed. The adsorption length at 25 °C temperature was equal to 0.5, 1, 3, 5, 15, 25, 60, and 120 min.

To determine the stability of synthetic gyrolite during the experiment, 1 g of synthetic gyrolite was immersed in 100 ml of the pure solution of EDTA (600 mg/l) and kept for 2 h. The obtained solution was characterized by AAS analysis (Fig. 4.60), while synthetic gyrolite was dried and characterized by XRD and STA analysis (Fig. 4.61).

We established that synthetic gyrolite is not stable in EDTA solution because it releases a significant amount of Ca^{2+} ions (Fig. 4.60). The liquid medium analysis results indicated that ~100 mg of Ca^{2+} ions was released from 1 g of synthetic gyrolite after 1 min (Fig. 4.60). By extending the duration to 60 min, the amount of the released ions more than doubled, and, at the end of the experiment (120 min), 1 g of synthetic gyrolite released 230 mg of Ca^{2+} ions. It should be highlighted that 1 g of pure synthetic gyrolite contains ~275 mg of Ca^{2+} ions; thus, synthetic gyrolite released more than 83% of calcium ions from its structure into a liquid medium.

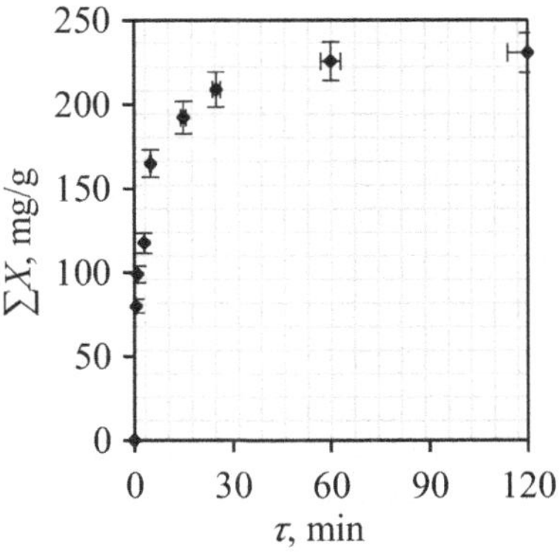

Fig. 4.60 Integral kinetic curve of the variation of Ca^{2+} ions concentration in EDTA solution

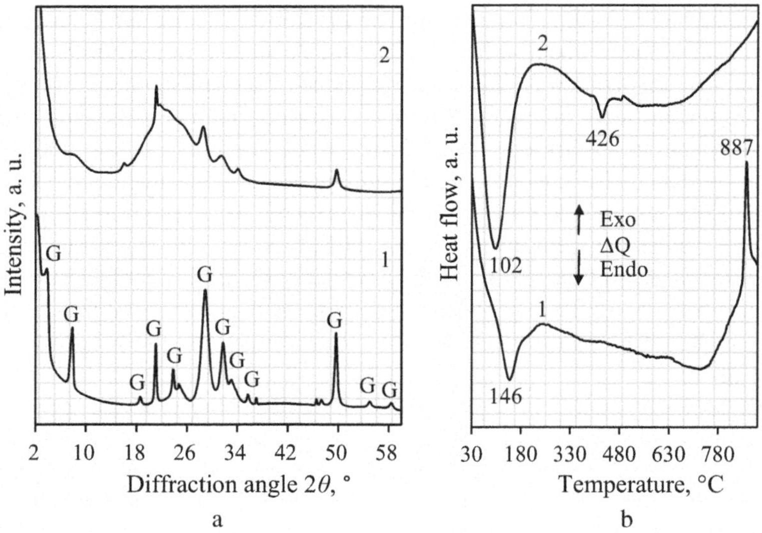

Fig. 4.61 X-ray diffraction patterns (**a**) and DSC curves (**b**) of synthetic gyrolite (curve 1) and synthetic gyrolite after immersion in EDTA solution. *Indexes* G—gyrolite

The values obtained in the course of liquid medium analysis were corroborated by XRD and STA analysis (Fig. 4.61). It was established that synthetic gyrolite fully decomposed because the diffraction peaks characteristic of synthetic gyrolite could not be identified in the XRD pattern (Fig. 4.61a). Furthermore, the exothermic effect at ~890 °C characteristic of the recrystallization process of gyrolite to wollastonite was not identified in the DSC curve of the products obtained after the experiment (Fig. 4.61b). The absence of this thermal effect shows that even amorphous or semicrystalline calcium silicate hydrates were not formed in the product. Thus, synthetic gyrolite was completely destroyed by EDTA, which forms a strong calcium chelate.

In the second step of the experiment, zinc, copper, and cadmium nitrates were added to the solution of EDTA (600 mg/l). The concentration of metal ions was equal to 142.23 mg/l of Zn^{2+} ions, 25.8 mg/l of Cu^{2+} ions, and 2.17 mg/l of Cd^{2+} ions. The liquid medium pH value was equal to 3.04. To obtain a basic solution (pH = 10.86), aqueous solution of NH_4OH (c = 10%) was employed. During the adsorption, 1 g of synthetic gyrolite was immersed in 100 ml of the solution.

We establish that synthetic gyrolite adsorbed all cadmium ions (0.217 mg/g) in both solutions (acidic and alkaline) after 30 s of adsorption. Intensive adsorption of Cu^{2+} ions was also observed because more than 91% (2.36 mg/g) of these ions was adsorbed by synthetic gyrolite after 30 s (Fig. 4.62). When the duration of the experiment was extended, the amount of the adsorbed copper ions modestly increased. At the end of the process, it was equal to 2.40–2.43 mg Cu^{2+}/g, i.e. synthetic gyrolite adsorbed 93% of copper ions which were initially present in the liquid medium.

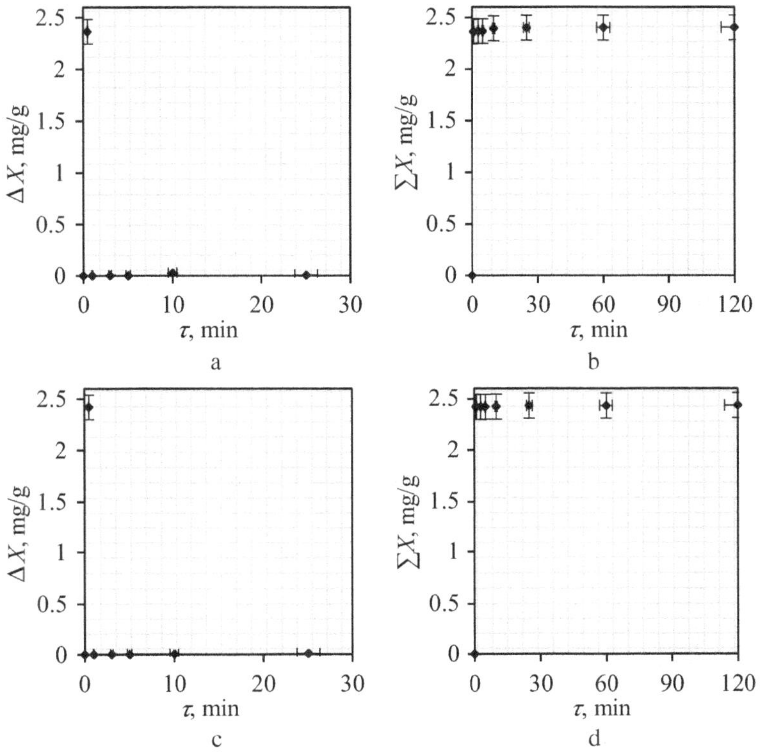

Fig. 4.62 Differential (**a**, **c**) and integral (**b**, **d**) kinetic curves of the adsorption of Cu^{2+} ions on synthetic gyrolite in EDTA solution with the pH value of the initial solution equal to 3.04 (**a**, **b**) and 10.68 (**c**, **d**)

Different results were observed in the changes in the adsorbed amount of Zn^{2+} ions (Fig. 4.63). During the initial period of the process (30 s), the adsorption of Zn^{2+} ions on synthetic gyrolite proceeded intensively because 1 g of synthetic gyrolite adsorbed ~6 mg (42%) of Zn^{2+} ions. However, the amount of the adsorbed ions increased slightly with the extension of the process duration. It was measured that, at the final stage of the process (120 min), synthetic gyrolite adsorbed only ~50% (~7.2 mg Zn^{2+}/g) of zinc ions that had been present in the acidic and alkaline solutions. These results allow to state that EDTA negatively affected the adsorption capacity of synthetic gyrolite for Zn^{2+} ions because, in a pure solution, the ratio of the adsorbed Zn^{2+} ions actually doubled (Figs. 4.50, 4.51 and 4.63).

We found that synthetic gyrolite was more stable in the EDTA solution with metal ions than in the pure EDTA solution (Figs. 4.60 and 4.64). It was measured that, at the end of the process in acidic and alkaline solutions, synthetic gyrolite released 12.66 Ca^{2+}/g and 5.76 mg Ca^{2+}/g, respectively (Fig. 4.64). Meanwhile, in the pure EDTA solution, synthetic gyrolite released more than 83% of calcium ions

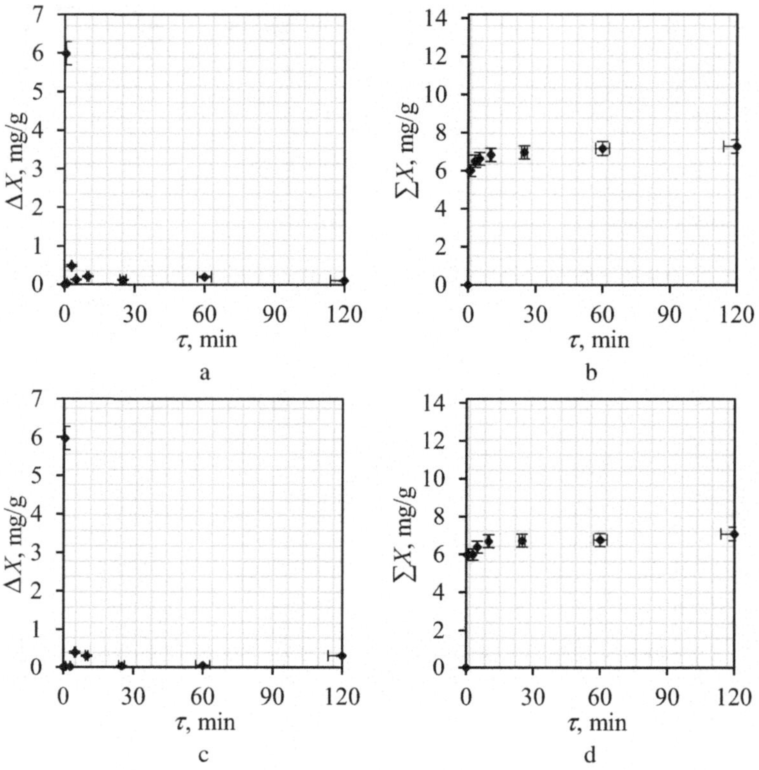

Fig. 4.63 Differential (**a**, **c**) and integral (**b**, **d**) kinetic curves of the adsorption of Zn^{2+} ions on synthetic gyrolite in EDTA solution with the pH value of the initial solution equal to 3.04 (**a**, **b**) and 10.68 (**c**, **d**)

and fully decomposed (Figs. 4.60 and 4.61). The adsorbed metal ions (Cd^{2+}, Cu^{2+}, Zn^{2+}) probably stabilized the structure of the adsorbent.

We also determined the irreversibility of the adsorption process as, after the leaching experiment (120 min, 25 °C), the concentration of metal ions in the liquid medium did not go beyond the value of the detection limit. It is worth emphasizing that calculations of the kinetic parameters according to the pseudo-first- or pseudo-second-order equations are inappropriate because the adsorption of Cu^{2+} and Cd^{2+} ions on synthetic gyrolite was finished after 30–60 s of adsorption.

In order to confirm that synthetic gyrolite remained stable after adsorption, it was dried and investigated by various methods of instrumental analysis. X-ray diffraction analysis results confirmed that synthetic gyrolite was stable in both investigated solutions (acidic and alkaline) because all the diffraction peaks characteristic of gyrolite were identified (Fig. 4.61a, curve 1 and Fig. 4.65a). Also, in the DSC curves of the adsorbent, all the thermal effects characteristics of synthetic gyrolite were observed (Fig. 4.65b). It should be specified that, due to the intercalation of metal ions into the structure of synthetic gyrolite, the intensity of the exothermic effect decreased,

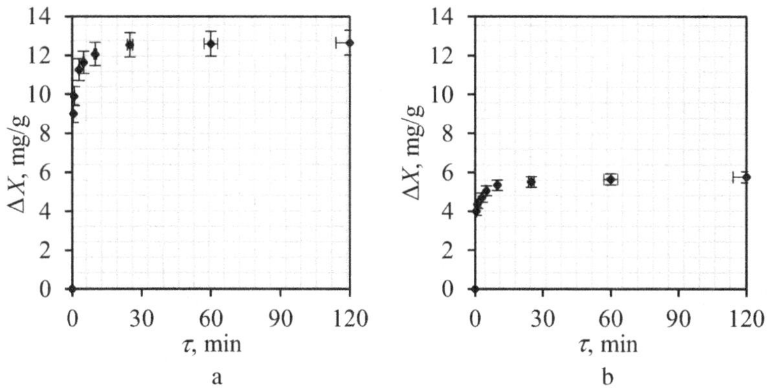

Fig. 4.64 Integral kinetic curve of the variation in the concentration of Ca^{2+} ions in EDTA solution with the pH value of the initial solution equal to 3.04 (**a**) and 10.68 (**b**)

and it shifted to a lower temperature range (Fig. 4.61b, curve 1, and Fig. 4.65b). Also, the carbonization of synthetic gyrolite occurred during the adsorption and/or drying process because the thermal effect at ~700 °C temperature characteristic of the decomposition of calcium carbonate was identified in the DSC curves (Fig. 4.65b).

XRD and STA analysis results were in line with the FT-IR data (Fig. 4.66). In the spectra of the adsorbent with intercalated metal ions, all the absorption bands characteristic of synthetic gyrolite were identified (Fig. 4.66). Furthermore, it was established that, due to the intercalation of heavy metal ions and the release of

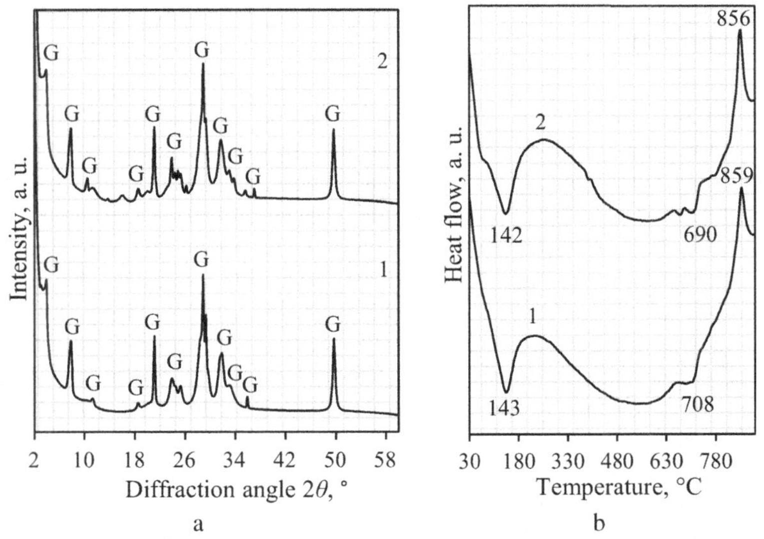

Fig. 4.65 X-ray diffraction patterns (**a**) and DSC curves (**b**) of synthetic gyrolite after adsorption in acidic (curve 1) and alkaline solutions (curve 2). *Indexes* G—gyrolite

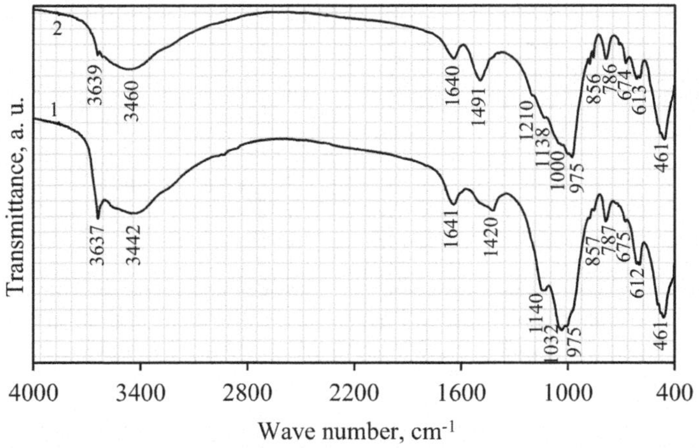

Fig. 4.66 FT-IR spectra of synthetic gyrolite (curve 1) and synthetic gyrolite after adsorption in alkaline solution (curve 2)

calcium and OH^- ions, the intensity of the absorption band at 3639 cm^{-1} decreased. This absorption band is characteristic of the vibrations of $\nu(OH^-)$ situated in the lattice. Meanwhile, due to carbonization, the intensity of the absorption bands (856 and 1491 cm^{-1}) characteristic of the CO_3^{2-} group vibrations increased. Also, the changes in the regions of δ(O–Si–O) and ν_s(O–Si–O) were observed. Probably, the adsorbed metal ions formed new functional groups and chemical bonds.

4.5 Summary

By summarizing the results of this Chapter, it can be stated that synthetic gyrolite is suitable for the adsorption of transition metal ions from wastewater and solutions of complexing compounds (EDTA). The adsorption rate and capacity depend on the parameters of adsorption (duration, pH, concentration) and the nature of the adsorbate. The best adsorption results can be achieved in solutions with alkaline pH values and when synthetic gyrolite of a lower crystallinity is used. The simplified scheme of the adsorption process on synthetic gyrolite is shown in Fig. 4.67.

According to the literature, when the adsorption capacity limit of the adsorbents has already been reached, usually, they get stored in landfills as a result of getting discarded, which eventually causes secondary pollution due to the risk of leaching and the subsequent release of non-biodegradable heavy metals [15, 16, 108, 110]. For these reasons, the research on novel adsorbents that can serve as additives in ordinary Portland cement or in the process of manufacturing alternative cementitious materials is in progress. Thus, in the further segment of the research, the influence of pure synthetic gyrolite and synthetic gyrolite with intercalated cadmium ions on the hydration of Portland cement shall be investigated in detail.

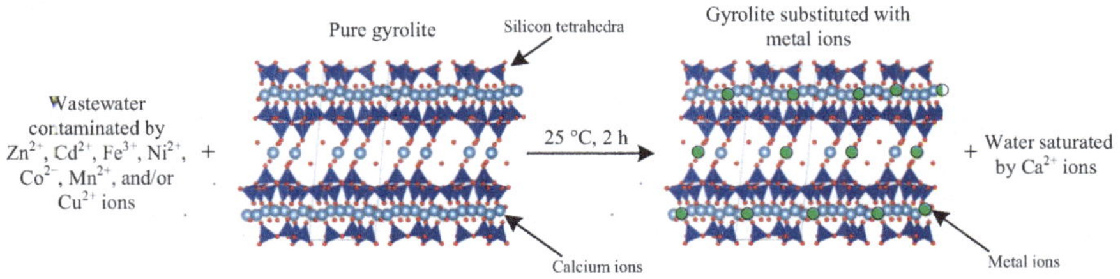

Fig. 4.67 Scheme of the intercalation of metal ions into the structure of synthetic gyrolite

References

1. The United Nations world water development report (2017) Wastewater: the untapped resource. United Nations Educational, Scientific and Cultural Organization, Paris
2. Droste RL, Gehr RL (2018) Theory and practice of water and wastewater treatment. Wiley
3. Liu DH, Lipták BG (2020) Wastewater treatment. CRC Press
4. Russell DL (2019) Practical wastewater treatment. Wiley
5. Shrestha R, Ban S, Devkota S, Sharma S, Joshi R, Tiwari AP, Kim HY, Joshi MK (2021) Technological trends in heavy metals removal from industrial wastewater: a review. J Environ Chem Eng 9:105688
6. Zhao M, Xu Y, Zhang C, Rong H, Zeng G (2016) New trends in removing heavy metals from wastewater. Appl Microbiol Biotechnol 100:6509–6518
7. Bradl HB (2005) Chapter 1 sources and origins of heavy metals. In: Heavy metals in the environment: origin, interaction and remediation. Interface Sci Technol 1–27
8. Singh R, Gautam N, Mishra A, Gupta R (2011) Heavy metals and living systems: an overview. Indian J Pharmacol 43:246–253
9. De Vries W, Römkens P, Van Leeuwen T, Bronswijk J (2002) Heavy metals. CABI Books
10. Zhou Q, Yang N, Li Y, Ren B, Ding X, Bian H, Yao X (2020) Total concentrations and sources of heavy metal pollution in global river and lake water bodies from 1972 to 2017. Glob Ecol Conserv 22:e00925
11. Gautam RK, Sharma SK, Mahiya S, Chattopadhyaya MC (2014) Chapter 1: contamination of heavy metals in aquatic media: transport, toxicity and technologies for remediation. In: Heavy metals in water: presence, removal and safety. The Royal Society of Chemistry
12. Mohammed AS, Kapri A, Goel R (2011) Heavy metal pollution: source, impact, and remedies. In: Biomanagement of metal-contaminated soils. Springer
13. Iakovleva E, Sillanpää M (2013) The use of low-cost adsorbents for wastewater purification in mining industries. Environ Sci Pollut Res 20:7878–7899
14. Crini G, Lichtfouse E, Wilson LD, Morin-Crini N (2019) Conventional and non-conventional adsorbents for wastewater treatment. Environ Chem Lett 17:195–213
15. Lata S, Singh PK, Samadder SR (2015) Regeneration of adsorbents and recovery of heavy metals: a review. Int J Environ Sci Technol 12:1461–1478
16. Baskar AV, Bolan N, Hoang SA, Sooriyakumar P, Kumar M, Singh L, Jasemizad T, Padhye LP, Singh G, Vinu A, Sarkar B, Kirkham MB, Rinklebe J, Wang S, Wang H, Balasubramanian R, Siddique KHM (2022) Recovery, regeneration and sustainable management of spent adsorbents from wastewater treatment streams: a review. Sci Total Environ 822:153555
17. Yahya N, Aziz F, Jamaludin NA, Mutalib MA, Ismail AF, W. Salleh WN, Jaafar J, Yusof N, Ludin NA (2018) A review of integrated photocatalyst adsorbents for wastewater treatment. J Environ Chem Eng 6:7411–7425
18. Manikandan S, Karmegam N, Subbaiya R, Karthiga Devi G, Arulvel R, Ravindran B, Kumar Awasthi M (2021) Emerging nano-structured innovative materials as adsorbents in wastewater treatment. Bioresour Technol 320:124394

19. Naiya TK, Bhattacharya AK, Das SK (2009) Adsorption of Cd(II) and Pb(II) from aqueous solutions on activated alumina. J Colloid Interface Sci 333:14–26
20. Mor S, Ravindra K, Bishnoi NR (2007) Adsorption of chromium from aqueous solution by activated alumina and activated charcoal. Bioresour Technol 98:954–957
21. Bhattacharya AK, Mandal SN, Das SK (2006) Adsorption of Zn(II) from aqueous solution by using different adsorbents. Chem Eng J 123:43–51
22. Tripathy SS, Bersillon J, Gopal K (2006) Removal of fluoride from drinking water by adsorption onto alum-impregnated activated alumina. Sep Purif Technol 50:310–317
23. Jha B, Singh DN (2016) Basics of zeolites. In: Jha B, Singh DN (eds) Fly ash zeolites: innovations, applications, and directions. Springer, pp 5–31
24. Gottardi G, Galli E (2012) Natural zeolites. Springer
25. Barakan S, Aghazadeh V (2021) The advantages of clay mineral modification methods for enhancing adsorption efficiency in wastewater treatment: a review. Environ Sci Pollut Res 28:2572–2599
26. Otunola BO, Ololade OO (2020) A review on the application of clay minerals as heavy metal adsorbents for remediation purposes. Environ Technol Innov 18:100692
27. Vedenyapina MD, Kurmysheva AY, Kulaishin SA, Kryazhev YG (2021) Adsorption of heavy metals on activated carbons (a review). Solid Fuel Chem 55:83–104
28. Sultana M, Rownok MH, Sabrin M, Rahaman MH, Alam SMN (2022) A review on experimental chemically modified activated carbon to enhance dye and heavy metals adsorption. Clean Technol 6:100382
29. Sosa JA, Laines JR, García DS, Hernández R, Zappi M, Monteros E, Alejandra E (2022) Activated carbon: a review of residual precursors, synthesis processes, characterization techniques, and applications in the improvement of biogas. Environ Eng Res 28:220100
30. Maneerung T, Liew J, Dai Y, Kawi S, Chong C, Wang C (2016) Activated carbon derived from carbon residue from biomass gasification and its application for dye adsorption: kinetics, isotherms and thermodynamic studies. Bioresour Technol 200:350–359
31. Garbev K (2004) Struktur, Eigenschaften und quantitative Rietveldanalyse von hydrothermal kristallisierten Calciumsilikathydraten (CSH-Phasen). Karlsruhe Research Center, Karlsruhe
32. Dambrauskas T, Baltakys K, Eisinas A, Siauciunas R (2017) A study on the thermal stability of kilchoanite synthesized under hydrothermal conditions. J Therm Anal Calorim 127:229–238
33. Hong S, Glasser FP (2004) Phase relations in the $CaO–SiO_2–H_2O$ system to 200 °C at saturated steam pressure. Cem Concr Res 34:1529–1534
34. Yanagisawa K, Hu X, Onda A, Kajiyoshi K (2006) Hydration of β-dicalcium silicate at high temperatures under hydrothermal conditions. Cem Concr Res 36:810–816
35. Meller N, Hall C, Phipps JS (2005) A new phase diagram for the $CaO–Al_2O_3–SiO_2–H_2O$ hydroceramic system at 200 °C. Mater Res Bull 40:715–723
36. Meller N, Kyritsis K, Hall C (2009) The mineralogy of the $CaO–Al_2O_3–SiO_2–H_2O$ (CASH) hydroceramic system from 200 to 350 °C. Cem Concr Res 39:45–53
37. Komarneni S, Roy DM, Fyfe CA, Kennedy GJ (1987) Naturally occurring 1.4 nm tobermorite and synthetic jennite: characterization by 27Al and 29Si MASNMR spectroscopy and cation exchange properties. Cem Concr Res 17:891–895
38. Komarneni S, Breval E, Roy DM, Roy R (1988) Reactions of some calcium silicates with metal cations. Cem Concr Res 18:204–220
39. Komarneni S, Roy DM (1983) Tobermorites: a new family of cation exchangers. Science 221:647–648
40. Komarneri S, Roy DM, Roy R (1982) Al-substituted tobermorite: shows cation exchange. Cem Concr Res 12:773–780
41. Tsuji M, Komarneni S, Malla P (1991) Substituted tobermorites: ^{27}Al and ^{29}Si MASNMR, cation exchange, and water sorption studies. J Am Ceram Soc 74:274–279
42. Tsuji M, Komarneni S (1989) Alkali metal ion exchange selectivity of Al-substituted tobermorite. J Mater Res 4:698–703
43. Komarneni S, Tsuji M (1989) Selective cation exchange in substituted tobermorites. J Am Ceram Soc 72:1668–1674

44. Shrivastava OP, Komarneni S, Breval E (1991) Mg^{2+} uptake by synthetic tobermorite and xonotlite. Cem Concr Res 21:83–90
45. Shrivastava OP, Komarneni S (1994) Cesium selectivity of (Al+Na)-substituted tobermorite. Cem Concr Res 24:573–579
46. Zhao Z, Wei J, Li F, Qu X, Shi L, Zhang H, Yu Q (2017) Synthesis, characterization and hexavalent chromium adsorption characteristics of aluminum- and sucrose-incorporated tobermorite. Materials 10:597
47. Coleman NJ (2005) Synthesis, structure and ion exchange properties of 11Å tobermorites from newsprint recycling residue. Mater Res Bull 40:2000–2013
48. Shrivastava OP, Shrivastava R (2001) Sr^{2+} uptake and leachability study on cured aluminum-substituted tobermorite–OPC admixtures. Cem Concr Res 31:1251–1255
49. Komarneni S, Roy R, Roy DM (1986) Pseudomorphism in xonotlite and tobermorite with Co^{2+} and Ni^{2+} exchange for Ca^{2+} at 25 °C. Cem Concr Res 16:47–58
50. Al-Wakeel E, El-Korashy SA, El-Hemaly S, Rizk MA (2001) Divalent ion uptake of heavy metal cations by (aluminum + alkali metals)—substituted synthetic 1.1 nm-tobermorites. J Mater Sci 36:2405–2415
51. Wang Z, Huang Z, Zheng S, Wu D, Cai W (2021) Effective removal of Pb^{2+} and Cd^{2+} from wastewater by mesoporous tobermorite synthesized from alumina-extracted fly ash. Desalin Water Treat 235:209–220
52. Zou J, Guo C, Zhou X, Sun Y, Yang Z (2018) Sorption capacity and mechanism of Cr^{3+} on tobermorite derived from fly ash acid residue and carbide slag. Colloids Surf Physicochem Eng Aspects 538:825–833
53. Zhou W, Tang W, Xin J (2020) Functionalization of xonotlite composite with amidoxime groups for the sorption of Cu(II) ion. Water Air Soil Pollut 231:406
54. Tang W, Zeng R, Gu B, Zhang Z, He H (2019) Preparation of sodium-substituted xonotlite from eggshell and its adsorption behavior for cadmium (II). Desalin Water Treat 174:279–290
55. Liu L, Liu S, Peng H, Yang Z, Zhao L, Tang A (2020) Surface charge of mesoporous calcium silicate and its adsorption characteristics for heavy metal ions. Solid State Sci 99:106072
56. Niuniavaite D, Baltakys K, Dambrauskas T, Eisinas A (2019) Cu^{2+}, Co^{2+} and Cr^{3+} adsorption by synthetic dibasic calcium silicate hydrates and their thermal stability in a 25–1000 °C temperature range. J Therm Anal Calorim 138:2241–2249
57. Niuniavaite D, Baltakys K, Dambrauskas T (2018) The adsorption kinetic parameters of Co^{2+} ions by α-C_2SH. Buildings 8:10
58. Niuniavaite D, Baltakys K, Dambrauskas T (2020) The adsorption kinetic parameters of Cu^{2+} and Cr^{3+} ions by α-C_2SH. Desalin Water Treat 185:175–184
59. Baltušnikas A, Šiaučiūnas R, Lukošiūtė I, Baltakys K, Eisinas A, Kriūkienė R (2015) Crystal structure refinement of synthetic pure gyrolite. Medziagotyra 21:111–116
60. Merlino S (1988) Gyrolite: its crystal structure and crystal chemistry. Mineral Mag 52:377–387
61. Meyer JW, Jaunarajs KL (1961) Synthesis and crystal chemistry of gyrolite and reyerite. Am Min 46:913–933
62. Garbev K, Black L, Beuchle G, Stemmermann P (2002) Inorganic polymers in cement based materials. Wasser-und Geotechnologie 1:19–30
63. Richardson IG (2008) The calcium silicate hydrates. Cem Concr Res 38:137–158
64. El-Korashy SA (2002) Synthetic crystalline calcium silicate hydrate (I): cation exchange and caesium selectivity. Monatsh Chem 133:333–343
65. El-Korashy SA, Kishar EA (1998) Hydrothermal synthesis XE "Hydrothermal synthesis" of calcium silicate hydrates in the presence of 3d-ferromagnetic cations. Indian J Chem 37A:1098–1101
66. El-Korashy SA (2004) Cation exchange of alkali metal hydroxides with some hydrothermally synthesized calcium silicate compounds. J Ion Exch 15:2–9
67. Miyake M, Iwaya M, Suzuki T, Kakehi H, Mitsuda T (1990) Aluminum-substituted gyrolite as cation exchanger. J Am Ceram Soc 73:3524–3527

68. Winters MA, Richter JD, Sagar SL, Lee AL, Lander RJ (2003) Plasmid DNA purification by selective calcium silicate adsorption of closely related impurities. Biotechnol Progress 19:440–447
69. Agnew UM, Slesinger TL (2002) Zinc toxicity. StatPearls Publishing
70. UNEP, WHO, ILO, International Programme on Chemical Safety, Inter-Organization Programme for the Sound Management of Chemicals (2001) Zinc. Published under the joint sponsorship of the United Nations Environment Programme, the International Labour Organisation, and the World Health Organization, and produced within the framework of the Inter-Organization Programme for the Sound Management of Chemicals. Available via: https://apps.who.int/iris/handle/10665/42337
71. Vengris T, Binkienė R, Sveikauskaitė A (2001) Nickel, copper and zinc removal from waste water by a modified clay sorbent. Appl Clay Sci 18:183–190
72. Lu C, Chiu H (2006) Adsorption of zinc(II) from water with purified carbon nanotubes. Chem Eng Sci 61:1138–1145
73. Mishra PC, Patel RK (2009) Removal of lead and zinc ions from water by low cost adsorbents. J Hazard Mater 168:319–325
74. Ali I, Burakov AE, Melezhik AV, Babkin AV, Burakova IV, Neskomornaya MEA, Galunin EV, Tkachev AG, Kuznetsov DV (2019) Removal of Copper(II) and Zinc(II) ions in water on a newly synthesized polyhydroquinone/graphene nanocomposite material: kinetics, thermodynamics and mechanism. Chem Select 4:12708–12718
75. Ayawei N, Ebelegi AN, Wankasi D (2017) Modelling and interpretation of adsorption isotherms. J Chem 2017:3039817
76. Qiu H, Lv L, Pan B, Zhang Q, Zhang W, Zhang Q (2009) Critical review in adsorption kinetic models. J Zhejiang Univ Sci A 10:716–724
77. Sen Gupta S, Bhattacharyya KG (2011) Kinetics of adsorption of metal ions on inorganic materials: a review. Adv Colloid Interface Sci 162:39–58
78. Stumm A, Garbev K, Beuchle G, Black L, Stemmermann P, Nüesch R (2005) Incorporation of zinc into calcium silicate hydrates, Part I: formation of C-S-H(I) with C/S=2/3 and its isochemical counterpart gyrolite. Cem Concr Res 35:1665–1675
79. Thommes M, Kaneko K, Neimark AV, Olivier JP, Rodriguez-Reinoso F, Rouquerol J, Sing KS (2015) Physisorption of gases, with special reference to the evaluation of surface area and pore size distribution (IUPAC Technical Report). Pure Appl Chem 87:1051–1069
80. Rouquerol J, Rouquerol F, Llewellyn P, Maurin G, Sing KS (2013) Adsorption by powders and porous solids: principles, methodology and applications. Academic Press
81. Webb PA, Orr C (1997) Analytical methods in fine particle technology. Micromeritics Instrument Corporation
82. Lowell S, Shields JE, Thomas MA, Thommes M (2006) Characterization of porous solids and powders: surface area, pore size and density. Springer
83. Naderi M (2015) Surface area: Brunauer–Emmett–Teller (BET). In: Progress in filtration and separation. Academic Press, pp 585–608
84. Abbaspour N, Hurrell R, Kelishadi R (2014) Review on iron and its importance for human health. J Res Med Sci 19:164–174
85. Britton RS, Leicester KL, Bacon BR (2002) Iron toxicity and chelation therapy. Int J Hematol 76:219–228
86. Yuen H, Becker W (2021) Iron toxicity. In: StatPearls [Internet]. StatPearls Publishing
87. Gryschko R, Kuhnle R, Terytze K, Jö B, Stahr K (2005) Soil extraction of readily soluble heavy metals and as with 1 M NH_4NO_3-solution—evaluation of DIN 19730 (6 pp). J Soils Sed 5:101–106
88. Lo IMC, Yang XY (1999) EDTA extraction of heavy metals from different soil fractions and synthetic soils. Water Air Soil Pollut 109:219–236
89. Pouresmaieli M, Ataei M, Forouzandeh P, Azizollahi P, Mahmoudifard M (2022) Recent progress on sustainable phytoremediation of heavy metals from soil. J Environ Chem Eng 10:108482

90. Lin H, Wang Z, Liu C, Dong Y (2022) Technologies for removing heavy metal from contaminated soils on farmland: a review. Chemosphere 305:135457
91. Yang X, Liu L, Tan W, Liu C, Dang Z, Qiu G (2020) Remediation of heavy metal contaminated soils by organic acid extraction and electrochemical adsorption. Environ Pollut 264:114745
92. Haladu SA, Mu'azu ND, Jarrah N, Zubair M, Ali SA (2021) Investigation of biodegradable polyaspartate as an effective chelant for washing of lead from soil: response surface methodology approach. Int J Environ Anal Chem 101:2679–2696
93. Liang F, Guo Z, Men S, Xiao X, Peng C, Wu L, Christie P (2019) Extraction of Cd and Pb from contaminated-paddy soil with EDTA, DTPA, citric acid and $FeCl_3$ and effects on soil fertility. J Cent South Univ 26:2987–2997
94. Ke X, Zhang FJ, Zhou Y, Zhang HJ, Guo GL, Tian Y (2020) Removal of Cd, Pb, Zn, Cu in smelter soil by citric acid leaching. Chemosphere 255:126690
95. Wang G, Pan X, Zhang S, Zhong Q, Zhou W, Zhang X, Wu J, Vijver MG, Peijnenburg WJGM (2020) Remediation of heavy metal contaminated soil by biodegradable chelator–induced washing: efficiencies and mechanisms. Environ Res 186:109554
96. Yoo J, Lee C, Yang J, Baek K (2013) Extraction characteristics of heavy metals from marine sediments. Chem Eng J 228:688–699
97. Dojindo EU GmbH. Metal chelates. Available via: https://www.dojindo.eu.com/images/Product%20Photo/Chelate_Table_of_Stability_Constants.pdf. Accessed 25 July 2023
98. Martell AE, Hancock RD (2013) Metal complexes in aqueous solutions. Springer
99. Regulation EC (2008) Regulation (EC) No 1272/2008 of the European Parliament and of the Council of 16 December 2008 on classification, labelling and packaging of substances and mixtures, amending and repealing Directives 67/548/EEC and 1999/45/EC, and amending Regulation (EC) No 1907/2006. Available via: https://eur-lex.europa.eu/legal-content/EN/TXT/?uri=celex%3A32008R1272. Accessed 20 July 2023
100. Nörtemann B (1999) Biodegradation of EDTA. Appl Microbiol Biotechnol 51:751–759
101. Chen L, Liu T, Ma C (2010) Metal complexation and biodegradation of EDTA and S, S-EDDS: a density functional theory study. J Phys Chem A 114:443–454
102. Rekab K, Lepeytre C, Goettmann F, Dunand M, Guillard C, Herrmann J (2015) Degradation of a cobalt(II)–EDTA complex by photocatalysis and H2O2/UV-C. Application to nuclear wastes containing 60Co. J Radioanal Nucl 303:131–137
103. Chen T, Macauley E, Hong A (1995) Selection and test of effective chelators for removal of heavy metals from contaminated soils. Can J Civ Eng 22:1185–1197
104. Peters RW (1999) Chelant extraction of heavy metals from contaminated soils. J Hazard Mater 66:151–210
105. Efligenir A, Mohamed MA, Fievet P, Fatin-Rouge N (2013) Reusing chelating agents to wash metal-contaminated soils. J Environ Chem Eng 1:448–452
106. Menoud P, Cavin L, Renken A (2000) New regeneration process of heavy-metal-loaded chelating resins. Chem Eng Technol 23:441–447
107. Gylienė O, Nivinskienė O, Razmutė I (2006) Copper(II)–EDTA sorption onto chitosan and its regeneration applying electrolysis. J Hazard Mater 137:1430–1437
108. Vakili M, Deng S, Cagnetta G, Wang W, Meng P, Liu D, Yu G (2019) Regeneration of chitosan-based adsorbents used in heavy metal adsorption: a review. Sep Purif Technol 224:373–387
109. Zhang K, Dai Z, Zhang W, Gao Q, Dai Y, Xia F, Zhang X (2021) EDTA-based adsorbents for the removal of metal ions in wastewater. Coord Chem Rev 434:213809
110. Gkika DA, Mitropoulos AC, Kyzas GZ (2022) Why reuse spent adsorbents? The latest challenges and limitations. Sci Total Environ 822:153612

Chapter 5
Utilization of Synthetic Gyrolite in Ordinary Portland Cement

The fifth chapter of the monograph starts on the review of literature on the environmental challenges posed by ordinary Portland cement production and possible solutions. One of the possible solutions is the application of synthetic pure gyrolite and synthetic gyrolite with intercalated cadmium ions as additives for ordinary Portland cement. The use of adsorbents in cement prevents secondary pollution caused by metal ions during landfilling. Furthermore, it was proven that both pure and contaminated synthetic gyrolite positively affects the early stage hydration of ordinary Portland cement as well as its compressive strength. Also, the hydration kinetics by using different models was investigated. The chapter is concluded with technological recommendations for the production and application of synthetic gyrolite.

5.1 Environmental Impact of Ordinary Portland Cement

Section 3.2.1. *Solid Wastes and Their Utilization* discusses the necessity of reusing wastes in an environmentally friendly way. Unfortunately, the reuse and/or recycling of spent adsorbents contaminated with heavy metal ions is a challenging process (Fig. 5.1) [2–4]. Quite commonly, adsorbents and pollutants (heavy metal ions) form strong chemical interaction (chemisorption), and thus, the regeneration of spent adsorbents is impossible or, in the best case scenario, very complex. For these reasons, a lot of spent adsorbents are landfilled [3–5]. Some spent adsorbents contaminated with heavy metal ions (e.g. carbon, organic adsorbents, etc.) can be incinerated by reducing the volume and increasing the concentration of metal ions [6–8]. Later, incineration residues (ash) are stored as hazard wastes, or, in some cases, they can be reused. However, this method is harmful to the environment because the incineration process requires high temperatures, a huge amount of greenhouse gas is emitted,

K. Baltakys and T. Dambrauskas, *Synthetic Gyrolite*,
https://doi.org/10.1007/978-3-031-49259-4_5

Fig. 5.1 Adsorbent and spent adsorbent

and it also requires underground mines or landfills to store the ash. It should be kept in mind that the storage of ashes contaminated with heavy metal ions can cause secondary pollution due to the interaction between ashes and the atmospheric and/or ground water.

Another way is the stabilization/solidification technique (Fig. 5.2) [9–13]. In this case, metal ions are strongly bonded with an adsorbent or converted into a less soluble form (stabilization), which leads to a reduction in their mobility. In the next step, the waste (spent adsorbent) contaminated with metal ions is encapsulated by creating a durable matrix (solidification). During the stabilization/solidification process, adsorbents or wastes contaminated with metal ions can be mixed with various binders (ordinary Portland cement [OPC], lime, belite-rich cement, etc.) to form an inert and stable matrix which can be landfilled safely. The purpose of this method is to keep metal ions within the permitted limits (according to the regulations of each particular contaminant), to transfer metal ions from liquid waste or sludge to a solid body, to improve the physical properties as well as the handling characteristics of the spent adsorbent, and to reduce the mobility of metal ions in contact with other materials and leaching fluids [10, 11, 14]. This method is not only more environmentally friendly but also cheaper compared to other alternative treatment techniques.

However, the spent adsorbents are still considered as wastes. Thus, the generally preferred way is the application of spent adsorbents in the production of products

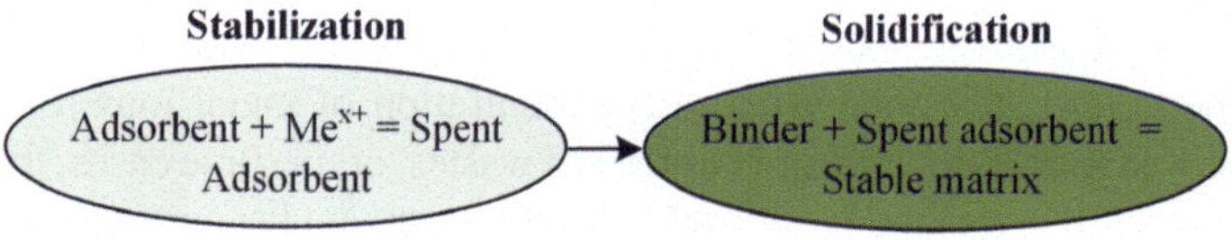

Fig. 5.2 Stabilization/solidification process

bearing some commercial value [15–17]. The obtained products must be non-toxic and insoluble. For these reasons, the solidification of spent adsorbents in the construction industry is in great demand. The influence of spent adsorbents on the properties of binding materials and hardened products, the mechanism of hydration, as well as the effectiveness of solidification have been widely studied and discussed in many reports [10–12, 18–20]. The possibility of using spent adsorbents in the production of construction materials is mainly determined by CSH phases—the main components of hardened OPC (the most commonly [essentially, worldwide] used synthetic compound). The CSH phases feature a high adsorption capacity and a low permeability, which allows reaching high immobilization of metal ions. Also, the structural features of CSH phases prevent the migration of liquids and metal ions, which will increase the immobilization potential. The influence of each spent adsorbent on the properties of hardened products depends on the nature of the adsorbent, the involved metal ions, and the binding material. Thus, the possibility to use any specific spent adsorbent in the production should be evaluated.

The usage of spent adsorbents as additives in OPC cement or concrete allows not only reusing spent adsorbents but can also contribute to reducing the negative impact of OPC production. Every year, about 1 m^3 of concrete is used per person, which is almost 10 times more than all other man-made materials combined [18, 21–24]. The main binder for concrete is Portland cement whose production exceeds 4 billion tons per year (Fig. 5.3). According to various sources, it is estimated that the production of OPC will increase up to 4.8 billion tons per year by 2030 [25–28].

The huge production significantly affects the environment. To produce 1 ton of OPC, a considerable quantity of energy is used (4.7–6.3 GJ/t, i.e. about 3% of the world's total energy production) [25–28]. Also, the cement production process is considered as one of the main sources of CO_2 emissions from industrial activities. The huge emissions of CO_2 in the production of OPC are caused due to [22–24, 26, 27]:

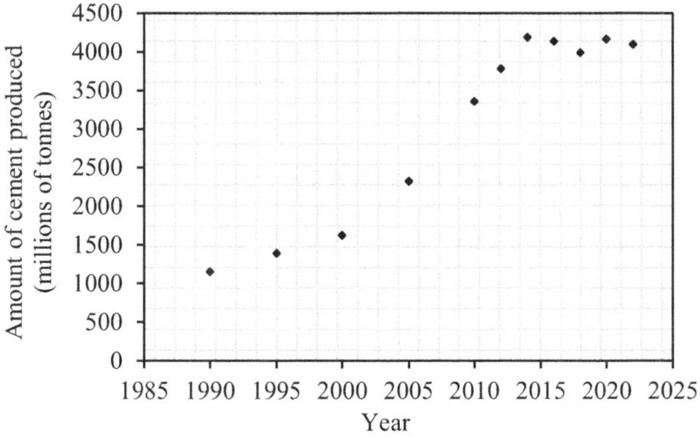

Fig. 5.3 Global production of Portland cement

- The high synthesis temperature of OPC (1450 °C) which requires combustion of large quantities of fossil fuels (mainly coal). The CO_2 produced during the synthesis accounts for around 40% of the total CO_2 emissions.
- The decomposition of $CaCO_3$ into CaO and CO_2, since limestone is one of the main components of the charge. The CO_2 produced during the decomposition process accounts for around 60% of the total CO_2 emissions.

With the steady increase in human needs, the demand for cement-based materials will be maintained or even further increased [25]. However, the increase in the production volumes and the use of conventional technologies will lead to an unacceptable increase in CO_2 emissions contributing to climate change [25–28]. Extensive research is being carried out worldwide to deploy carbon capture, storage, and reuse technologies across a wide range of industrial sectors, including the cement industry [21–24]. In the cement industry, it is possible to significantly reduce CO_2 emissions associated with the production and use of cement/concrete in the following ways (but not limited to those):

- 3D printing.
- Big data and smart materials.
- Alternative cementitious materials.
- Computationally designed composites.
- Supplementary cementitious materials (SCMs).
- More efficient use of clinker in mortars and concretes.
- Other.

In modern cements, about 20 wt% of OPC clinker is replaced by supplementary cementitious materials, mainly limestone, granulated blast furnace slag, fly ash, and natural pozzolans (Fig. 5.4) [29–32]. Furnace slag and fly ash are limited worldwide; thus, it is possible to replace about 15–25 wt% of clinker, and this number is unlikely to increase. It is estimated that the 40 wt% replacement rate of cement clinker by SCM could reduce CO_2 emissions by up to 400 million tons per year. According to some scientists, it is possible to reduce CO_2 emissions by up to 70–90%, by replacing the maximal possible amount of clinker in cements and their products [33–36]. However, it is limited by hardness regarding the necessity to ensure the operational properties of concrete structures.

A similar effect (~40%) can be achieved by optimizing the composition of mortars and concretes, i.e. the amount of fine and coarse fractions of the aggregates should be optimized by achieving the densest packing [22, 23, 26, 27, 37, 38]. In this case, the dispersants and additives take the crucial role. Also, cement consumption can be reduced by using the required strength class and preprepared (dry) concrete and mortar mixtures [22, 23, 38, 39]. In the long term, the consumption of cement, aggregates, and additives can be reduced by extending the lifetime of concrete structures and components; however, it requires the development of more efficient and innovative design and production methods.

Scientists and members of industry are interested in the development of alternative (environmentally friendly) cements and are looking for economically viable and

Fig. 5.4 Limestone (**a**) (Photo ID: 20766668, *Source* Dreamstime.com), blast furnace slag (**b**) (Photo ID: 196012136, *Source* Dreamstime.com), pumice quarry (**c**) (Photo ID: 19471988, *Source* Dreamstime.com)

attractive ways of their production [21–27]. The usage and production of environmentally friendly cements will allow the cement industry to reach climate neutrality because CO_2 emissions can potentially be reduced by up to 90–100%. Even more, some environmentally friendly cements exhibit negative CO_2 emission; i.e. they bind more CO_2 during the process of curing than it is released during the production [40, 41]. However, currently, only a small amount of environmentally friendly cements is being produced in the industrial way because they cannot offer yet a cost-effective alternative to OPC clinker [21–27]. Alternative proposals to cement can be found in the scientific literature; however, only some of them manage to attract great interest. One of them is belite–ye'elimite–ferrite clinker which has significantly lower CO_2 emissions than OPC clinker [42, 43]. The main production difference between OPC and belite–ye'elimite–ferrite clinkers is the proportions of raw materials. In both cases, limestone is a main raw material which ensures the required content of calcium oxide in the load. However, the production of belite–ye'elimite–ferrite clinkers requires 20–30 wt% less limestone than OPC, and thus, CO_2 emissions get reduced by up to 35% [42–45]. Still, the production of belite–ye'elimite–ferrite clinkers is limited by the comparatively more expensive raw materials than OPC.

The significant reduction of CO_2 can be achieved by using cements hardened in CO_2 atmosphere [22–24, 40, 41]. These cements can be produced by using calcium silicate hydrates with a CaO/SiO_2 molar ratio from 1 till 1.5 (kilchoanite, rankinite,

Fig. 5.5 Carbonized wollastonite samples

wollastonite) [46–48]. The calcination temperatures of these cements can be as low as 1200 °C, which will make it easier to use alternative fuels with relatively low heating values; meanwhile, the clinkers can also be easier to grind than ordinary Portland clinkers. These cements are hardened in a CO_2-rich atmosphere at ambient gas pressures and at moderate temperatures (20–60 °C) (Fig. 5.5). This technology allows reducing CO_2 emissions by up to 70%. It is worth noting that, by using natural calcium silicates or magnesium silicates rather than limestone in the production of this type of cements, CO_2 emissions may even be negative [40, 41]. Unfortunately, the use of these technologies is limited because they can only be applied to products produced in the factory and are restricted to a localized production site near the cement factory [21–27].

Based on the discussed data and the available scientific literature, it can be assumed that OPC-based cements, mortars, and concretes will continue to dominate in the near future and will be replaced by other materials only after a century or even later [21, 23, 25, 26, 49, 50]. For this reason, the research on the reduction of the OPC content in concrete is in great demand. This alternative (SCM application) does not require any technological investment regarding the production of OPC and allows utilizing wastes (ash, slag, sorbents, etc.) in concretes.

5.2 Supplementary Cementitious Materials and Their Influence on Ordinary Portland Cement Hydration

Different authors propose various classifications of supplementary cementitious materials because SCM can be classified according to their activity, origin, mechanism of interaction with cement, and some other parameters [20, 29, 30, 32, 33, 50, 51]. One of the most universal classifications was proposed by Massazza and modified by Snellings et al. [50] (Fig. 5.6). This classification is based on the origin of SCM. Thus, all SCMs can be divided into natural and artificial ones. Natural SCMs may be volcanic or sedimentary in origin, while artificial SCMs are classified according to the production process. It is worth highlighting that most artificial SCMs are wastes from other processes.

According to the activity, supplementary cementitious materials can be divided into four groups [20, 29, 30, 32, 33, 50, 51]:

- Hydraulically active additives, for example, ashes and slags with binding properties.
- Additives with a latent hydraulic effect, for example, glass or amorphous materials containing a high amount of SiO_2, CaO, and Al_2O_3.
- Artificial and natural pozzolans, for example, fly ash, silica fume, diatomaceous earth, calcined clays, etc. These compounds react with calcium hydroxide at ambient temperature and humidity to form calcium silicate hydrates and/or aluminates.
- Inert additives or fillers, for example, ground limestone and dolomite. These compounds change the rheological properties of cement or concrete and/or act as crystallization/reaction centers during hydration.

The most commonly used (together > 800 Mt/year) and most important SCM in cement is blast furnace slag, fly ash, silica fume, natural pozzolans, limestone, and calcined clay [20, 29, 30, 32, 33, 50, 51]. The usage of SCMs has been increasing every year; however, the supply of high-quality SCM is limited. Thus, the future reduction of CO_2 emissions requires expanded use of other SCMs. The future SCM materials can be spent adsorbent, spent catalyst, wastes, or other worthless materials [20, 29, 33, 50, 52, 53]. However, the usage of these materials is limited due to their amount or still unexamined effects on OPC hydration.

The hydration of ordinary Portland cement is the interaction between OPC minerals and water, which involves various exothermic reactions [54–56]. This interaction or binding is simply called the hydration of OPC, while the newly formed solid compounds are called hydration products or hydrates. Cement consists of different minerals, see Table 5.1, and different-sized particles. For these reasons, the hydration of OPC is a complex process, in which cement particles react with water at different rates, and the hydration of the minerals overlaps. Also, the formed hydrates react with unreacted minerals and other hydrates; thus, the final composition of the hydrates may vary.

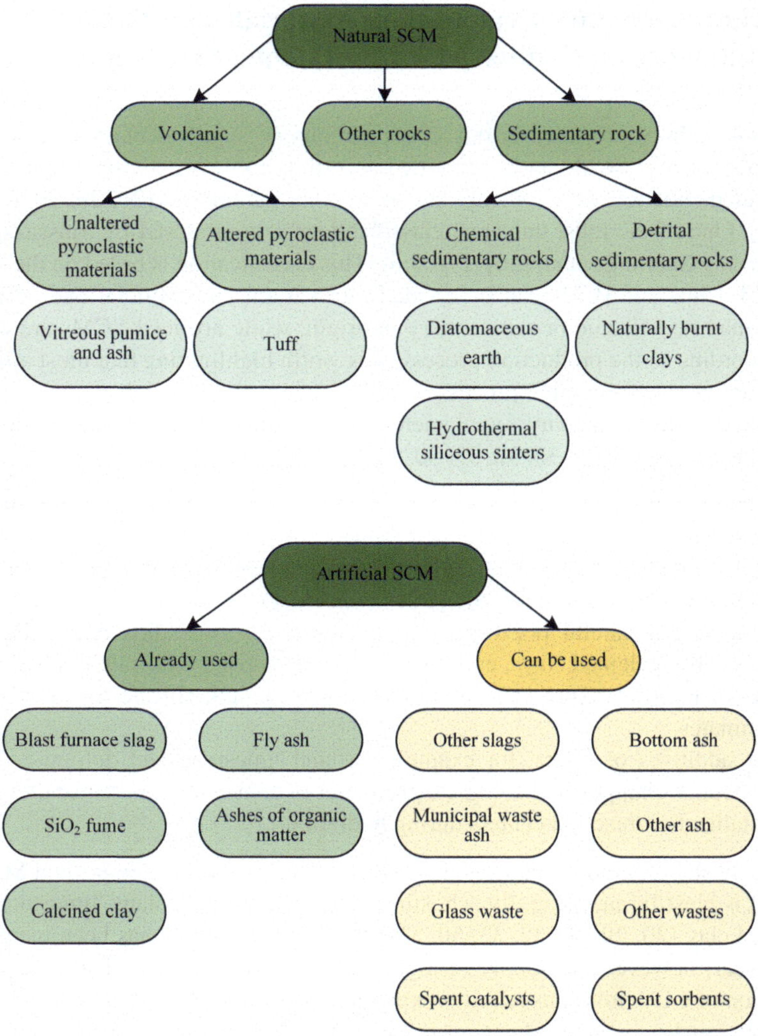

Fig. 5.6 Classification of supplementary cementitious materials

OPC consists of tricalcium silicate (alite, C_3S), dicalcium silicate (belite, C_2S), tricalcium aluminate (C_3A), and tetracalcium aluminoferrite (brownmillerite, C_4AF) (Table 5.1) [54–56]. Since OPC mainly (>65 wt%) consists of calcium silicates, the main hydration products are semi-amorphous calcium silicate hydrates whose CaO/SiO_2 molar ratio changes continuously throughout the hydration. Due to the hydration of other cement phases, calcium hydroxide, calcium monosulfoaluminate, calcium carboaluminate, calcium hydroaluminates, and ettringite are also formed.

The hydration of Portland cement involves various exothermic reactions; thus, the hydration process can be shown as a function of the released heat versus the

Table 5.1 Characteristics of the main cement minerals

Compound	Abbreviation	Released heat during hydration, J/g	Concentration in OPC, wt%	Molar mass, g/mol	Density, g/cm^3
Tricalcium silicate	C_3S	517	37–60	228	3.21
Dicalcium silicate	C_2S	262	15–37	172	3.28
Tricalcium aluminate	C_3A	1144	6–15	270	3.03
Tetracalcium aluminoferrite	C_4AF	725	10–18	486	3.73

duration [54–56]. The heat released during the initial hydration (72 h) depends on the composition of OPC and the fineness of particles. This value usually varies from 200 to 250 J/g. Microcalorimetric analysis is the most popular method to measure the dependence of the released heat (J/g) and the heat flow (W/g) on the duration of hydration.

The hydration of OPC includes five periods: initial, induction, acceleration, deceleration, and slow continued reaction (Fig. 5.7) [54–56]. These periods describe the initial and subsequent hydration processes.

First period—initial (Fig. 5.7) [54–58]. Cement particles are wetted, and the solution is saturated with various ions: Na^+, K^+, Ca^{2+}, OH^-, SiO_4^{4-}, SO_4^{2-}, etc. After saturation, the crystallization of portlandite and the interaction between SO_4^{2-} ions and tricalcium aluminate proceeds. The latter interaction results in the formation of a film of needle-like crystals of ettringite. From the rheological point of view, the

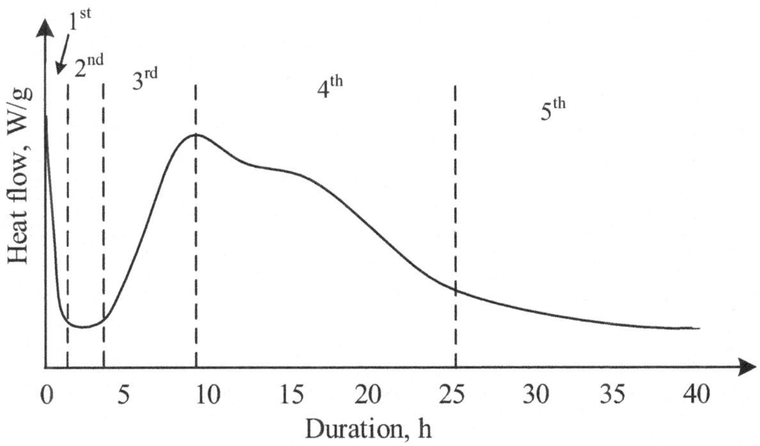

Fig. 5.7 Microcalorimetric analysis curve of OPC

formation of the ettringite film is one of the most important reactions as it determines the beginning of cement setting ($\geq$ 60 min). Also, in this period, a small amount (1–5 wt%) of calcium silicate gel is formed. After a few minutes of hydration, cement particles are covered with a layer of hydration products which prevents the penetration of water into the deeper layers of the non-hydrated cement particles. At this stage, the reaction stops, and the heat release is significantly reduced.

Second period—induction [54–58]. This period lasts for several hours (1–4 h), during which, hydration reactions slow down considerably, and a small amount of heat is released. During this period of hydration, the slow diffusion of water through the hydrate layer to the non-hydrated cement particles proceeds, which leads to an increase in the C–S–H gel layer. At the end of the period, the critical thickness of the latter layer leads to its destruction.

Third period—acceleration [54–58]. During this period, the cement paste loses its rheological properties and forms cement stone which can no longer be worked upon. The beginning of the acceleration period can be explained by the disintegration of the initial crystalline layer (film) of cement hydrates. As the latter layer grows, the surface tension of the film and the osmotic pressure of the water in the film increase, which leads to the cracking of the film and to a reaction between the deeper layers of cement particles and water. During the latter process, the intensive hydration of C_3S and the further reactions of C_3A and minerals with water proceed, which results in the formation of CSH and portlandite as well as the recrystallization of ettringite.

Fourth period—deceleration [54–58]. During this period, a cement stone matrix is formed. It consists mainly of portlandite, ettringite and calcium silicate hydrates. These hydrates grow from the surface of the cement particles toward the empty space (pores) and connect with each other. As a result, the porosity of the cement stone begins to decrease while its strength is increasing. Although the cement particles and hydrates are bounded with each other, the cement does not yet possess a significant strength. The hardening and hydration processes of the cement slow down because the diffusion of water toward the non-hydrated cement particles becomes more difficult. Thus, the hydration rate mostly depends on water diffusion.

Fifth period—slow continued reaction [54–58]. During this period, the hydration of cement depends on the diffusion of water, the water content in the pores, and the size of these pores. The hydrated cement stone consists of short-fiber calcium silicate hydrates, portlandite, calcium aluminates, and other compounds formed on non-hydrated cement particles. The hydration of non-hydrated cement particles and the recrystallization processes of the hydrates lead to a further increment of cement stone.

Thus, the microcalorimetric curve and its characteristics describe the process and mechanism of cement hydration. This curve is also used to calculate the total released heat (J/g) during hydration (Fig. 5.8). This is useful to assess the influence of each period or any additive on the overall hydration process [54–58]. For example, from the curve presented in Fig. 5.8, it is possible to determine the cumulative heat

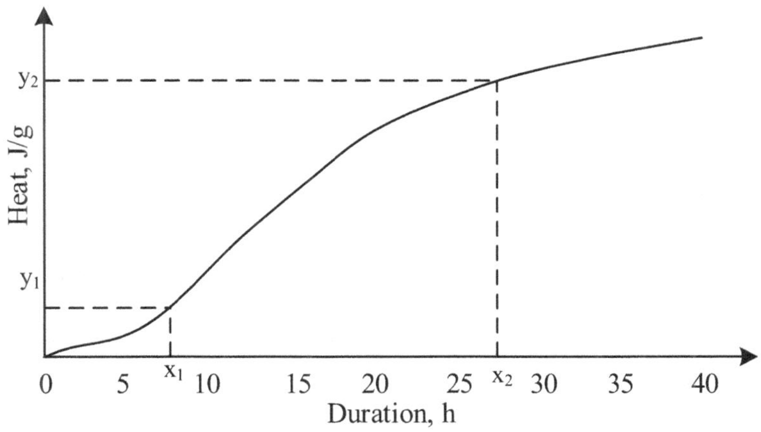

Fig. 5.8 Cumulative heat curve of OPC

released during the initial 40 h of hydration, or the heat released ($\Delta y = y_1 - y_2$) within a specific range ($x_1 - x_2$).

The above-discussed literature shows that the chemical and mineralogical composition of cement determines the heat flow and the final composition of cement stone. Therefore, the investigation of the influence of SCM on the kinetics of cement hydration and the mineral composition of a hardened sample is challenging [30, 32, 34, 35, 55, 59, 60].

The hydration of cement with SCM is influenced by pozzolanic and/or hydraulic reactions [57, 59–65]. For this reason, the kinetics of the hydration reactions, the composition, structure, and morphology of the hydrates, and the properties of the fresh and hardened cement paste differ compared to pure cement samples. The chemical and mineralogical composition as well as the quantity of used SCM is the main factors affecting hydration. When the amount of SCM and water required for hydration is properly selected, the hydration of cement is accelerated; i.e. the initial period is prolonged, while the induction period is shortened. The usage of SCM requires to determine the beginning of the binding process. In contrast, SCM-dominated (slag) cement reacts with water more slowly than OPC; thus, it is suitable for use in concrete for massive structures (dams, reservoirs, bridges, etc.).

According to the scientific literature [57, 59–65], despite the wide variety of SCMs, there are relatively few compounds formed during pozzolanic, hydraulic, or hydration reactions of SCM. This can be explained by a similar/related chemical composition of SCM and cement. Also, some SCMs participate in the hydration reactions, while others act only as crystallization centers. Thus, only a limited number of hydrates are thermodynamically stable during hydration. During hydration, the reactive silicon dioxide, which is released from SCM, reacts with portlandite (formed during cement hydration) by forming calcium silicate hydrates [59, 61, 66]. If the SCM contains active compounds with aluminum ions, the dissolved alumina intercalates into calcium silicate hydrates and/or reacts with calcium ions to form calcium aluminates [59, 61, 67–69]. The molar ratio of calcium silicate hydrates depends on the duration of hydration and the properties of SCM (solubility, composition, quantity, etc.). The latter ratio is higher in the cement with SMC because the pore solution is more saturated with Si^{2+} and Al^{3+} ions compared to pure cement samples. Meanwhile, the structure of the formed calcium silicate hydrates is closer to that of tobermorite than to that of jennite.

In systems with alumina and in the absence of sulfates, carbonates, and chlorides, C_4AH_{13} is the main calcium aluminate hydrate formed during hydration [59, 61, 67–69]. If the concentration of calcium ions in the pore solution decreases (due to a pozzolanic reaction), the calcium ions that are present in the structure of C_4AH_{13} can be replaced by aluminum and silicon to form stratlingite—C_2ASH_8. The prolonged duration of hydration, the higher water to solid ratio, or the presence of alkalis leads to the formation of hydrogrossular $[Ca_3Al_2(SiO_4)_{3-x}(OH)_{4x}]$.

Those soluble sulfates which are present in SCM (fly ash, furnace slags, etc.) may react with portlandite to form calcium sulfoaluminates: ettringite, monosulfoaluminate, or both [59, 63, 70–72]. The mineralogical composition of the formed compounds depends on the ratio of the sulfate to aluminum. The presence of sulfates in the system accelerates the pozzolanic reaction; i.e. the formation of ettringite from dissolved sulfates is compensated by an increase in the hydroxide concentration and pH, which accelerates the dissolution of the pozzolans and the progress of the pozzolanic reaction. When all the soluble sulfate has been consumed, ettringite recrystallizes to monosulfoaluminate.

The formation and stability of ettringite and monosulfoaluminate are strongly affected by the presence of CO_2 in the environment or cement matrix [54, 56, 57, 73–76]. When the CO_2 concentration in an environment and cement matrix (featuring very little added carbonate) is low, calcium aluminumhydrates (C_4AH_{13}) react with carbon dioxide to form semi-carboaluminate. In the case of a higher concentration of CO_2 in an environment, or if more calcium carbonate is added, monocarboaluminate is formed. The carbonation of calcium aluminates inhibits the recrystallization of ettringite to monosulfoaluminate. For this reason, in the presence of a higher concentration of CO_2, ettringite is stabilized in hardened cement samples, which leads to an increase in the compressive strength.

Thus, the mineralogical composition of hardened cement with or without SCM is very similar [57, 59–65]; however, SCMs influence the hydration process of individual compounds and the concentration of hydrates. Each SCM has a different effect on the course and rate of cement hydration and on the mineral properties of the hydrated stone. Therefore, in order to use calcium silicate hydrates with intercalated metal ions as SCMs, it is necessary to determine their influence on the hydration mechanism and the compressive strength values of cement stone.

5.3 Effect of Synthetic Gyrolite Additive on Portland Cement Hydration

It is well known that cement hydration involves numerous chemical reactions occurring simultaneously, and is, therefore, a highly complex process. The system becomes even more complicated when an adsorbent serves as a cement substitute in the cement–water reaction system. New hydration products can form, which will not only affect the kinetics of the hydration process but will also change not only the mineral composition, but also the properties of hardened cement. Most likely, the additive of pure synthetic gyrolite and/or synthetic gyrolite with intercalated heavy metal ions can affect the hydration of ordinary Portland cement. If we want to understand the impact of synthetic gyrolite with intercalated metal ions on the hydration of OPC, the first step is to determine the influence of pure synthetic gyrolite on this process. Thus, 2.5, 5.0, and 7.5 wt% of cement was replaced with synthetic gyrolite. For the analysis, standard samples (prisms 4 × 4 × 16 cm) were produced according to Standard EN 196–1. The ratio of cement and sand equaled to 1:3 by weight, while the water-to-cement ratio (w/c) was equal to 0.5.

5.3.1 *Influence of Synthetic Gyrolite on the Heat Flow of Cement Paste*

Since the cement hydration reactions are exothermic, isothermal calorimetry is one of the most precise methods of monitoring the global reaction process by observing the heat production rate. To compare the heat evolution rate of each sample and avoid the influence of a reduced amount of OPC in the samples, all the results were calculated based on the unit weight of ordinary Portland cement in the sample. The heat evolution rate (W/g OPC) and the cumulative heat of hydration (J/g OPC) data of the binary blended pastes are shown in Figs. 5.9, 5.11 and 5.12. The five stages of the hydration reaction (the initial reaction, the induction period, the acceleration period, the deceleration period, and the period of slow continued reaction) were observed in the microcalorimetry curves of the pure OPC sample and the samples with a synthetic gyrolite additive. The obtained numbers align fairly well with the data presented in scientific literature [54–58]. At the beginning of hydration, the wetting of cement particles and the hydration of calcium oxide take place; thus, the rapid increment of the heat with a peak after ~5 min was observed (Fig. 5.9). Also, Ca^{2+}, OH^-, $H_2SiO_4^{4-}$, SO_4^{2-}, and alkali ions were released from the cement compounds during the initial reaction of hydration. It was established that the additive of synthetic gyrolite accelerated the initial reaction because the maximum of the first peak increased by increasing the amount of synthetic gyrolite (Fig. 5.9).

Most likely, the additive of synthetic gyrolite increases the speed of the early stage of hydration of OPC because synthetic gyrolite acts as an adsorbent. As a result, both alkaline and calcium ions were adsorbed on it. The adsorption process leads to the

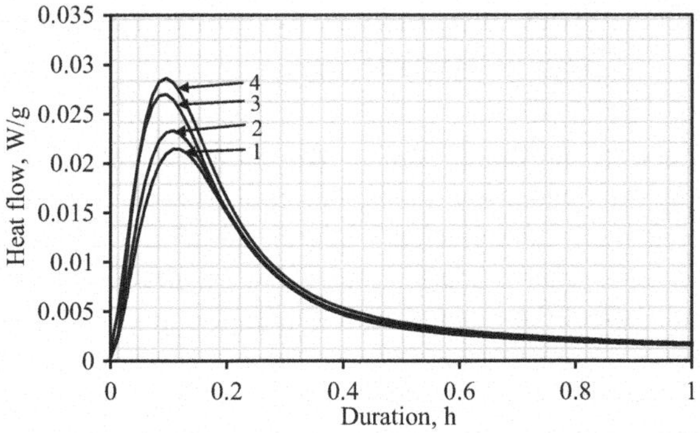

Fig. 5.9 Heat evolution rate of OPC samples. Amount of synthetic gyrolite additive, wt%: 1—0.0; 2—2.5; 3—5.0; 4—7.5

decrement of Ca^{2+} ions in the solution, which boosts the C_3S dissolution rate. These results conform well to the values suggested by the dissolution theory proposed by Juilland et al. [77]; as long as the concentration of Ca^{2+} is kept undersaturated, a higher rate of dissolution is enabled.

In order to prove that synthetic gyrolite adsorbs calcium ions, a mixture of synthetic gyrolite and portlandite with C/S molar ratio equal to 1.75, which corresponds to the main hydration product (calcium silicate hydrate gel), was prepared and poured into distilled water (w/s = 10). Adsorption experiments were carried out at 25 °C in a thermostatic absorber while stirring 0.5 g of the solid mixture in distilled water for 1, 2, 3, or 7 min. The initial mixture was calculated to contain a 25.62% excess of calcium ions according to $Ca(OH)_2$ solubility (1.13 g CaO/1000 g H_2O). Scientific data suggests that portlandite can exist in crystalline and amorphous forms during the hydration of cement minerals. A number of scientists have proven that thermal analysis techniques are highly reliable for calculations of the portlandite content in samples of hardened Portland cement [78–80]. Thus, simultaneous thermal analysis was used to determine the concentration of portlandite in the mixture at different times of adsorption. The adsorption curves of Ca^{2+} ions adsorption on synthetic gyrolite are presented in Fig. 5.10. It was determined that the amount of adsorbed Ca^{2+} ions on synthetic gyrolite increased when extending the length of the experiment. As in the case of heavy metal ions adsorption (Chap. 4), the intercalation of calcium ions takes place intensively at the beginning of the process. It was calculated that, after 1 min, the number of intercalated Ca^{2+} ions in the crystal lattice of synthetic gyrolite equals to 182 mg/g (Fig. 5.10a). Meanwhile, at the end of the process (7 min), the amount of the adsorbed ions increased only to 248 mg Ca^{2+}/g (Fig. 5.10b).

The leaching experiment was performed with the objective to establish the nature of the adsorption process (i.e. whether that is chemical or physical adsorption): hence, synthetic gyrolite obtained after adsorption was dried and later immersed in distilled water. The analysis of the liquid medium showed that Ca^{2+} ions are not released from the crystal lattice of synthetic gyrolite into the solution during desorption. Thus, it can be stated that adsorption reactions are irreversible and specific to chemisorption processes.

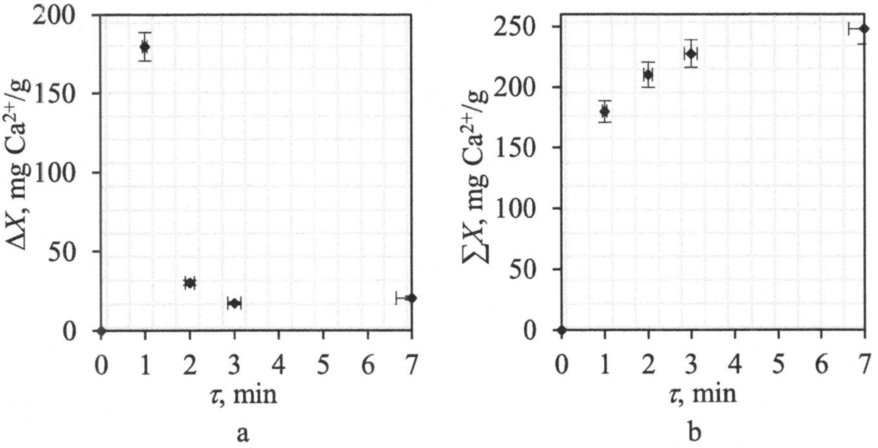

Fig. 5.10 Differential (**a**) and integral (**b**) kinetic curves of Ca^{2+} ions adsorption on synthetic gyrolite

According to the scientific literature and the obtained nitrogen adsorption–desorption results, different pores (micro, meso, and macro) are present in the structure of synthetic gyrolite (see Sect. 2.4. *Effect of hydrothermal synthesis duration at 200 °C temperature on the properties of gyrolite*). Also, the crystal lattice of gyrolite is layered [81, 82]. Thus, in the course of adsorption, calcium ions can easily migrate through the synthetic gyrolite pores and the layers of its crystal lattice until the equilibrium of the C/S ratio has been achieved. However, we should emphasize that the adsorption is fairly rapid, yet the solubility rate of the initial compounds limits this process. The same tendency was observed during the hydration of OPC; i.e. at the beginning of the process, synthetic gyrolite adsorbed the calcium ions that had been contained in the liquid medium; consequently, the solubility of C_3S was increased.

The data obtained in STA analysis showed that synthetic gyrolite remained stable while adsorption was taking place because the area of the endothermic effect at ~140 °C does not alter with the duration of the adsorption, and its value equals to ~30.4 J/g (according to the linear method) (Table 5.5). As expected, the area of the endothermic effect typical of portlandite decomposition (~420 °C) went down from 2328.2 J/g (before the experiment) to 238.7 J/g (after 7 min of adsorption) (Table 5.2). The results of STA and adsorption allow to state that synthetic gyrolite is stable under all investigated adsorption conditions, i.e. the temperature is equal to 25 °C, the duration is equal to 7 min, and the molar ratio is equal to 1.75.

The influence of synthetic gyrolite additive on the released heat during the induction period, and the second exothermic reaction is presented in Figs. 5.11 and 5.12. The induction period of the pure OPC sample lasts for ~2.5 h, while it was shortened to 20–30 min for the sample with 7.5 wt% of synthetic gyrolite.

Thus, in the samples with the synthetic gyrolite additive, the induction period is significantly shortened as synthetic gyrolite serves as a nucleation site (for growing CSH and CH on the surface of the particles of synthetic gyrolite) for the formation

Table 5.2 Main characteristics of thermal effects manifested by synthetic gyrolite and portlandite

Duration of adsorption, min	Endothermic effect at ~135 °C		Endothermic effect at ~450 °C	
	Area, J/g	Mass loss, %	Area, J/g	Mass loss, %
0	30.9	3.61	2328.2	24.34
1	30.3	3.55	327.5	3.41
2	29.8	3.79	285.9	3.00
3	30.1	3.50	264.6	2.77
5	30.6	3.61	251.2	2.63
7	30.8	3.53	238.7	2.50

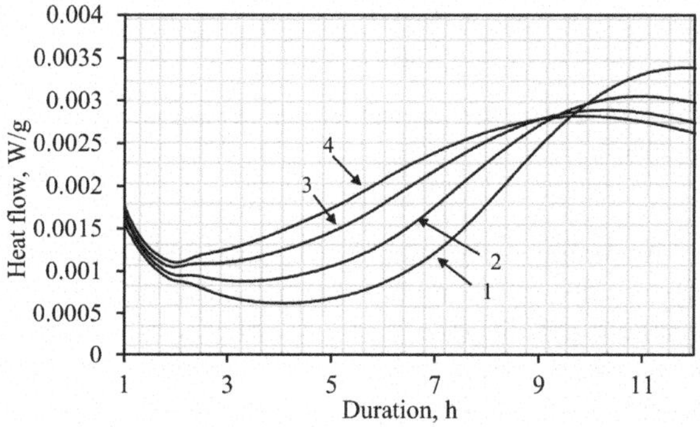

Fig. 5.11 Heat evolution rate of OPC samples. Amount of synthetic gyrolite additive, wt%: 1—0.0; 2—2.5; 3—5.0; 4—7.5

of hydration products. However, these results are only in partial agreement with the numbers provided by Thomas et al. [83]. These authors discussed a general OPC hydration theory by which fine additives, such as silica fumes, colloidal silica, and especially pure CSH, stimulate the nucleation and growth kinetics by seeding the hydration reaction. In general, such additives promote the main hydration reactions, for instance, increasing the released heat (area and intensity) during the second exothermic reaction (stage II) (Fig. 5.12).

It was determined that the acceleration period in the sample with 7.5 wt% of synthetic gyrolite and the pure OPC sample starts after ~2 h and ~4.5 h, respectively (Fig. 5.11). However, the heat evolution rate of the second exothermic reaction, i.e. the intensity of the second exothermic effect, decreased by increasing the amount of the additive (Fig. 5.12).

The diminished rate of heat evolution during the second stage of hydration can be related to the destruction of the crystal lattice of synthetic gyrolite (Fig. 5.13). To confirm the data of the experiments and the theoretical hypothesis, we performed thermodynamic calculations according to the methodology developed by Babushkin et al. [84]. Theoretically, the amount of the released heat during OPC sample hydration is sufficient to destroy the crystal lattice of synthetic gyrolite as the enthalpy of this compound equals to $\Delta H_f = -4914.42$ kJ/mol. Furthermore, the obtained thermodynamic results show that synthetic gyrolite remained stable at the beginning

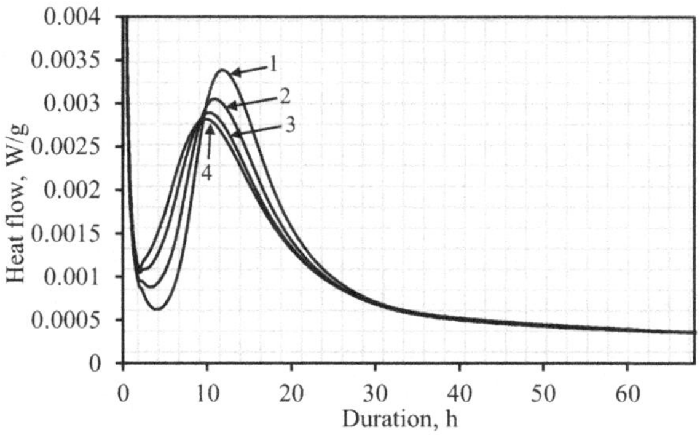

Fig. 5.12 Heat evolution rate of OPC samples. Amount of synthetic gyrolite additive, wt%: 1—0.0; 2—2.5; 3—5.0; 4—7.5

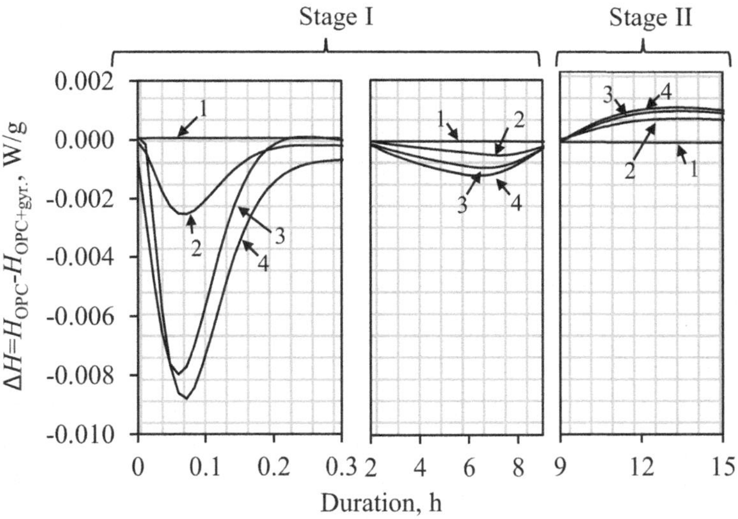

Fig. 5.13 Differences between heat evolution of pure OPC sample and OPC sample with the additive. Amount of synthetic gyrolite additive, wt%: 1—0.0; 2—2.5; 3—5.0; 4—7.5

of the process of hydration. At the same time, this compound recrystallizes into CSH with a higher molar ratio of CaO and SiO_2 after 8–9 h of hydration.

In order to confirm the previous results, a synthetic gyrolite–portlandite mixture (CaO/SiO_2 = 1.75) was produced and kept in water (w/s = 10) for 10 h at 25 °C temperature. The changes in the CaO/SiO_2 ratio of the reaction products are presented in Fig. 5.14a. It was established that, by extending the length of the experiment, the molar ratio of the products increases, and, at the end of the process (10 h), it is equal to 1.23. This data is in line with the obtained XRD results (Fig. 5.14b). Furthermore, it was established that synthetic gyrolite is not stable during its interaction with portlandite because the main diffraction peak (*d*—2.231 nm) characteristic of synthetic gyrolite could not be identified after 10 h of the experiment. The main reaction product was semicrystalline calcium silicate hydrates with a CaO/SiO_2 molar ratio equal to 0.8–2.0. These results confirm a decreased rate of heat evolution (stage II) during the hydration of OPC samples containing the synthetic gyrolite additive (Fig. 5.13).

As already established, synthetic gyrolite significantly affects the cumulative heat of the samples at the very beginning of hydration (till 20 h) (Fig. 5.15). Meanwhile, at the later stages of hydration, the cumulative heat increased monotonically in all the samples, and the difference between the amounts of released heat is insignificant. Thus, synthetic gyrolite affects OPC hydration as well as the usual pozzolanic additives (see Sect. 5.2. *Supplementary Cementitious Materials and Its Influence on OPC Hydration*).

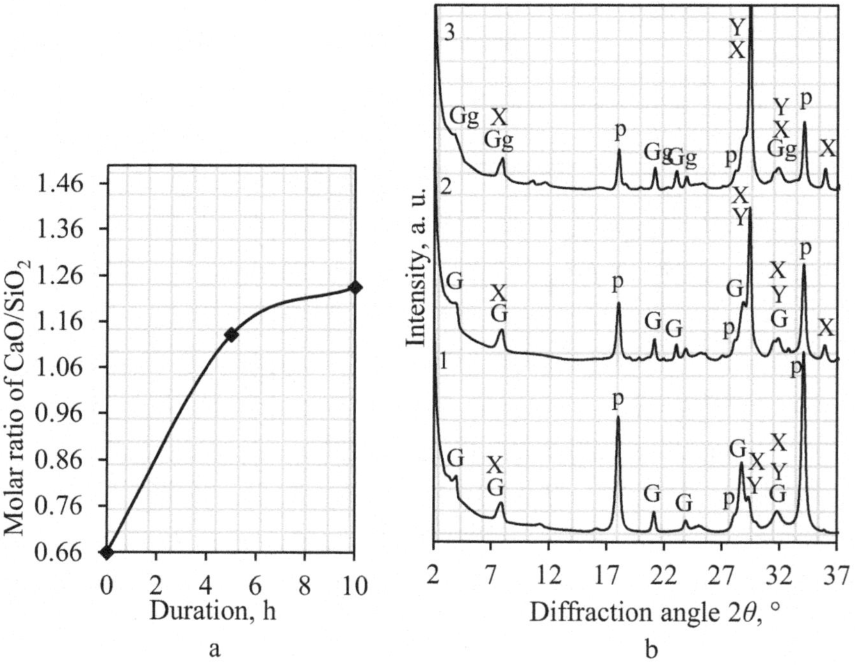

Fig. 5.14 Variation of CaO/SiO_2 molar ratio (**a**) and XRD pattern (**b**) of synthetic gyrolite and portlandite mixture at initial molar ratio of CaO/SiO_2 = 1.75, w/s = 10.0, t = 25 °C, and the following lengths of hydration, h: 1—0 (after mixing); 2—5; 3—10. *Indexes* G—gyrolite; Gg—gyrolite gel; P—portlandite; X—calcium silicate hydrates (CaO/SiO_2 = 0.8–1.3); Y—calcium silicate hydrates (CaO/SiO_2 = 1.3–2.0)

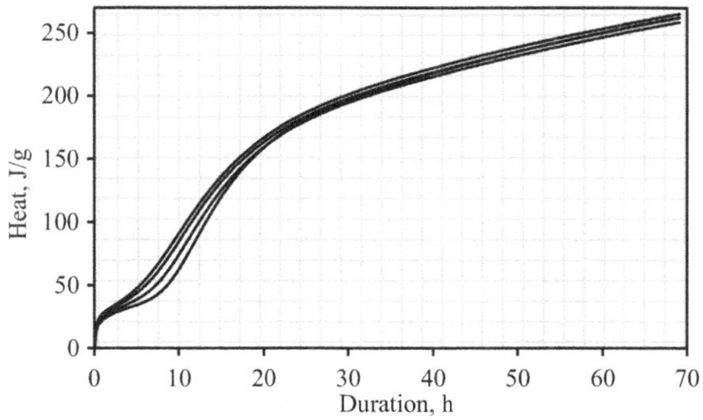

Fig. 5.15 Cumulative heat of OPC samples. Amount of synthetic gyrolite additive, wt%: 1—0.0; 2—2.5; 3—5.0; 4—7.5

5.3.2 Influence of Synthetic Gyrolite on Portlandite Quantity in Hydration Products

To prove the positive effect of synthetic gyrolite on the formation of portlandite, the quantity of the latter compound was measured in the hydration products. Thus, microcalorimetric analysis experiments were repeated in a thermostat at 25 °C temperature with the reaction times of 2, 5, 9, and 12 h, w/c ratio equal to 0.5, and hydration was stopped by using acetone. Afterward, the samples were pulverized, dried at 50 ± 5 °C, and sieved (80 μm mesh was used). The selected duration of the halted reactions coincided with the time interval of the second exothermic reaction when using various compositions of cement samples. The DSC, TG, and DTG curves of pure OPC samples and OPC samples with the synthetic gyrolite additive after 2 h of hydration (i.e. beginning of the acceleration period in OPC samples with 7.5 wt% synthetic gyrolite) are given in Fig. 5.16. Summarized data of the thermal analysis is presented in Table 5.3. We thus established that the quantity of portlandite formed in the hydration products depends on both the duration of hydration and the amount of the additive (Table 5.3). All the investigated samples showed the thermal effect at 400–475 °C temperature characteristic of the decomposition of portlandite. According to the mass losses and the decomposition reaction of portlandite, the concentration of the latter compound can be calculated in the hydration products. We found out that, after 2 h of hydration, portlandite concentration in the hydration products increased from 1.11 to 3.66 wt% by increasing the amount of the synthetic gyrolite additive (Fig. 5.16). Quantitative DSC analysis data strongly confirmed these findings (Table 5.3).

Although the TG curves look somewhat similarly shaped, major differences in both DSC and DTG profiles were observed. The thermal effect characteristic of portlandite is broader in the samples with the additive than in the pure sample (Fig. 5.16, curve 1). Also, the peaks of DSC and DTG curves shift to higher temperatures in the samples with the synthetic gyrolite additive. Most likely, this trend depends on the morphology of the portlandite crystals.

This tendency was also detected after 5 h of hydration (Table 5.3). In that particular case, the concentration of portlandite went up from 2.47 wt% (in the pure OPC

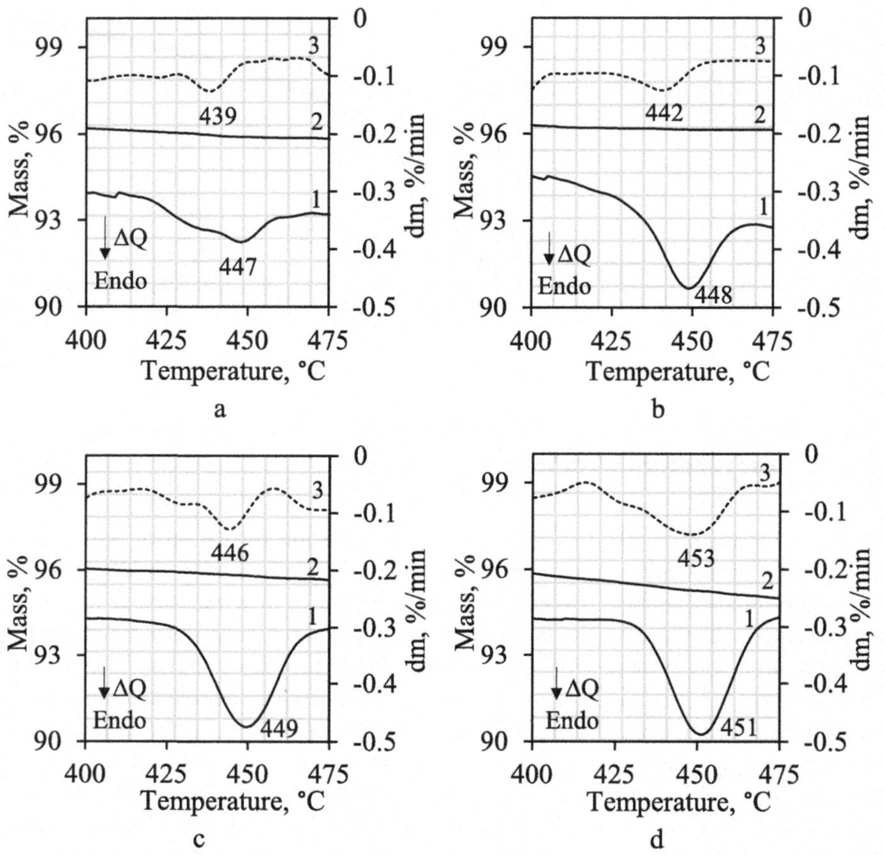

Fig. 5.16 DSC (1), TG (2), and DTG (3) curves of pure OPC sample and OPC sample with the additive after 2 h of hydration. Amount of synthetic gyrolite additive, wt%: **a** 0.0; **b** 2.5; **c** 5.0; **d** 7.5

samples) to 5.76 wt% (in the sample with 7.5 wt% of the synthetic gyrolite additive). Also, the shapes of the DSC and DTG curves of all the samples manifest a trend similar to that after 2 h of hydration; i.e. when raising the amount of the synthetic gyrolite additive, we observe a shift of the peak of portlandite to higher temperatures (DSC_{max} from 449 to 452 °C; DTG_{max} from 438 to 442 °C). The data obtained in the thermal analysis was confirmed by the data of XRD (Fig. 5.17). XRD results showed that, by increasing the quantity of the additive in the OPC sample, the intensity of the diffraction peak (*d*—0.489 nm) characteristic of portlandite increased significantly (Fig. 5.17).

Further analysis indicated that the concentration of portlandite was growing when extending the length of hydration to 9 and 12 h (Table 5.3). However, the differences between the samples decreased with the duration of 9 h, especially, after passing the mark of 12 h, these differences diminish and thus become largely insignificant.

Table 5.3 Quantity of portlandite in cement paste during hydration

Duration, h	Amount of additive, wt%	TGA method-based data		DSC method-based data	
		Mass losses at 400–475 °C, %	$W_{Ca(OH)_2}$, wt%	Peak area at 400–475 °C, mVs/mg	$W_{Ca(OH)_2}$, wt%
2	0.0	0.27	1.11	7.87	1.24
	2.5	0.45	1.85	11.62	1.83
	5.0	0.57	2.34	14.54	2.29
	7.5	0.89	3.66	23.37	3.68
5	0.0	0.60	2.47	15.94	2.51
	2.5	0.82	3.37	21.46	3.38
	5.0	1.16	4.77	29.78	4.69
	7.5	1.40	5.76	36.64	5.77
9	0.0	1.24	5.09	31.94	5.03
	2.5	1.55	6.38	40.70	6.41
	5.0	1.57	6.44	40.83	6.43
	7.5	1.65	6.77	42.48	6.69
12	0.0	1.78	7.31	46.54	7.33
	2.5	1.81	7.44	46.80	7.37
	5.0	1.82	7.49	47.75	7.52
	7.5	1.88	7.73	49.33	7.77

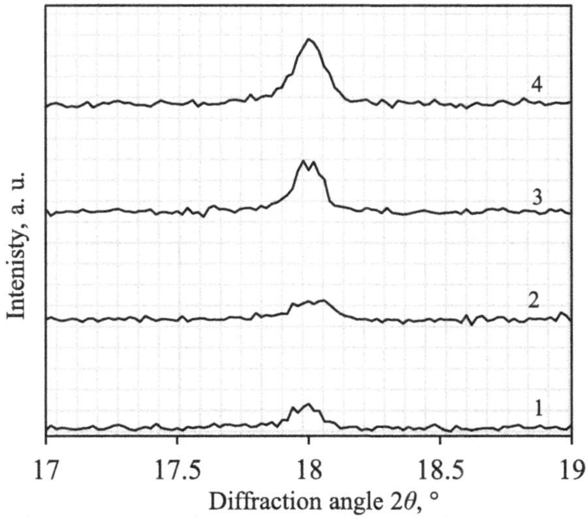

Fig. 5.17 Change of the main diffraction peak intensity of portlandite (*d*-spacing—0.489 nm) after 5 h of hydration. Amount of synthetic gyrolite additive, wt%: 1—0.0; 2—2.5; 3—5.0; 4—7.5

5.3.3 Kinetics and Modeling of the Hydration Process

The kinetics of cement hydration is related to the relationships between the hydration degree α and the age τ as well as with the influencing factors [54–58]. Theoretically, α can be calculated for an individual phase or OPC. Since cement is a mixture of phases reacting at different rates, it is difficult to determine and define the value of α for the entirety of cement. For this reason, α was calculated by adding up the amounts of the reacted C_3S. To determine the amount of the reacted C_3S, the samples obtained during the hydration experiments in the thermostat were used. The diffraction peak (*d*-spacing—0.2604 nm) characteristic of C_3S was used for the calculations. It should be highlighted that this peak does not coincide with the other peaks characteristic of the primary compounds or their hydration products. Each calculation was repeated five times, and the deviation of the values was below 5%. It was accepted that the C_3S quantity was 100% in all the samples (pure and with the additive) before the hydration. The amount of unreacted C_3S was established by considering the changes in the intensity of the diffraction peak with a *d*-spacing of 0.260 nm (Table 5.4 and Fig. 5.18).

Table 5.4 Quantity of unreacted C_3S

Duration of hydration, h	Amount of synthetic gyrolite additive, wt%			
	0.0	2.5	5.0	7.5
	Amount of reacted C_3S from the initial concentration, %			
0	0	0	0	0
2	8	30	32	36
5	12	32	35	37
9	37	37	40	40
12	44	45	46	47

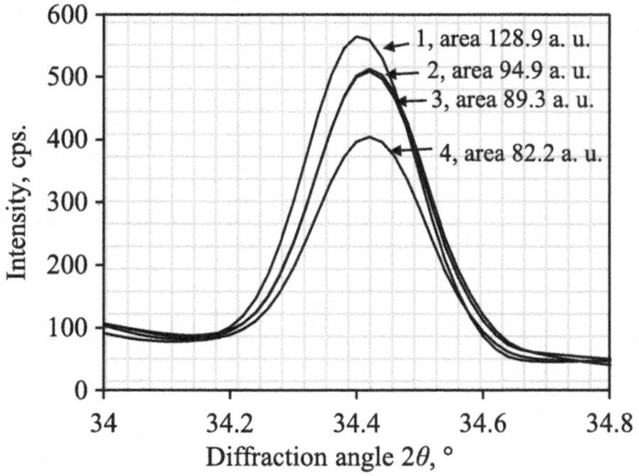

Fig. 5.18 Change of the main diffraction peak intensity of C_3S (*d*-spacing—0.260 nm) after 2 h of hydration. Amount of synthetic gyrolite additive, wt%: 1—0.0; 2—2.5; 3—5.0; 4—7.5

It was determined that only 8% of C_3S reacted in the pure OPC samples after 2 h of hydration (Fig. 5.18). By raising the length of hydration to 12 h, the amount of reacted C_3S reached 44%. Meanwhile, the additive of synthetic gyrolite significantly promoted the earliest hydration of C_3S because, after 2 h of the experiment in the sample with 7.5 wt% of the additive, 36% of this compound had already reacted (Fig. 5.18). However, in the further hydration stages, the influence of synthetic gyrolite decreased, and after 12 h of the experiment, the amount of the reacted C_3S was similar in all the samples (Table 5.4).

We employ the kinetic equation of the chemical process [85] and determine that the reaction rate $W_{\text{app.}}$ increases due to $f(c)^n$ variation:

$$W = \frac{\mathrm{d}x}{\mathrm{d}\tau} = k\big(f(c)^n\big), \tag{5.1}$$

where k is the constant of the reaction rate, $f(c)^n$ is the influence of the concentration of reagents on the reaction rate W.

The linear form of Eq. 5.1 is:

$$\ln W = \ln k_{\text{app}} + n_{\text{app}} \cdot \ln(X_0 - X_\tau), \tag{5.2}$$

where $W = \frac{\mathrm{d}x}{\mathrm{d}\tau}$ is the reaction rate as the reacted amount of C_3S, k_{app} is the apparent reaction rate constant, n_{app} is the apparent order of the reaction, X_0 is the initial amount of C_3S, and X_τ is the residual amount of C_3S at reaction time τ.

The apparent kinetic parameters of the total process of C_3S hydration were calculated by using Eq. 5.2 and the data presented in Table 5.4. The obtained results are presented in Fig. 5.19.

We determined that the process rate as a function of $(X_0 - X_\tau)$ in a logarithmic plot yields straight lines with correlation coefficients higher than 0.97 (Fig. 5.20 and Table 5.5).

By using the slope and intercept of plots presented in Fig. 5.20, the values of n_{app} and $\ln k_{\text{app}}$ can be calculated. The values of the reaction order and the rate constants are presented in Table 5.5. It is clearly seen that the additive of synthetic gyrolite strongly affects the apparent kinetic parameters of the C_3S hydration reaction rate (Table 5.5). It was calculated that, in the pure sample, the reaction order $n_{\text{app.}}$ and the reaction rate k_{app} are equal to 0.288 and 3.881, respectively (Table 5.5). The usage of the synthetic gyrolite additive leads to the increment of the presently discussed values, and the sample with 7.5 wt% of the synthetic gyrolite additive is characterized by a 2.5 times higher value of n_{app} (0.716) and by more than 2 times higher value of k_{app} (8.112).

The obtained apparent kinetic parameters were fitted in the common hydration models to describe the mechanism of the OPC sample hydration. The most appropriate model of OPC hydration is the one in which the equation produces a straight line in the linear form. The hydration OPC process occurs by nucleation and growth, as well as diffusion [86–91]. Therefore, the early stage hydration rate for C_3S was

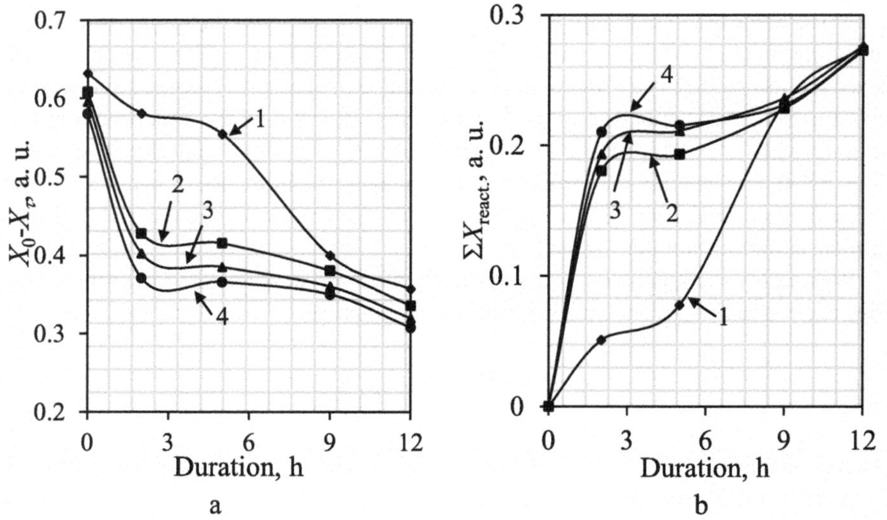

Fig. 5.19 Kinetic curves of the residual amount of C_3S ($X_0 - X_\tau$) (**a**) and the amount of reacted C_3S (**b**) during OPC hydration. Amount of synthetic gyrolite additive, wt%: 1—0.0; 2—2.5; 3—5.0; 4—7.5

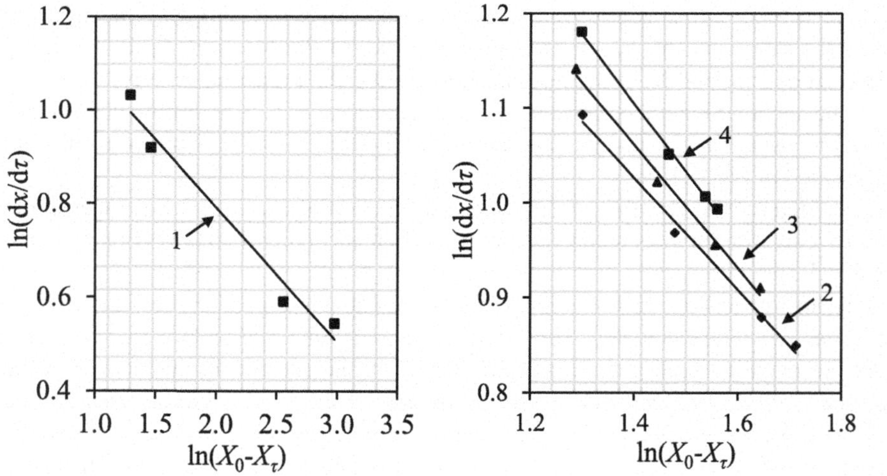

Fig. 5.20 Reaction rate of C_3S as a function of ($X_0 - X_\tau$) in a logarithmic plot during OPC hydration. Amount of synthetic gyrolite additive, wt%: 1—0.0; 2—2.5; 3—5.0; 4—7.5

Table 5.5 Apparent kinetic parameters of reacted C_3S during OPC hydration

Amount of additive, wt%	0.0	2.5	5.0	7.5
k_{app}	3.881	6.284	7.074	8.112
n_{app}	0.288	0.588	0.650	0.716
R^2	0.972	0.993	0.993	0.997

determined on the basis of the Avrami (otherwise known as the 'JMAK' model) nucleation and growth model and the Jander diffusion model.

It is possible to write the general form of the Avrami equation in the following way:

$$-\ln(1-a)^{n_a} = k_a(\tau - \tau_0), \tag{5.3}$$

where a is the reacted fraction, k is the constant of the reaction rate, τ is duration, τ_0 is the duration of the induction period, and n is a constant. The linear form of the Avrami equation is:

$$\ln(-\ln(1-a)) = -\frac{\ln k_a}{1} - \frac{1}{n_a} \cdot \ln(\tau - \tau_0). \tag{5.4}$$

The curves obtained by using the linear form of the Avrami equation are presented in Fig. 5.21. The slope of the curves was used to calculate the reaction order n_a (Table 5.6). When knowing n_a, the mean value for k_a can be determined by establishing the reaction rate constant during OPC hydration with different amounts of the synthetic gyrolite additive (Table 5.6).

Based on Jander's model, the hydration reaction can be written as follows:

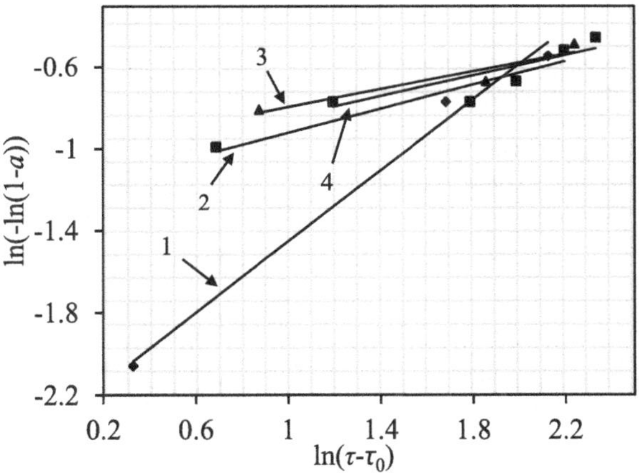

Fig. 5.21 Determination of n and k by using Avrami model in logarithmic scale during pure OPC hydration. Amount of synthetic gyrolite additive, wt%: 1—0.0; 2—2.5; 3—5.0; 4—7.5

Table 5.6 Apparent kinetic parameters of reacted C_3S during OPC hydration obtained by using Avrami model

Amount of additive, wt%	0.0	2.5	5.0	7.5
k_a	0.069	0.016	0.009	0.013
n_a	0.866	0.292	0.217	0.253
R^2	0.990	0.901	0.888	0.893

$$\left(1-(1-a)^{1/3}\right)^{n_j} = k_j\tau, \tag{5.5}$$

where a is the degree of hydration, k_j is the constant of the reaction rate, τ is the duration of hydration, and n_j varies in relation with the type of the reaction governing the process.

The value of n is calculated by plotting $\ln\left(1-(1-a)^{1/3}\right)$ versus $\ln(t)$ and establishing the slope ($1/n_j$). When $n_j < 1$, the process occurs through nucleation; when $n_j = 1$, the process is based on phase boundary kinetics, and when $n_j = 2$, the process is run by diffusion. The results obtained by using the Jander equation are presented in Fig. 5.22, while the summarized data of the calculated parameters is given in Table 5.7.

We discovered that the Avrami equation only fits for the pure OPC sample due to the fact that the value of the correlation coefficient (R^2) equaled to 0.99 (Table 5.6). Yet, in the samples with the synthetic gyrolite additive, the value of the latter coefficient was lower than 0.91. The calculations revealed that the Jander equation fits all the investigated samples as the correlation coefficient values exceeded the scores obtained by using the Avrami equation (Tables 5.6 and 5.7). It was estimated that the pure OPC sample is characterized by constant n_j which is less than 1. This shows that the reaction occurs by nucleation kinetics (Table 5.7). This fact was also confirmed by the linear dependency of the results obtained by using the Avrami equation. Different results were produced in the samples with the synthetic gyrolite additive

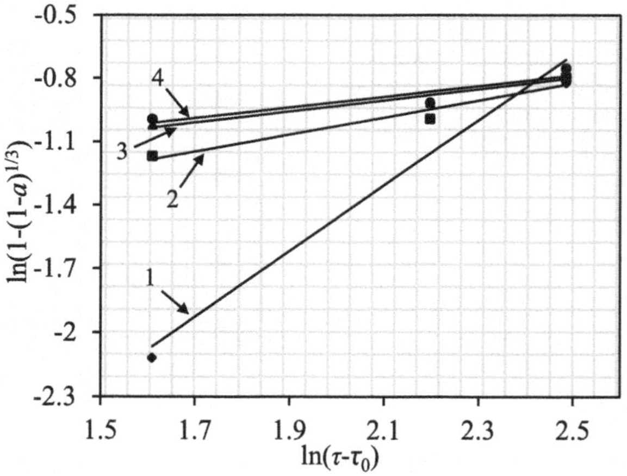

Fig. 5.22 Determination of n and k by using Jander model in logarithmic scale during pure OPC hydration. Amount of synthetic gyrolite additive, wt%: 1—0.0; 2—2.5; 3—5.0; 4—7.5

Table 5.7 Apparent kinetic parameters of reacted C_3S during OPC hydration determined by using Jander model

Amount of additive, wt%	0.0	2.5	5.0	7.5
k_j	0.007	0.389	0.875	0.959
n_j	0.646	2.455	3.772	3.961
R^2	0.958	0.950	0.926	0.917

because these samples showed the values of n_j higher than 2, which means that the process is controlled by diffusion processes (Table 5.7). Thus, we can state that the synthetic gyrolite additive strongly affects the hydration mechanism of OPC.

5.3.4 Peculiarities of the Hydration Process Mechanism

As discussed in the previous chapters, synthetic gyrolite adsorbs calcium ions during the OPC hydration; however, cement consists of various chemical elements which can also be released into the liquid medium. Therefore, iron and aluminum ions may penetrate into the crystal lattice of synthetic gyrolite within the initial minutes of hydration (stage I). The adsorption process leads to the decrement of the concentration of the presently discussed ions in the liquid medium and the faster dissolution of the individual phases of cement. In order to prove the adsorption of Al^{3+}, Ca^{2+}, and Fe^{3+} ions on synthetic gyrolite during OPC hydration, the following experiment was performed: the OPC samples were mixed with water (w/s = 0.5) and maintained for 1, 3, 5, 7, and 10 min under environmental conditions (t ~23 °C). After the selected duration, the liquid medium of the samples was separated and collected by vacuum filtration. The obtained liquid medium was analyzed by atomic absorption spectroscopy. We established that only a minor amount of aluminum ions was released into the liquid medium at the very beginning of hydration (1 min) (Table 5.8). Further analysis showed that the concentration of the ions was lower than the detecting limit of the spectrometer. A similar tendency was observed in the changes in the concentration of iron ions in the liquid medium. In both cases, after 1 min of hydration, the iron concentration in the liquid medium did not even reach the value of 0.045 mg/dm^3 (Table 5.8). It was thus established that the concentration of iron ions in the liquid medium increased after 7–10 min of hydration of the pure OPC sample. Meanwhile, these ions were not detected in the sample with the synthetic gyrolite additive. An interesting tendency was observed in the changes of the concentration of calcium ions in the liquid medium because the concentration of calcium ions decreased till 7 min of hydration, whereas, after 10 min, it slightly increased. Meanwhile, in the sample with the synthetic gyrolite additive, the concentration of calcium ions equaled to ~580 mg/dm^3. Such a difference can be related to the excellent adsorption properties of synthetic gyrolite for calcium ions (Fig. 5.10). Thus, the obtained results demonstrate that, at the beginning of hydration, the ions released from OPC compounds into the liquid medium intercalate into the structure of synthetic gyrolite.

By extending the duration of hydration, the additive of synthetic gyrolite starts to act as a nucleation site for the hydration products. However, synthetic gyrolite is not a stable compound during the later stages of hydration, and it recrystallizes into CSH of a higher basicity (C/S > 0.8) (Fig. 5.14). The summarized suggested mechanism of OPC hydration with the synthetic gyrolite additive is schematically presented in Fig. 5.23.

Table 5.8 Dependence of Al^{3+}, Ca^{2+}, and Fe^{3+} ion concentrations in the liquid medium on the duration of hydration

Duration, min	Pure OPC sample			OPC sample with 7.5 wt% synthetic gyrolite additive		
	Ion, mg/dm^3					
	Al^{3+}	Ca^{2+}	Fe^{3+}	Al^{3+}	Ca^{2+}	Fe^{3+}
1	< 0.005	660	< 0.045	< 0.005	585	< 0.045
3	Not detected	581	Not detected	Not detected	575	Not detected
5	Not detected	496	Not detected	Not detected	577	Not detected
7	Not detected	485	0.134	Not detected	586	Not detected
10	Not detected	536	0.108	Not detected	594	Not detected

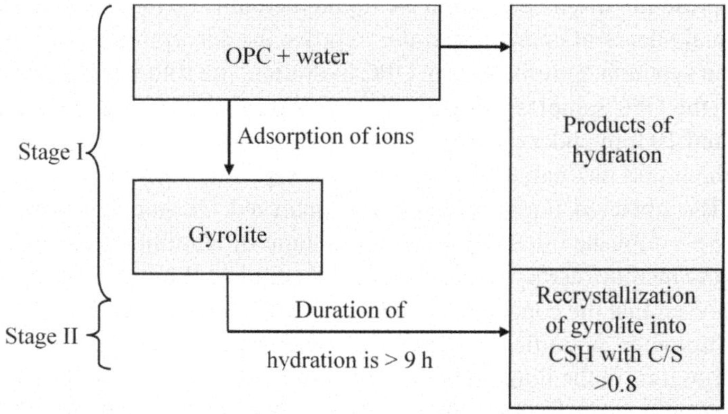

Fig. 5.23 Peculiarities of hydration mechanism of OPC with additive of synthetic gyrolite

5.3.5 *Influence of Synthetic Gyrolite on the Compressive Strength and Mineralogical Composition of Hardened OPC*

Although the influence of additives on the kinetics of the hydration process is significant, the main characteristic of cement and concrete is their compressive strength. Thus, to establish the influence of the synthetic gyrolite additive on this property, samples (prisms $4 \times 4 \times 16$ cm) were produced according to Standard EN 196-1 (with cement/sand ratio 1:3 by weight). Synthetic gyrolite was added as partial replacement of OPC by weight, and the water/cement ratio of all the samples equaled to 0.5. The samples were kept in molds at 20 ± 1 °C and 100% air humidity during the first day of the experiment. After 24 h of formation, the samples were transferred into distilled water and stored there for 6, 27, and 83 days at 20 ± 1 °C. The data of the compressive strength test was derived as the arithmetic mean of six individual

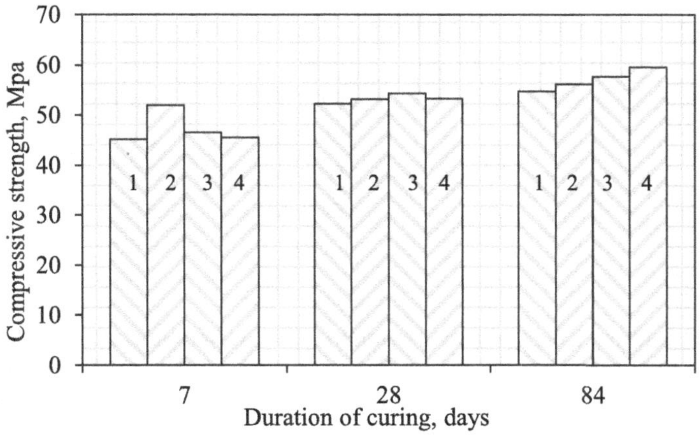

Fig. 5.24 Dependence of compressive strength of OPC samples on the duration of curing. Amount of synthetic gyrolite additive, wt%: 1—0.0; 2—2.5; 3—5.0; 4—7.5

results. Finally, the samples (for instrumental analysis) were pulverized, dried at 50 ± 5 °C, and sieved by employing a mesh of 80 μm width.

The impact of synthetic gyrolite on the compressive strength of OPC is presented in Fig. 5.24. As shown previously, synthetic gyrolite promotes the hydration of C_3S; however, synthetic gyrolite is not a cementitious material; i.e. it has no binding properties. We established that after seven days of hydration, the sample with 2.5 wt% of synthetic gyrolite additive was characterized by the highest value of compressive strength (52 MPa) (Fig. 5.24). Meanwhile, other samples showed similar values of strength (~45 MPa). By prolonging the duration of hardening to 28 days, the compressive strength of all the samples increased, but samples containing synthetic gyrolite could be singled out for slightly higher values of the compressive strength than the pure sample. Finally, after 84 days, the compressive strength of the samples increased when raising the amount of the synthetic gyrolite additive. Such dependence can be attributed to a higher rate of hydration.

To fully understand the changes in the compressive strength, the samples after hydration were characterized by XRD and STA analysis. Quantitative XRD analysis (QXRD) of the phases was implemented according to the Rietveld method by using *Autoquan* software. The accuracy of QXRD analysis is outlined in Table 5.9.

The XRD curves and QXRD results of the pure OPC sample and the OPC sample with 5 wt% of the synthetic gyrolite additive after 28 days of hardening are presented in Figs. 5.25 and 5.26. It can be observed that synthetic gyrolite does not have any impact on the mineralogical composition of the hydration products as, in both samples, the peaks characteristic of the hydration products (portlandite, calcite, ettringite, and calcium silicate hydrates) were identified (Figs. 5.25a and 5.26a). Also, under these hardening conditions, the peaks characteristic of unreacted alite and belite were identified. It is of importance to emphasize that the diffraction peaks characterizing the presence of gyrolite were not observed in the XRD pattern

Table 5.9 Errors of quantitative XRD analysis

Phase	Typical accuracy, %	Phase	Typical accuracy, %
Amorphous	± 3.0	Calcite	± 0.7
C_2S	± 1.7	Ettringite	± 1.1
C_3A	± 0.7	Periclase	± 0.4
C_3S	± 1.0	Portlandite	± 0.7
C_4AF	± 1.1	Quartz	± 0.3

(Fig. 5.26a). Although the additive of synthetic gyrolite does not affect the mineralogical composition, it slightly changes the quantity of the formed hydration products (Figs. 5.25b and 5.26b). It was established that in the sample with the additive of synthetic gyrolite, smaller amounts of C_4AF, C_3A, and C_3S were present, and as a result, a higher amount of amorphous phases was formed. The obtained results are well-aligned with the data of compressive strength; the sample with the additive of synthetic gyrolite was characterized by a slightly higher value of strength (Fig. 5.24).

The data of the thermal analysis of the samples after 28 days of hardening is presented in Fig. 5.27 and Table 5.10. Three endothermic peaks are identified in all the DSC curves of the hydration products (Fig. 5.27). The first endothermal peak within the temperature interval from 50 to 150 °C can be attributed to the dehydration of calcium silicate hydrates and aluminate phases (ettringite, etc.). The second peak in the higher temperature range (400–500 °C) corresponds to portlandite dehydration, while the third peak at 650–800 °C temperatures characterizes calcium carbonate

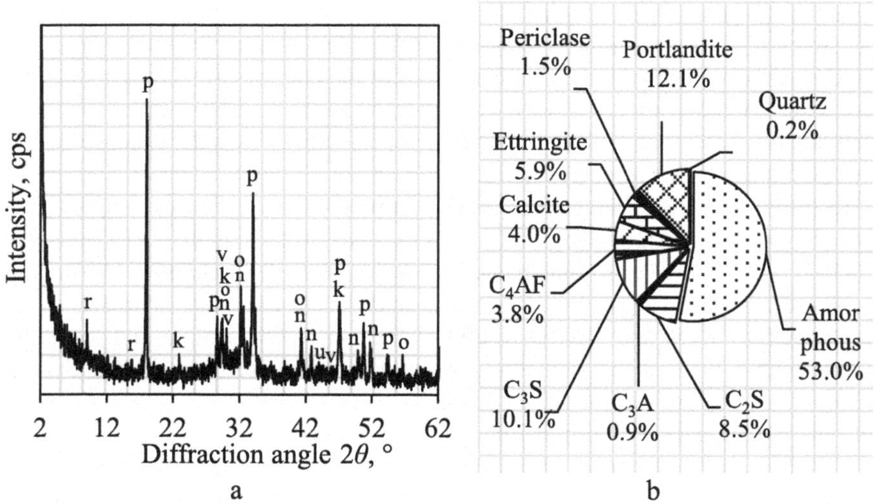

Fig. 5.25 XRD pattern (**a**) and Rietveld quantitative analysis (**b**) of pure OPC sample hardened for 28 days. *Indexes* n—alite; o—belite; v—calcium silicate hydrates; r—ettringite; k—calcite; p—portlandite; u—periclase

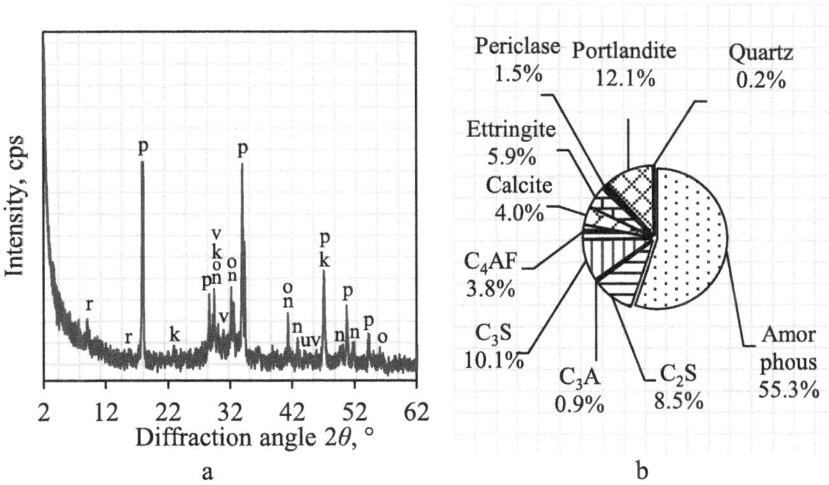

Fig. 5.26 XRD pattern (**a**) and Rietveld quantitative analysis (**b**) of OPC sample with 5.0 wt% synthetic gyrolite additive sample hardened for 28 days. *Indexes* n—alite; o—belite; v—calcium silicate hydrates; r—ettringite; k—calcite; p—portlandite; u—periclase

decomposition. The loss on ignition during the presently discussed effects is given in Table 5.10.

We established that the loss on ignition during the decomposition of calcium silicate hydrates and aluminates, and also the decomposition of portlandite, depends on the duration of hardening and the amount of the additive (Table 5.10). After seven days of hydration, the highest values of the loss on ignition were obtained in the OPC sample with 2.5 wt% of the synthetic gyrolite additive. Meanwhile, the pure

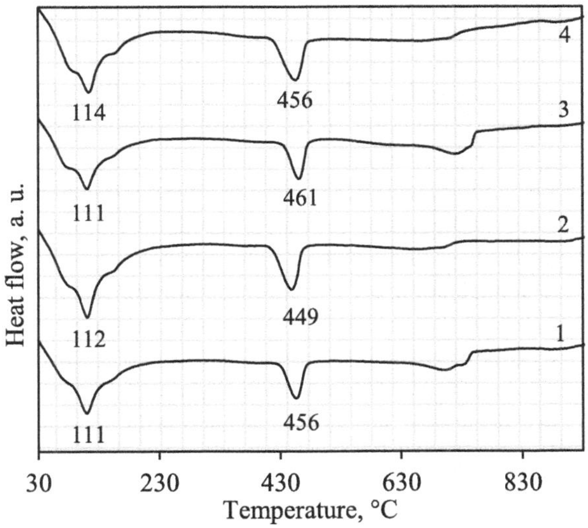

Fig. 5.27 DSC curves of OPC samples after 28 days of hydration. Amount of synthetic gyrolite additive, wt%: 1—0.0; 2—2.5; 3—5.0; 4—7.5

Table 5.10 Loss on ignition of OPC samples after 7 and 28 days of hardening

Duration of hardening, days	Amount of additive, wt%	Loss on ignition, %	
		50–150 °C	400–500 °C
7	0.0	7.4	2.6
	2.5	8.2	3.4
	5.0	7.6	2.7
	7.5	7.7	2.7
28	0.0	12.4	2.9
	2.5	12.6	3.7
	5.0	14.8	3.7
	7.5	13.2	3.2

OPC samples lost the lowest amount of their mass. These results showed that more hydration products are formed in the samples with the additive of synthetic gyrolite, and it aligns well with the compressive strength data (Fig. 5.24). Similar trends were determined in the samples after 28 days of hydration. The main difference is that the highest value of the loss on ignition was obtained in the sample with 5.0 wt% of the synthetic gyrolite additive (Table 5.10). Also, this sample had the highest value of the compressive strength.

By summarizing the results of Sects. 5.1–5.3, the following conclusions can be made:

- Synthetic gyrolite acts as chemisorbent during the hydration of OPC samples because synthetic gyrolite adsorbs aluminum, iron, and calcium ions. The adsorption experiments demonstrated that after 7 min of adsorption, 248 mg of Ca^{2+} ions is intercalated into the crystal lattice of synthetic gyrolite.
- Because of the adsorption reactions (consequently, concentration of Ca^{2+} ions in the liquid medium decreases), the additive of synthetic gyrolite increases the speed of the early stage of hydration of the OPC samples; thus, samples featuring this additive have the induction period eliminated. These results were confirmed by the calculations of the apparent kinetic parameters of hydration. As a result, it was estimated that the hydration reaction of the pure OPC sample proceeds by nucleation kinetics; meanwhile, samples containing the synthetic gyrolite additive have the hydration process controlled by diffusion.
- At the beginning of hydration, synthetic gyrolite acts as an adsorbent and serves as a nucleation site for the formation of hydration products (stage I). Afterward, synthetic gyrolite becomes unstable and gets recrystallized to calcium silicate hydrates with a molar ratio higher than 0.8. During the recrystallization, energy is consumed, and, as a result, the heat evolution rate of the second exothermic reaction decreases. The values obtained in thermodynamic calculations were approved by experimental data.
- The additive of synthetic gyrolite does not impact the mineralogical composition of the OPC sample but increases the compressive strength. It was obtained that

the samples with the additive of synthetic gyrolite are characterized by a higher compressive strength under all experimental conditions. After 84 days of hardening, the compressive strength of samples increased by raising the amount of the additive. The increase of the strength is connected with the higher amount of the hydration products.

5.4 Utilization of Synthetic Gyrolite with Incorporated Cd^{2+} Ions in Ordinary Portland Cement

As discussed in Chap. 4, the elimination of heavy metal ions from water is one of the most critical environmental concerns. Various treatment processes can be used for the elimination of toxic elements (chemical precipitation, ion exchange, foam floatation, filtration, etc.); however, most of these technologies suffer from essential disadvantages: complex operation, incomplete removal, high-energy requirements, secondary pollution, etc. [92–95]. One of the most acceptable ways to remove heavy metal ions from water is adsorption; however, when adsorbents get to their adsorption capacity limit, in most cases, they are stored in a landfill and essentially abandoned, which eventually leads to secondary pollution [3–5]. Thus, in the subsequent part of this research, synthetic gyrolite with intercalated heavy metal ions was utilized in OPC samples.

5.4.1 Synthetic Gyrolite Application for Cd^{2+} Ions Adsorption Process

Cadmium is a toxic chemical element, even in low concentrations, with a biological half-life in the body from 10 to 33 years [96–98]. The toxicity of cadmium depends on both the concentration and the duration of exposure. Short periods of exposure can cause nausea, coughing, muscle cramps, vomiting, liver injury, etc., while extended periods of exposure can cause kidney, bone, lung, and blood damage. This element enters the environment through both natural and anthropogenic sources. Thus, it is very important to control the concentration of cadmium in water even in low concentrations.

During the adsorption experiment, 1 g of synthetic gyrolite was immersed in 100 ml cadmium nitrate solution. The concentration of cadmium ions was equal to 1000 mg/dm^3, while the duration of the experiment at 25 °C temperature was equal to 1, 3, 5, 60, and 120 min. Ammonium hydroxide ($c = 10\%$) was employed to keep the liquid medium pH value equal to 8.70.

We determined that initially, i.e. in the early stages of the experiment, the adsorption of cadmium ions on synthetic gyrolite proceeds intensively since ~87% of cadmium ions was adsorbed (87.54 mg Cd^{2+}/g) (Fig. 5.28). After 5 min of adsorption, the amount of the adsorbed ions reached 91.18 mg Cd^{2+}/g, while, with the further

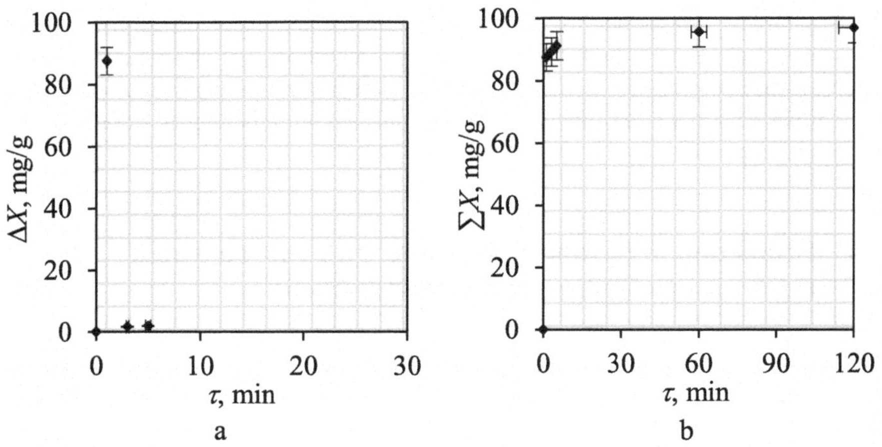

Fig. 5.28 Differential (**a**) and integral (**b**) kinetic curves of Cd^{2+} ions adsorption on synthetic gyrolite

prolongation of the experiment duration, the amount of the adsorbed ions went up insignificantly; i.e. the value reached 96.96 mg Cd^{2+}/g.

It was obtained that part of Ca^{2+} ions is released from the crystal lattice of synthetic gyrolite (Fig. 5.29). At the beginning of the experiment (1 min), 5.64 mg of Ca^{2+} ions from 1 g of synthetic gyrolite were released into a liquid medium (Fig. 5.29a). The latter value was doubled (13.01 mg Ca^{2+}/g) by prolonging the length of the experiment to 60 min (Fig. 5.29 b). Also, after this experiment period, the equilibrium was reached because after 120 min, the change in Ca^{2+} ions concentration in the liquid medium was negligible (Fig. 5.29). We highlight that the amount of the released calcium ions was more than 7 times lower than the amount of the adsorbed cadmium ions.

Since the pH value of the primary solution was kept at 8.70 by using ammonium water and the amount of released calcium ions (together, OH^- anions are released) was not significant, the pH value of the liquid medium did not change in the course of the adsorption experiment (Fig. 5.30).

We calculated that the exchange of Ca^{2+} ions to Cd^{2+} is non-stoichiometric because 1 g (3.75×10^{-4} mol) of synthetic gyrolite adsorbs 8.66×10^{-4} mol of Cd^{2+} ions and releases 3.28×10^{-4} mol of Ca^{2+} ions. Thus, ~38% of Cd^{2+} ions were involved in the substitution reaction, i.e. cadmium ions replaced calcium ions, while the remaining part of Cd^{2+} ions was adsorbed on synthetic gyrolite according to the addition reaction. Therefore, the general chemical formula of synthetic gyrolite including the intercalated Cd^{2+} ions can be represented in the following way: $Ca_{15.1}Cd_{2.3}Si_{24}O_{60}(OH)_8 \cdot 14H_2O$. It should be specified that the amount of OH^- anion in the formula can be slightly different because, together with calcium ions, part of OH^- anions is also released to the liquid medium.

For the sake of determining the reversibility of the adsorption process, we performed leaching experiments. The synthetic gyrolite obtained after adsorption

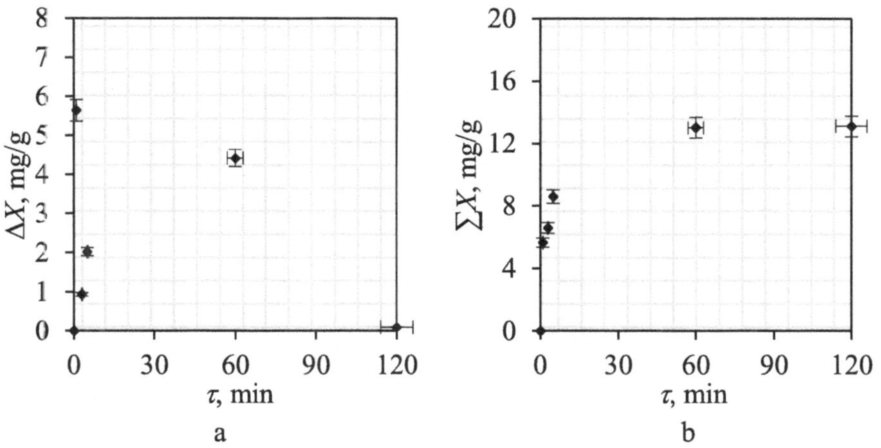

Fig. 5.29 Differential (**a**) and integral (**b**) kinetic curves of the concentration of released Ca^{2+} ions

Fig. 5.30 Variation of liquid medium pH value

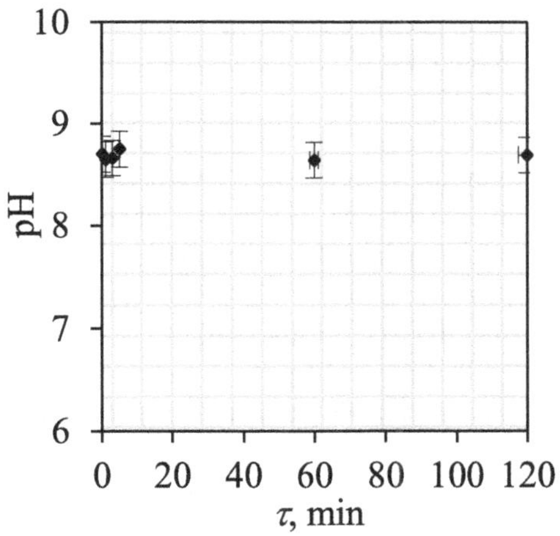

was dried and subsequently immersed in the solutions with different pH values (6.11, 8.70, and 11.57) for 60 min at 25 °C temperature. The changes in the concentration of cadmium ions in synthetic gyrolite are presented in Fig. 5.31. After 1 min of desorption, the concentration of cadmium ions in the liquid medium was equal to 0.006–0.03 mg/l. By increasing the duration of desorption, the concentration of cadmium ions in the liquid medium went down, and, after 60 min, it was lower than 0.001 mg/l in all the investigated cases. At the beginning of the experiments, a small amount of cadmium ions was probably released from the surface of the adsorbent and later bounded chemically.

To prove the nature of the adsorption process, the pseudo-first- and pseudo-second-order models were used. Calculations showed that the value of R^2 was relatively high in both cases (Fig. 5.32). Despite the high value of R^2, the pseudo-first-order kinetic

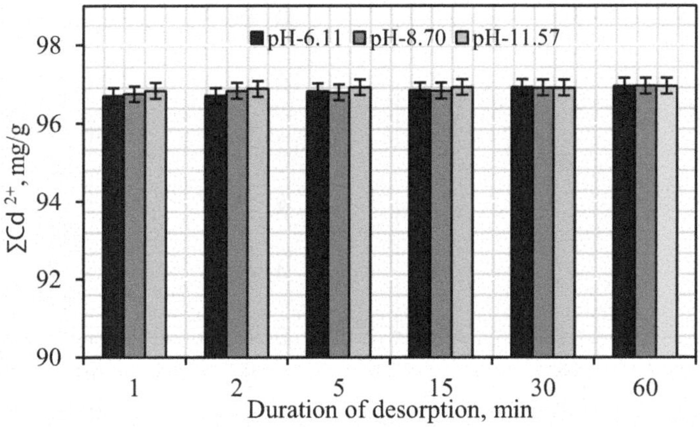

Fig. 5.31 Concentration of Cd^{2+} ions in synthetic gyrolite after desorption process at various pH levels

equation did not conform to the adsorption mechanism of cadmium ions as the calculated $X_{e(cal)}$ values were largely different from the experimental $X_{e(exp)}$ values (Fig. 5.33).

As expected, when using the pseudo-second-order kinetic equation, $X_{e(cal)}$ values were in line with the $X_{e(exp)}$ values (Fig. 5.33). Thus, it can be stated that this model is suitable for the description of the adsorption process. These results confirmed the experimental data that cadmium ions were chemisorbed on synthetic gyrolite. Calculations showed that the adsorption rate (k_2) of Cd^{2+} ions on synthetic gyrolite is equal to 0.025 g/(mg min).

Synthetic gyrolite was found to be stable during Cd^{2+} ions adsorption because all the diffraction peaks defining this compound could be identified in the XRD

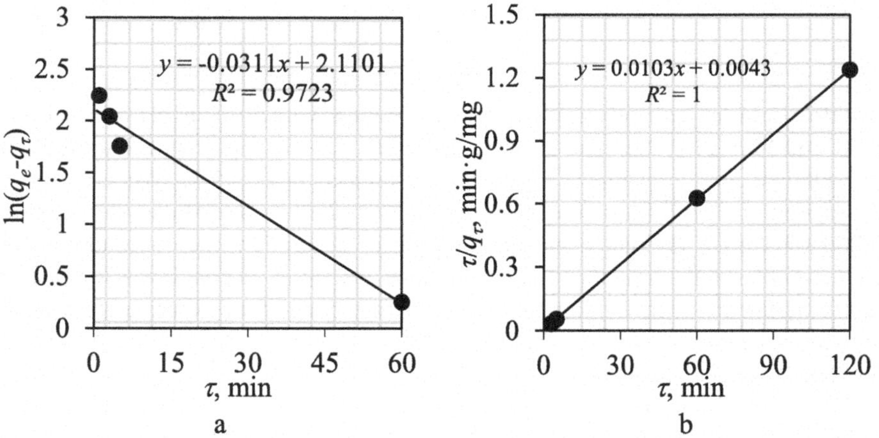

Fig. 5.32 Pseudo-first-order (**a**) and pseudo-second-order (**b**) kinetic plots

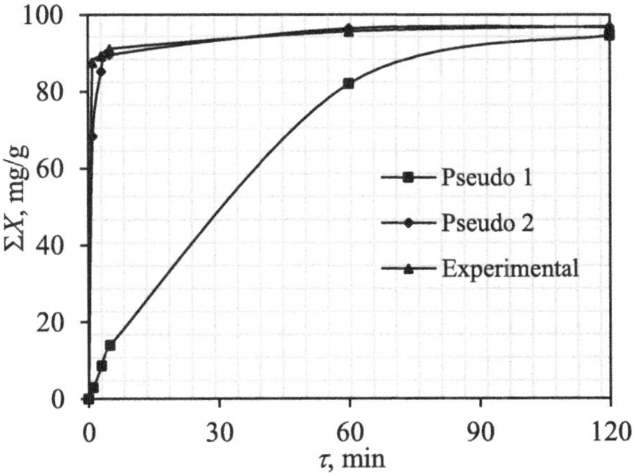

Fig. 5.33 Calculated and experimental differential kinetic curves of Cd^{2+} ions adsorption on synthetic gyrolite

pattern (Fig. 5.34a). It needs to be specified that, due to the large quantity of Cd^{2+} ions intercalated into the structure of the adsorbent, the intensity of some diffraction peaks decreased; however, in the XRD pattern, no other phases could be detected. The intercalation of Cd^{2+} ions was also proven by EDS analysis because the peak (3.1 keV) typical of cadmium (L_α) was observed in the EDS spectra of the adsorbent after adsorption (Fig. 5.34b).

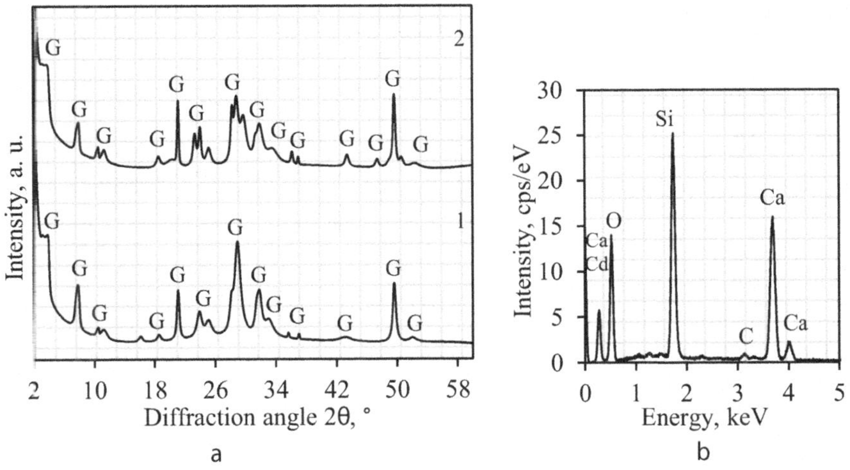

Fig. 5.34 XRD patterns (**a**) and EDS curve (**b**) and of synthetic gyrolite (curve 1) and synthetic gyrolite after adsorption (curve 2). *Indexes* G—gyrolite

5.4.2 Influence of Synthetic Gyrolite with Incorporated Cd^{2+} Ions on Ordinary Portland Cement Hydration

Section 5.3.1 showed that synthetic gyrolite strongly affects the hydration kinetics of ordinary Portland cement; thus, if we seek to evaluate the effect of synthetic gyrolite with intercalated Cd^{2+} ions on the early stage hydration of OPC, microcalorimetric analysis was performed. For the experiments, 0.0, 5.0, and 7.5 wt% of cement were replaced with synthetic gyrolite. With the objective to compare the rate of heat evolution of each sample and avoid the influence of a reduced amount of OPC in the samples, all the results were determined on the grounds of the unit weight of OPC in the sample. The rate of heat evolution and the cumulative heat of hydration data of the samples are visualized in Figs. 5.35, 5.36 and 5.37.

As expected, the additive of synthetic gyrolite with intercalated Cd^{2+} ions affects the hydration kinetics of OPC samples. Still, it does not influence the number of the hydration reaction stages because all the samples showed five stages of hydration in their microcalorimetric curves (Figs. 5.35, 5.36 and 5.37).

We established that the additive of synthetic gyrolite with Cd^{2+} ions promoted the hydration of OPC during the initial reaction because the samples with the additive

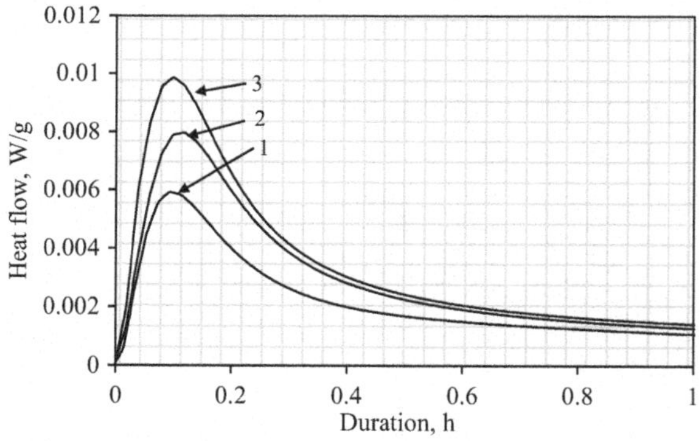

Fig. 5.35 Heat evolution rate of OPC samples during the initial reaction of hydration. Amount of additive, wt%: 1—0.0; 2—5.0; 3—7.5

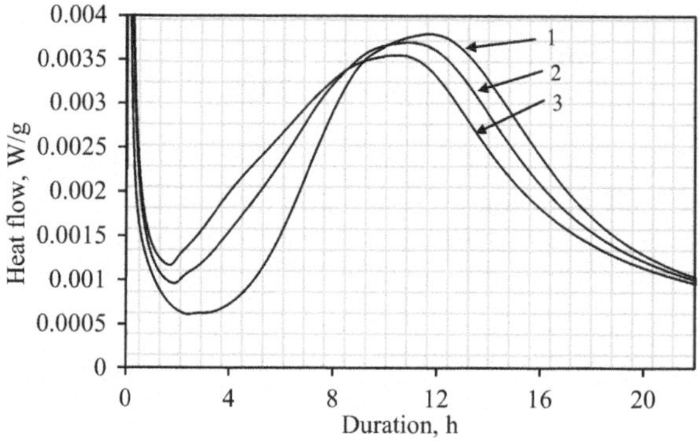

Fig. 5.36 Heat evolution rate of OPC samples during the induction and acceleration period of hydration. Amount of additive, wt%: 1—0.0; 2—5.0; 3—7.5

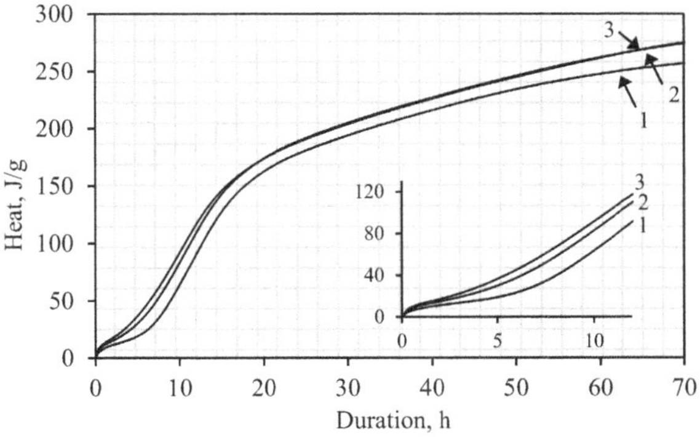

Fig. 5.37 Cumulative heat of OPC samples. Amount of synthetic gyrolite additive, wt%: 1—0.0; 2—5.0; 3—7.5

showed a significantly higher peak of the heat flow compared with the pure OPC sample (Fig. 5.35). It was measured that the intensity of the first exothermic peak was ~1.67 times higher in the sample with 7.5 wt% of the additive (0.00986 W/g) in comparison with the pure sample (0.00592 W/g). The increase of the released heat is associated with the higher solubility of the primary OPC compounds because of the good adsorption capacity of synthetic gyrolite regarding Al^{3+}, Fe^{3+}, Ca^{2+}, and other ions, which led to the decrement of the concentration of these ions in the liquid medium.

We determined that the induction period got shortened in the samples featuring the additive of synthetic gyrolite because, in the pure sample, it lasted for ~2.5 h, while in the sample with 7.5 wt% of the additive, it lasted only for as little as 20–30 min (Fig. 5.36). It must be emphasized that the same tendency was observed when the pure synthetic gyrolite additive (without Cd^{2+} ions) was used (Fig. 5.11). As expected, after the induction period, the acceleration and the deceleration periods started (the second endothermic effect) (Fig. 5.36). As in the previous case, the highest value of the heat flow (0.00379 W/g) was observed in the pure OPC sample, while the lowest value was registered in the sample with 7.5 wt% of the additive (Figs. 5.12 and 5.36). The decrease in the intensity of the second exothermic peak is related to the consumed energy (heat) for the destruction of the crystal lattice of synthetic gyrolite.

The data of cumulative heat confirmed the previously discussed results (Fig. 5.37). We established that the values of cumulative heat released prior to the end of the second exothermic reaction increased by increasing the amount of the additive (Fig. 5.37). Meanwhile, by extending the length from 20 to 72 h, the cumulative heat curves of the samples with the additive essentially coincided. It was measured that, at the end of the microcalorimetric experiment (72 h), the quantity of the released heat per 1 g of OPC was equal to 258 J/g in the pure OPC sample and ~277 J/g in the samples featuring additives.

In order to confirm the results of the microcalorimetric test and the theoretical assumptions, synthetic gyrolite and portlandite mixture (CaO/SiO_2 = 1.75) was performed and held in water (w/s = 10) for 7 and 9 h at 25 °C temperature. We found out that synthetic gyrolite with intercalated Cd^{2+} ions is stable till 7 h of the

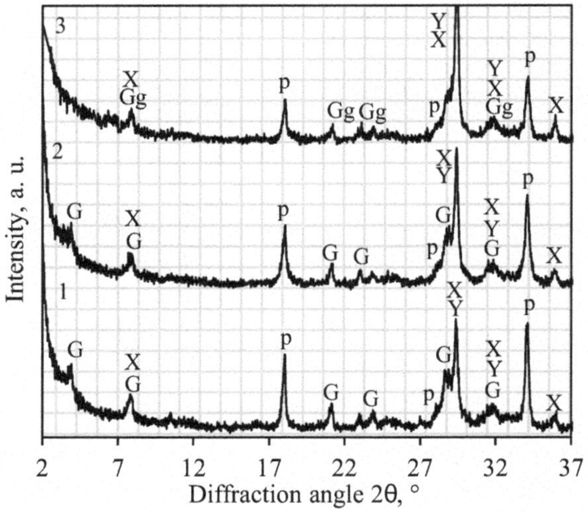

Fig. 5.38 XRD pattern of synthetic gyrolite substituted with Cd^{2+} ions and CH mixture at initial molar ratio $CaO/SiO_2 = 1.75$, w/s = 10.0, $t = 25$ °C, duration of hydration, h: 1—0 (after mixing); 2—7; 3—9. *Indexes* G—gyrolite; Gg – gyrolite gel; p—portlandite; X—calcium silicate hydrates ($CaO/SiO_2 = 0.8$–1.3); Y—calcium silicate hydrates ($CaO/SiO_2 = 1.3$–2.0)

experiment because all the diffraction peaks characteristic of synthetic gyrolite could be identified in the XRD pattern (Fig. 5.38). However, after the prolongation of the duration to 9 h, synthetic gyrolite became metastable and recrystallized to gyrolite gel and semicrystalline calcium silicate hydrates with a molar ratio equal to 0.8–2.0. During the recrystallization process, heat is consumed; thus, the intensity of the second exothermic reaction decreased (Fig. 5.36).

We established that the pure synthetic gyrolite additive strongly affected the kinetic parameters of the hydration of OPC samples (Sect. 5.3.3); thus, presumably, synthetic gyrolite with intercalated Cd^{2+} ions should also exert some impact on the kinetic parameters of the hydration reactions. The main parameter of the hydration process is the degree of hydration (α) which can be obtained by summing up the amount of the reacted C_3S. The microcalorimetric data was used for the calculations. All measurements and calculations were repeated five times. The main hydration reaction of C_3S can be written as follows:

$$C_3S + 3.9H \rightarrow C_{1.7}SH_{2.6} + 1.3CH. \tag{5.6}$$

The amount of the reacted C_3S (X_τ) can be calculated by using the results of microcalorimetry (the heat flow) and the following equation:

$$X_\tau = \frac{\Delta H_H}{\left(\Delta H_{C_3S} \cdot c_{C_3S}\right)/100\%} \cdot m, \tag{5.7}$$

where X_τ is the amount of the reacted C_3S, g; ΔH_τ is hydration enthalpy which was determined by employing the data of the heat flow, J/g_{OPC}; ΔH_{C_3S} is the enthalpy of C_3S, J/g_{C_3S}; c_{C_3S} is C_3S concentration in the primary sample, wt%; and m is the mass of the OPC sample, g.

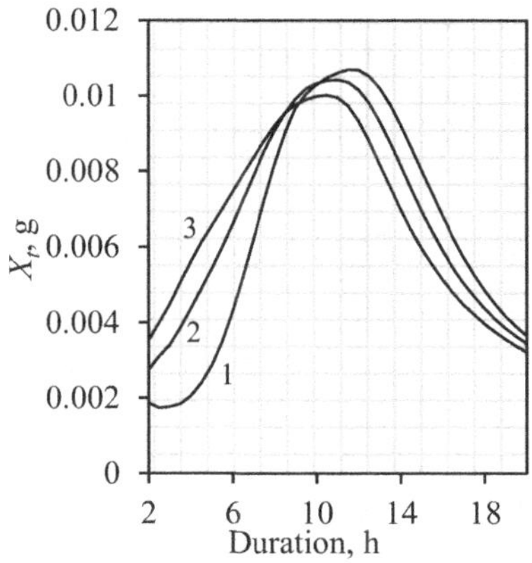

Fig. 5.39 Differential kinetic curves of C_3S reacting in 2–20 h of hydration. Amount of synthetic gyrolite additive, wt%: 1—0.0; 2—5.0; 3—7.5

The calculated amount of C_3S which reacted after 2–20 h (during the second exothermic peak) of the samples with various amounts of additive hydration is visually represented in Fig. 5.39. In addition, the obtained results of the reacted C_3S quantity according to Eq. 5.5 agree fairly well with the heat evolution rate data of OPC samples during the acceleration and deceleration periods of hydration (Fig. 5.36).

A linear apparent hydration rate of C_3S hydration can be obtained by plotting $\sum X_\tau$ as a function of τ. Linear dependencies with high values of R^2 (> 0.99) were obtained in the duration range from 2 to 12 h (Fig. 5.40 and Table 5.11). Therefore, the apparent rate (W_{app}) of Eq. 5.6 during the hydration of OPC samples can be calculated by using Eq. 5.1. Since the linear dependence in $\sum X_\tau$ versus τ coordination was observed, the apparent order of the reaction (n) starts approaching zero; thus, Eq. 5.1 may be rewritten in the following way:

$$W_{app.} = \frac{dx}{d\tau} = k_{app}, \ \text{g/h}. \tag{5.8}$$

The slope of the obtained linear plots determines the values $tg\alpha = k_{app.}$ (Fig. 5.40). The calculated values of the kinetic parameters along with the reaction rate constant within the duration range of 7–12 h are presented in Table 5.11, while the whole dependency of the reaction rate constant on the duration is given in Fig. 5.41.

We have established that the additive of synthetic gyrolite with intercalated Cd^{2+} ions strongly affects the reaction rate constant at the beginning of hydration (Fig. 5.41). It was calculated that 5.3% (0.034 g) of the primary amount of C_3S was hydrated in the pure sample after 7 h of hydration (Table 5.11). Meanwhile, after the same duration of hydration, the amount of the hydrated C_3S was two times higher in the sample with the synthetic gyrolite additive (10.5%, 0.066 g) than in the pure sample. The differences in the reacted amount of C_3S are related to a higher value of the reaction rate constant of the sample with the additive (Fig. 5.41). The main reason for the different reaction rate constant is that the particles of synthetic gyrolite serve as a nucleation site for the formation of the hydration products; i.e.

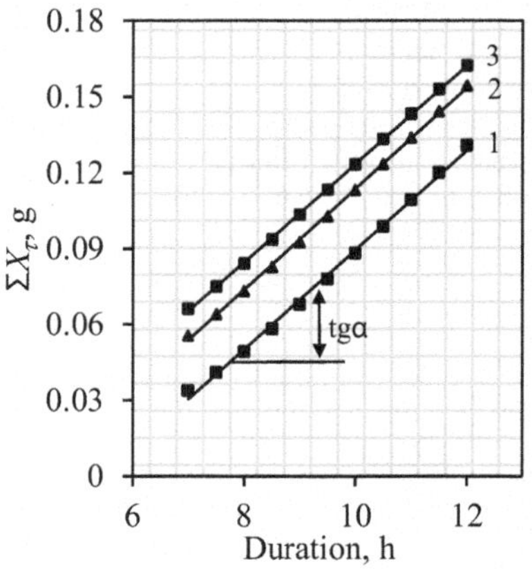

Fig. 5.40 Kinetic curves of reacted C_3S during OPC hydration. Amount of synthetic gyrolite additive, wt%: 1—0.0; 2—5.0; 3—7.5

Table 5.11 Kinetic parameters of reacted C_3S during OPC hydration within the duration range of 7–12 h

Amount of additive, wt%	X_0, g	X_τ, g	W_{app}, g/h	R^2
0.0	0.034	0.131	0.019	0.997
5.0	0.056	0.154	0.020	0.999
7.5	0.066	0.162	0.019	1.000

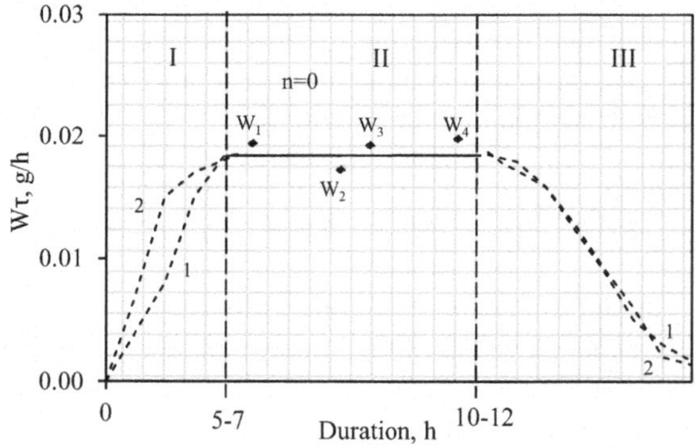

Fig. 5.41 Dependence of the reaction rate (W_τ) on hydration time (τ) of pure OPC sample (1) and sample with 7.5 wt% of additive (2). I, II, III—hypothetical macro stages of the process of hydration

synthetic gyrolite stimulates the formation of the hydration products on its surface (see Sect. 5.3). Further calculations showed that the additive of synthetic gyrolite with incorporated Cd^{2+} ions did not affect the hydration rate of C_3S during the later stage of hydration (Fig. 5.41). It was established that the values of W_{app} of all the investigated samples did not change within the duration range of 7–12 h and were equal to ~0.02 g/h (Table 5.11 and Fig. 5.41). Meanwhile, after 12 h of hydration, the hydration rate diminished in all the explored cases due to the high amount of the hydration products formed on the surfaces of non-hydrated particles. These results are in good agreement with the law of mass action which stipulates that the reaction rate depends upon the molar concentration of the reactants involved in any chemical reaction.

By summarizing the obtained data, it can be stated that synthetic gyrolite has good adsorption properties for Cd^{2+} ions because 1 g of synthetic gyrolite adsorbs near 100% of Cd^{2+} ions that are present in the liquid medium when the primary concentration of these ions is equal to 100 mg/l. Therefore, the adsorbent with the intercalated ions can be used as an additive for concrete production because it improves the hydration rate of OPC samples. However, it is necessary to prepare additional experiments such as long-term leaching, immobilization mechanism, mechanical properties, etc., which would lead to the full understanding of the effect of synthetic gyrolite with intercalated metal ions on the properties of concrete.

5.5 Technological Recommendations for the Production and Application of Synthetic Gyrolite

By summarizing the theoretical and experimental results, a sequence of technological operations for the production and application of synthetic gyrolite has been proposed (Figs. 5.42 and 5.44).

Synthetic gyrolite can be synthesized in the suspension (with water/solid ratio equaling 10) of materials containing the components calcium oxide and silicon dioxide. The duration and temperature of the synthesis depend on the chemical composition and activity of the raw materials; thus suggestions are given to use chemically active compounds, such as calcium oxide and amorphous silicon dioxide or silica gel waste. Quartz can also be used as a source of SiO_2; however, alkaline additives (NaOH, KOH, etc.) which would increase the solubility of quartz should be used. The molar ratio of calcium oxide and silicon dioxide in the primary mixture should correspond to the molar ratio of synthetic gyrolite, i.e. $CaO/SiO_2 = 0.66$.

The obtained experimental results showed that the length of hydrothermal synthesis in calcium oxide and amorphous silicon dioxide should be equal to no less than 32 h at 200 °C temperature. The formed synthetic gyrolite is stable to 336 h of hydrothermal treatment. To shorten the synthesis duration (16 h), silica gel waste as a SiO_2 source can be used; however, it should be noted that the formed product has traces of calcium fluoride and calcium aluminate.

In the principal technological scheme of synthetic gyrolite production, the raw materials are stored in silo *1*, and the necessary amount is weighed by the feeder (Fig. 5.42). Then, the powder compounds are transported to mixer *3* for homogenization. After homogenization, the required amount of water is supplied from water reservoir 2. The prepared mixture is then transported to autoclave *4* for hydrothermal treatment.

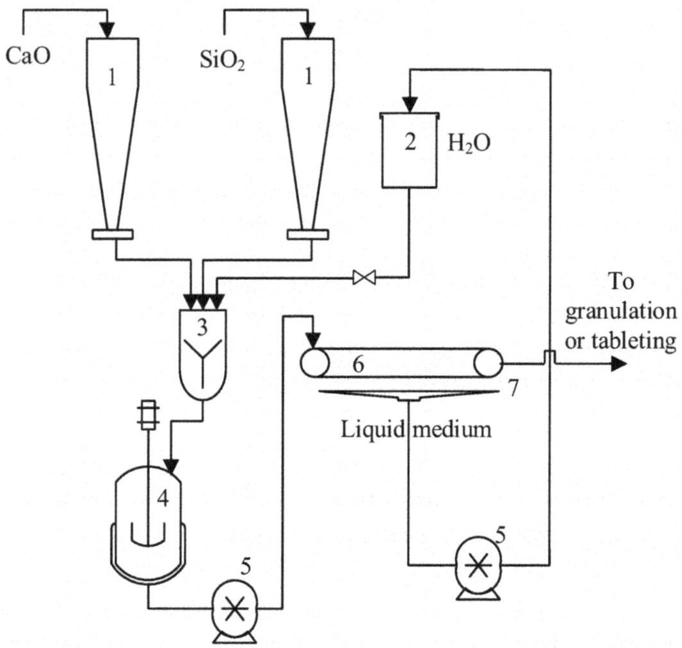

Fig. 5.42 Principal technological scheme of synthetic gyrolite production. 1—silo; 2—reservoir of water; 3—mixer; 4—autoclave; 5—pump; 6—vacuum belt filter; 7—collector of liquid medium

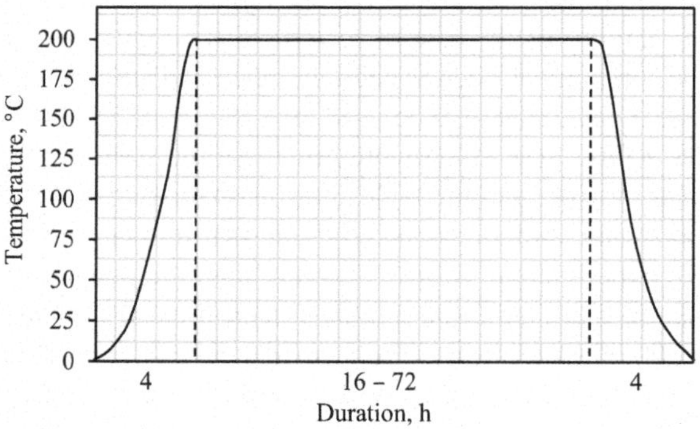

Fig. 5.43 Stages of hydrothermal treatment

The process of autoclave curing consists of the following steps: washing of the autoclave; pouring the suspension into the autoclave; closing the autoclave; increment of the temperature; isothermal treatment; transferring or releasing the steam to achieve ambient pressure; opening the autoclave; discharging the suspension. This sequence of all the stages forms one work cycle. The most important stages of the work cycle are (see Fig. 5.43):

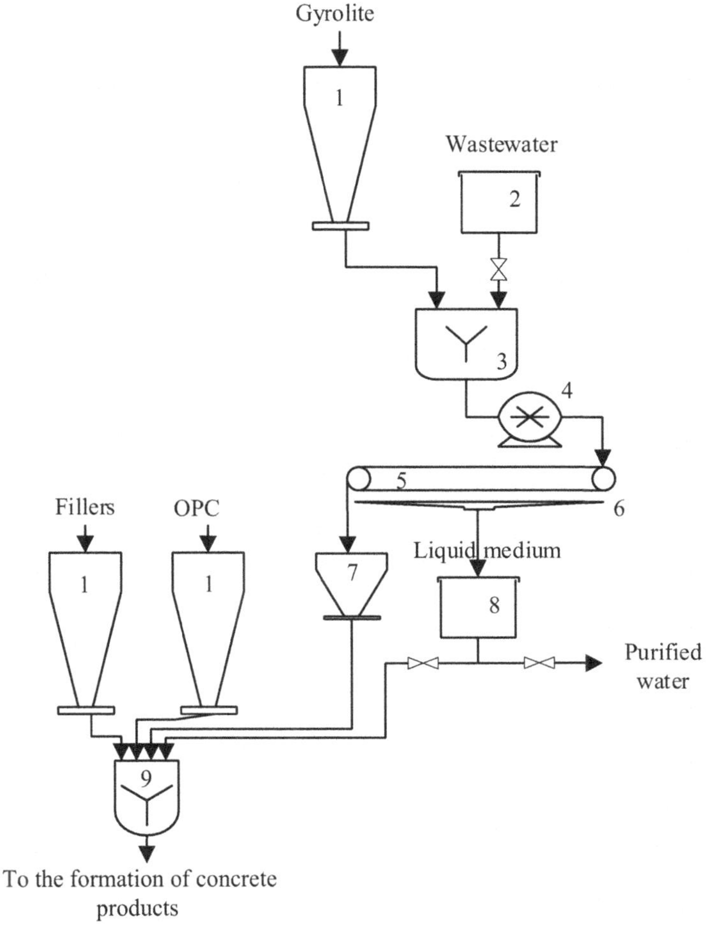

Fig. 5.44 Principal technological scheme of synthetic gyrolite application in adsorption and utilization in the production of concrete. 1—silo; 2—reservoir of wastewater; 3—bath adsorber; 4—pump; 5—vacuum belt filter; 6—collector of liquid medium; 7—bunker for synthetic gyrolite; 8—reservoir of purified water; 9—mixer for concrete

- The increment of temperature.
- Isothermal treatment.
- The decrement of temperature to the ambient level.

The synthesis product obtained after hydrothermal treatment contains a high amount of water (w/s = 10); thus, water should be separated by using the vacuum belt filter *6* or another apparatus (Fig. 5.42). When the amount of water has been reduced, synthetic gyrolite can be transported for granulation/tableting and/or drying. To avoid dust when transporting and physically using synthetic gyrolite, it is recommended to granulate or tablet it. Also, it is easier and more acceptable to use granules in the adsorption process than powder because granules can be used in a packed adsorption column, and, on top of that, separating the liquid medium from granules is easier. After the drying process, synthetic gyrolite can be transported to wastewater treatment plants.

The principal technological scheme of synthetic gyrolite application and utilization is presented in Fig. 5.42. Synthetic gyrolite is transported to the batch adsorber *3* or to a packed adsorption column, where it is mixed with wastewater (Fig. 5.44). Based on the experimental results, the optimal duration of adsorption should be 30–60 min, and 1 kg of the adsorbent can effectively adsorb 0.8–0.1 kg of heavy metal ions. The amount of wastewater that can be purified by synthetic gyrolite depends on the wastewater pollution by metal ions. After adsorption, the adsorbent is transported to separation from the liquid medium on the vacuum belt filter *5* (Fig. 5.44). The liquid medium is transported to the reservoir *8* while synthetic gyrolite is transferred to the bunker *7*. Part of the purified water can be used to produce concrete, while the rest of it can be released into the environment. Meanwhile, synthetic gyrolite with incorporated metal ions can be used as an additive in the production of concrete (Fig. 5.44). The experimental results showed that 5 wt% of ordinary Portland cement could be replaced by synthetic gyrolite. Such an amount of the additive positively affects the hydration kinetics and improves the mechanical properties of concrete.

5.6 Summary

The successful application of synthetic gyrolite in wastewater treatment and concrete production can benefit environmental sustainability and construction industry practices. It is evident that synthetic gyrolite can significantly contribute to these domains with its high adsorption capacity, positive impact on the hydration of OPC, and the strength properties of concrete.

The application of synthetic gyrolite as an adsorbent for wastewater or contaminated soils purification can lead to cleaner water resources and reduced pollution from heavy metal ions. In the next stage, the incorporation of synthetic gyrolite contaminated with heavy metal ions in concrete production not only improves the mechanical properties of concrete but also allows preventing secondary pollution with heavy metal ions during the landfilling of spent adsorbents.

With its dual functionality as an adsorbent for wastewater treatment and an additive for concrete, synthetic gyrolite can become one of the main components in sustainable development.

The significant impact of synthetic gyrolite continues to inspire further research and innovation in the fields of wastewater treatment and construction materials. However, some additional research must still be done in order to use synthetic gyrolite in practice. For example, applying synthetic gyrolite in real wastewater treatment lines would allow to evaluate the effectiveness and suitability of synthetic gyrolite for practical applications. Also, environmental impact and economic benefit assessments should be performed.

References

1. Eisinas A, Baltakys K, Siauciunas R (2012) The effect of gyrolite additive on the hydration properties of Portland cement. Cem Concr Res 42:27–38
2. Gkika DA, Mitropoulos AC, Kyzas GZ (2022) Why reuse spent adsorbents? The latest challenges and limitations. Sci Total Environ 822:153612
3. Baskar AV, Bolan N, Hoang SA, Sooriyakumar P, Kumar M, Singh L, Jasemizad T, Padhye LP, Singh G, Vinu A, Sarkar B, Kirkham MB, Rinklebe J, Wang S, Wang H, Balasubramanian R, Siddique KHM (2022) Recovery, regeneration and sustainable management of spent adsorbents from wastewater treatment streams: a review. Sci Total Environ 822:153555
4. Lata S, Singh PK, Samadder SR (2015) Regeneration of adsorbents and recovery of heavy metals: a review. Int J Environ Sci Technol 12:1461–1478
5. Crini G, Lichtfouse E, Wilson LD, Morin-Crini N (2019) Conventional and non-conventional adsorbents for wastewater treatment. Environ Chem Lett 17:195–213
6. Chaukura N, Gwenzi W, Tavengwa N, Manyuchi MM (2016) Biosorbents for the removal of synthetic organics and emerging pollutants: opportunities and challenges for developing countries. Environ Dev 19:84–89
7. Efaq AN, Rahman NN, Ab N, Nagao H, Al-Gheethi AA, Shahadat M, Kadir MA (2015) Supercritical carbon dioxide as non-thermal alternative technology for safe handling of clinical wastes. Environ Process 2:797–822
8. Kozyatnyk I, Yacout DMM, Van Caneghem J, Jansson S (2020) Comparative environmental assessment of end-of-life carbonaceous water treatment adsorbents. Bioresour Technol 302:122866
9. Jiang Q, He Y, Wu Y, Dian B, Zhang J, Li T, Jiang M (2022) Solidification/stabilization of soil heavy metals by alkaline industrial wastes: a critical review. Environ Pollut 312:120094
10. Navarro-Blasco I, Duran A, Sirera R, Fernández JM, Alvarez JI (2013) Solidification/stabilization of toxic metals in calcium aluminate cement matrices. J Hazard Mater 260:89–103
11. Latifi N, Vahedifard F, Siddiqua S, Horpibulsuk S (2018) Solidification–stabilization of heavy metal–contaminated clays using gypsum: multiscale assessment. Int J Geomech 18:04018150
12. Luo Z, Hao Y, Mu Y, Tang C, Liu X (2022) Solidification/stabilization of red mud with natural radionuclides in granular blast furnace slag based geopolymers. Constr Build Mater 316:125916
13. Soonthornwiphat N, Kobayashi Y, Toda K, Kuroda K, Islam CR, Otake T, Elakneswaran Y, Provis JL, Sato T (2020) Encapsulation of Sr-loaded titanate spent adsorbents in potassium aluminosilicate geopolymer. J Nucl Sci Technol 57:1181–1188
14. Hunce SY, Akgul D, Demir G, Mertoglu B (2012) Solidification/stabilization of landfill leachate concentrate using different aggregate materials. Waste Manage 32:1394–1400
15. Nthwane YB, Fouda-Mbanga B, Thwala M, Pillay K (2023) Synthesis and characterization of MC/TiO2 NPs nanocomposite for removal of Pb^{2+} and reuse of spent adsorbent for blood fingerprint detection. ACS Omega 8:26725–26738
16. Hooda S, Lanjewar R, Mondal P (2023) In-depth study of kinetics, thermodynamics, and reaction mechanism of catalytic pyrolysis of disposable face mask using spent adsorbent based catalysts. J Energy Inst 108:101247
17. Ali A, Khan S, Garg U, Luqman M, Bhagwath SS, Azim Y (2023) Chitosan-based hydrogel system for efficient removal of Cu[II] and sustainable utilization of spent adsorbent as a catalyst for environmental applications. Int J Biol Macromol 247:125805
18. Yang W, Kong Y, Yin H, Cao M (2023) Study on the adsorption performance of ZIF-8 on heavy metal ions in water and the recycling of waste ZIF-8 in cement. J Solid State Chem 326:124217
19. Nunes MABS, Fuentes DP, Mesquita JAFS, Romano RCO, Pileggi RG, Oliveira PV, Petri DFS (2023) Feasibility of sugarcane bagasse/polydopamine as sustainable adsorbents for Cr(VI) with reusability in cement composition. J Hazard Mater 12:100366
20. Paris JM, Roessler JG, Ferraro CC, DeFord HD, Townsend TG (2016) A review of waste products utilized as supplements to Portland cement in concrete. J Clean Prod 121:1–18
21. Rodrigues FA, Joekes I (2011) Cement industry: sustainability, challenges and perspectives. Environ Chem Lett 9:151–166

22. Ighalo JO, Adeniyi AG (2020) A perspective on environmental sustainability in the cement industry. Waste Dispos Sustain Energy 2:161–164
23. Schneider M, Romer M, Tschudin M, Bolio H (2011) Sustainable cement production—present and future. Cem Concr Res 41:642–650
24. Uwasu M, Hara K, Yabar H (2014) World cement production and environmental implications. Environ Dev 10:36–47
25. International Energy Agency (2018) Technology roadmap low-carbon transition in the cement industry. International Energy Agency, Paris
26. Guo Y, Luo L, Liu T, Hao L, Li Y, Liu P, Zhu T (2024) A review of low-carbon technologies and projects for the global cement industry. J Environ Sci 136:682–697
27. Farfan J, Fasihi M, Breyer C (2019) Trends in the global cement industry and opportunities for long-term sustainable CCU potential for Power-to-X. J Clean Prod 217:821–835
28. ICR Research (2022) Uncertain times. Available via: https://www.cemnet.com/Articles/story/171972/uncertain-times.html. Accessed 30 July 2023
29. Aprianti SE (2017) A huge number of artificial waste material can be supplementary cementitious material (SCM) for concrete production—a review part II. J Clean Prod 142:4178–4194
30. Duchesne J (2021) Alternative supplementary cementitious materials for sustainable concrete structures: a review on characterization and properties. Waste Biomass Valorization 12:1219–1236
31. von Greve-Dierfeld S, Lothenbach B, Vollpracht A, Wu B, Huet B, Andrade C, Medina C, Thiel C, Gruyaert E, Vanoutrive H, del Bosque S, Isabel F, Ignjatovic I, Elsen J, Provis JL, Scrivener K, Thienel K, Sideris K, Zajac M, Alderete N, Cizer Ö, Van den Heede P, Hooton RD, Kamali-Bernard S, Bernal SA, Zhao Z, Shi Z, De Belie N (2020) Understanding the carbonation of concrete with supplementary cementitious materials: a critical review by RILEM TC 281-CCC. Mater Struct 53:136
32. Juenger MCG, Snellings R, Bernal SA (2019) Supplementary cementitious materials: new sources, characterization, and performance insights. Cem Concr Res 122:257–273
33. Siddique R, Khan MI (2011) Supplementary cementing materials. Springer Science & Business Media
34. Shanmugasundaram N, Praveenkumar S (2021) Influence of supplementary cementitious materials, curing conditions and mixing ratios on fresh and mechanical properties of engineered cementitious composites—a review. Constr Build Mater 309:125038
35. Johari MM, Brooks JJ, Kabir S, Rivard P (2011) Influence of supplementary cementitious materials on engineering properties of high strength concrete. Constr Build Mater 25:2639–2648
36. Güneyisi E, Gesoğlu M (2008) A study on durability properties of high-performance concretes incorporating high replacement levels of slag. Mater Struct 41:479–493
37. Dils J, De Schutter G, Boel V (2012) Influence of mixing procedure and mixer type on fresh and hardened properties of concrete: a review. Mater Struct 45:1673–1683
38. AlShareedah O, Nassiri S (2021) Pervious concrete mixture optimization, physical, and mechanical properties and pavement design: a review. J Clean Prod 288:125095
39. Braga M, de Brito J, Veiga R (2014) Reduction of the cement content in mortars made with fine concrete aggregates. Mater Struct 47:171–182
40. Winnefeld F, Leemann A, German A, Lothenbach B (2022) CO_2 storage in cement and concrete by mineral carbonation. Curr Opin Green Sustain Chem 38:100672
41. Gartner E, Sui T (2018) Alternative cement clinkers. Cem Concr Res 114:27–39
42. Kumar R (2023) Recent progress in newer cementitious binders as an alternative to Portland cement: need for the 21st century. Recent Adv Struct Eng Constr Manage Select Proc ICSMC 2021 797–812
43. Bullerjahn F, Zajac M, Ben Haha M (2015) CSA raw mix design: effect on clinker formation and reactivity. Mater Struct 48:3895–3911
44. Sabbah A, Zhutovsky S (2023) The influence of gypsum content on the hydration and properties of Belite-Ye'elimite-Ferrite (BYF) clinker. In: International RILEM conference on synergising expertise towards sustainability and robustness of CBMs and concrete structures, pp 197–208

45. Schneider M (2019) The cement industry on the way to a low-carbon future. Cem Concr Res 124:105792
46. Meyer V, de Cristofaro N, Bryant J, Sahu S (2018) Solidia cement an example of carbon capture and utilization. Key Eng Mater 761:197–203
47. Sahu S, Meininger RC (2020) Sustainability and durability of solidia cement concrete. Concr Int 42:29–34
48. Siauciunas R, Hilbig H, Prichockiene E, Smigelskyte A, Takulinskas Z (2020) Accelerated carbonation of C_2SH based dense concrete. Ceram Int 46:29436–29442
49. Scrivener KL, John VM, Gartner EM (2018) Eco-efficient cements: potential economically viable solutions for a low-CO_2 cement-based materials industry. Cem Concr Res 114:2–26
50. Snellings R, Mertens G, Elsen J (2012) Supplementary cementitious materials. Rev Mineral Geochem 74:211–278
51. Habert G, Miller SA, John VM, Provis JL, Favier A, Horvath A, Scrivener KL (2020) Environmental impacts and decarbonization strategies in the cement and concrete industries. Nat Rev Earth Environ 1:559–573
52. Zhang J, Ye H, Gao X, Wu W (2022) Adsorption and desorption of polycarboxylate ether superplasticizer in fresh cementitious materials blended with mineral admixtures. J Mater Res Technol 17:1740–1751
53. Antiohos SK, Chouliara E, Tsimas S (2006) Re-use of spent catalyst from oil-cracking refineries as supplementary cementing material. China Particuology 4:73–76
54. John E, Lothenbach B (2023) Cement hydration mechanisms through time—a review. J Mater Sci 58:9805–9833
55. Zunino F, Dhandapani Y, Ben Haha M, Skibsted J, Joseph S, Krishnan S, Parashar A, Juenger MCG, Hanein T, Bernal SA, Scrivener KL, Avet F (2022) Hydration and mixture design of calcined clay blended cements: review by the RILEM TC 282-CCL. Mater Struct 55:234
56. Scrivener K, Ouzia A, Juilland P, Kunhi Mohamed A (2019) Advances in understanding cement hydration mechanisms. Cem Concr Res 124:105823
57. Bullard JW, Jennings HM, Livingston RA, Nonat A, Scherer GW, Schweitzer JS, Scrivener KL, Thomas JJ (2011) Mechanisms of cement hydration. Cem Concr Res 41:1208–1223
58. Taylor HF (1997) Cement chemistry. Thomas Telford, London
59. Ndahirwa D, Zmamou H, Lenormand H, Leblanc N (2022) The role of supplementary cementitious materials in hydration, durability and shrinkage of cement-based materials, their environmental and economic benefits: a review. Clean Mater 5:100123
60. Pacewska B, Wilińska I (2020) Usage of supplementary cementitious materials: advantages and limitations. J Therm Anal Calorim 142:371–393
61. Scrivener KL, Lothenbach B, De Belie N, Gruyaert E, Skibsted J, Snellings R, Vollpracht A (2015) TC 238-SCM: hydration and microstructure of concrete with SCMs. Mater Struct 48:835–862
62. Snellings R, Scrivener KL (2016) Rapid screening tests for supplementary cementitious materials: past and future. Mater Struct 49:3265–3279
63. Al-Shmaisani S, Kalina RD, Ferron RD, Juenger MCG (2022) Comparison of SCM reactivity to performance in cement-based mixtures. Mater Struct 55:241
64. De Belie N, Soutsos M, Gruyaert E (2018) Properties of fresh and hardened concrete containing supplementary cementitious materials. Springer
65. Skibsted J, Snellings R (2019) Reactivity of supplementary cementitious materials (SCMs) in cement blends. Cem Concr Res 124:105799
66. Mejdi M, Wilson W, Saillio M, Chaussadent T, Divet L, Tagnit-Hamou A (2019) Investigating the pozzolanic reaction of post-consumption glass powder and the role of portlandite in the formation of sodium-rich C-S-H. Cem Concr Res 123:105790
67. Snellings R, Almenares Reyes R, Hanein T, Irassar EF, Kanavaris F, Maier M, Marsh AT, Valentini L, Zunino F, Alujas Diaz A (2022) Paper of RILEM TC 282-CCL: mineralogical characterization methods for clay resources intended for use as supplementary cementitious material. Mater Struct 55:149

68. Taylor-Lange SC, Lamon EL, Riding KA, Juenger MCG (2015) Calcined kaolinite–bentonite clay blends as supplementary cementitious materials. Appl Clay Sci 108:84–93
69. Chappex T (2012) The role of aluminium from supplementary cementitious materials in controlling alkali-silica reaction. Dissertation, EPFL
70. Deschner F, Winnefeld F, Lothenbach B, Seufert S, Schwesig P, Dittrich S, Goetz-Neunhoeffer F, Neubauer J (2012) Hydration of Portland cement with high replacement by siliceous fly ash. Cem Concr Res 42:1389–1400
71. Palomo A, Fernández-Jiménez A, Kovalchuk G, Ordoñez LM, Naranjo MC (2007) Opc-fly ash cementitious systems: study of gel binders produced during alkaline hydration. J Mater Sci 42:2958–2966
72. Dittrich S, Neubauer J, Goetz-Neunhoeffer F (2014) The influence of fly ash on the hydration of OPC within the first 44h—a quantitative in situ XRD and heat flow calorimetry study. Cem Concr Res 56:129–138
73. Pelletier L, Winnefeld F, Lothenbach B (2010) The ternary system Portland cement–calcium sulphoaluminate clinker–anhydrite: hydration mechanism and mortar properties. Cem Concr Compos 32:497–507
74. Moffatt ETG, Thomas MD (2019) Effect of carbonation on the durability and mechanical performance of ettringite-based binders. ACI Mater J 116:95–102
75. Plank J, Zhimin D, Keller H, Hössle F, Seidl W (2010) Fundamental mechanisms for polycarboxylate intercalation into C_3A hydrate phases and the role of sulfate present in cement. Cem Concr Res 40:45–57
76. Baquerizo LG, Matschei T, Scrivener KL (2016) Impact of water activity on the stability of ettringite. Cem Concr Res 79:31–44
77. Juilland P, Gallucci E, Flatt R, Scrivener K (2010) Dissolution theory applied to the induction period in alite hydration. Cem Concr Res 40:831–844
78. Zhou W, Li L, Liu S, Vinh TND, Liu X (2017) Hydration properties and thermal analysis of cement-based materials containing limestone powder. J Cent South Univ 24:2932–2939
79. Villagrán-Zaccardi YA, Egüez-Alava H, De Buysser K, Gruyaert E, De Belie N (2017) Calibrated quantitative thermogravimetric analysis for the determination of portlandite and calcite content in hydrated cementitious systems. Mater Struct 50:179
80. Gaviria X, Borrachero MV, Payá J, Monzó JM, Tobón JI (2018) Mineralogical evolution of cement pastes at early ages based on thermogravimetric analysis (TG). J Therm Anal Calorim 132:39–46
81. Baltušnikas A, Lukošiūtė I, Levinskas R, Grybėnas A, Baltakys K, Eisinas A (2012) Analysis of rietveld method application for gyrolite crystal structure refinement. Medziagotyra 18:379–384
82. Merlino S (1988) Gyrolite: its crystal structure and crystal chemistry. Mineral Mag 52(366):377–387
83. Thomas JJ, Jennings HM, Chen JJ (2009) Influence of nucleation seeding on the hydration mechanisms of tricalcium silicate and cement. J Phys Chem C 113:4327–4334
84. Babushkin VI, Matveev GM, Mchedlov-Petrosyan OP (1962) Thermodynamics of silicates. Gosstoiizdat, Moscow
85. Levenspiel O (1998) Chemical reaction engineering. Wiley
86. Bishnoi S, Scrivener KL (2009) Studying nucleation and growth kinetics of alite hydration using μic. Cem Concr Res 39:849–860
87. Bullard JW (2008) A determination of hydration mechanisms for tricalcium silicate using a kinetic cellular automaton model. J Am Ceram Soc 91:2088–2097
88. Dove PM, Han N, De Yoreo JJ (2005) Mechanisms of classical crystal growth theory explain quartz and silicate dissolution behavior. Proc Natl Acad Sci 102:15357–15362
89. Thomas JJ, Biernacki JJ, Bullard JW, Bishnoi S, Dolado JS, Scherer GW, Luttge A (2011) Modeling and simulation of cement hydration kinetics and microstructure development. Cem Concr Res 41:1257–1278
90. Provis JL (2016) On the use of the Jander equation in cement hydration modelling. RILEM Tech Lett 1:62–66

91. Khawam A, Flanagan DR (2006) Solid-state kinetic models: basics and mathematical fundamentals. J Phys Chem B 110:17315–17328
92. Shrestha R, Ban S, Devkota S, Sharma S, Joshi R, Tiwari AP, Kim HY, Joshi MK (2021) Technological trends in heavy metals removal from industrial wastewater: a review. J Environ Chem Eng 9:105688
93. Zhao M, Xu Y, Zhang C, Rong H, Zeng G (2016) New trends in removing heavy metals from wastewater. Appl Microbiol Biotechnol 100:6509–6518
94. Liu DH, Lipták BG (2020) Wastewater treatment. CRC Press
95. Droste RL, Gehr RL (2018) Theory and practice of water and wastewater treatment. Wiley
96. Wang M, Chen Z, Song W, Hong D, Huang L, Li Y (2021) A review on cadmium exposure in the population and intervention strategies against cadmium toxicity. Bull Environ Contam Toxicol 106:65–74
97. Rahimzadeh MR, Rahimzadeh MR, Kazemi S, Moghadamnia A (2017) Cadmium toxicity and treatment: an update. Caspian J Intern Med 8:135–145
98. Genchi G, Sinicropi MS, Lauria G, Carocci A, Catalano A (2020) The effects of cadmium toxicity. Int J Environ Res Public Health 17:3782

Author Index

A
Anderson, T., 41–43
Armbruster, T., 12

B
Babushkin, V.I., 48, 226
Baltakys, K., 1, 41, 50, 79, 82, 83, 139, 211
Baltušnikas, A., 80
Bernal, J.D., 2
Blakeman, E.A., 80
Blanc, P., 48, 49
Burri, G., 12

C
Cann, J.R., 45
Chan, C.F., 49
Christie, P., 43
Clarke, F.W., 42, 43
Currie, J., 42

D
Dambrauskas, T., 1, 10, 13, 41, 79, 95, 139, 211

E
Eberhard, E., 45
El-Korashy, S.A., 148
Esteoule, J., 49

F
Fleischmann, F., 43
Flint, E.P., 43, 48, 54
Franke, B., 48

G
Galuskina, I.O., 13
Galuskin, E.V., 13, 14, 16
Gard, J.A., 50, 56
Glasser, F.P., 3
Grice, J.D., 13
Groves, G.W., 4
Grudemo, Ä., 3
Gundlach, H., 50

H
Hamid, 45
Hara, N., 49
Harker, R.I., 48
Heddle, M.F., 5, 42
How, H., 42, 43
Howison, J.W., 3

I
Inoue, N., 49

J
Jadhav, S.V., 92
Jander, W., 48, 235, 236
Jauberthie, R., 49
Jaunarajs, K.L., 45
Juilland, P., 224

K. Baltakys and T. Dambrauskas, *Synthetic Gyrolite*, https://doi.org/10.1007/978-3-031-49259-4

K
Kalousek, G.L., 48
Kaminskas, R., 95
Komarneni, S., 145
Koritnig, S., 44
Krivenko, P., 94
Krysztafkiewicz, A., 95
Kurczyk, H.G., 3

L
Lachowski, E.E., 56
Laquerbe, M., 49
Lazic, B., 13
Lottisnatham, S.J., 13

M
Mackay, A.L., 43, 45, 48, 54
Mandeikyte, N., 93
Massazza, 217
Merlino, S., 45, 46
Meyer, V., 45
Micheelsen, H., 45
Mitsuda, T., 10, 48, 49
Miyake, M., 80
Moertel, H., 49

N
Nelson, E.B., 48
Nocun-Wczelik, W., 81, 83

O
Orr, C., 25, 26, 29, 67, 129, 166

P
Perez, D., 13
Petrovic, J., 48

R
Richardson, I.G., 1, 4
Roy, D.M., 1, 4, 10, 48
Różycka, A., 81

S
Sarkar, R., 95
Sarp, H., 12
Schwiete, H.E., 3
Seager, A.F., 43
Shaw, S., 50, 81
Shpynova, L.G., 2
Shrivastava, O.P., 145, 146
Shrivastava, R., 146
Siauciunas, R., 50, 81–83
Smith, A., 13
Snellings, R., 217
Speakman, K., 10
Stade, H., 3
Stevula, L., 48
Strunz, H., 45
Stumm, A., 81, 155

T
Taylor, H.F.W., 1, 3, 4, 10, 43, 45, 48
Temimi, M., 49
Thomas, M.A., 226
Toraya, H., 19

V
Vaičiukynienė, D., 94

W
Wieker, W., 3

Z
Zhao, X., 146

Subject Index

A

Acceleration, 219, 220, 223, 226, 229, 248, 249, 251
Acicic solution, 150, 155–159, 175, 176, 180–184, 190–195
Activated alumina, 142, 143
Activated carbon, 142–145
Addition reaction, 162, 173, 186, 194, 195, 244
Adsorbents, 2, 10, 29, 57, 59, 60, 63, 86, 93, 99, 139–145, 147, 149, 154–156, 159, 161, 162, 166, 169, 171, 173, 177, 178, 180, 185, 190, 199, 203, 204, 211–213, 223, 242, 243, 247, 253, 256
Adsorption, 10, 18, 22–25, 27, 29, 30, 47, 57–64, 67, 68, 91–93, 96, 99, 125–129, 131, 139, 141–143, 145–160, 162, 164–166, 168, 169, 171, 173, 175–191, 193–196, 199–204, 213, 223–226, 237, 242–244, 246, 247, 253, 255, 256
Adsorption capacity, 151, 152, 171, 184, 187, 190, 201, 204, 249
Adsorption experiments, 139, 150, 154, 155, 159, 166, 185, 190
Adsorption isotherm, 23, 27, 30, 57, 60, 164
Adsorption properties, 145, 190, 237
Adsorption rate, 158, 162, 180, 182–184, 187, 204
Adsorption selectivity, 141, 154, 182
Afwillite, 2, 7, 11, 15
Alkaline solution, 155–157, 159, 184, 190, 192–196, 201, 203, 204
α-C_2SH, 7, 11, 12, 15, 22, 49, 82, 146
Aluminum fluoride, 87, 88, 90, 91, 94
Aluminum ions, 43, 46, 79, 99, 104, 115, 145, 146, 221, 237
Anions, 2, 79, 93, 99, 121, 125–127, 129–131, 143, 145, 153, 173, 185, 193, 244
Apophyllite, 41–44
Autoclave curing, 254
Average diameter of pores, 166
Avrami equation, 235, 236

B

Basaltic rock, 10, 42
Biodegradation, 198, 199
Brunauer-Emmett-Teller, 24, 61, 128, 166

C

Cadmium, 140, 190, 195, 200, 204, 211, 243–247
Calcination, 13, 18, 19, 22, 27, 91, 95, 97, 115–117, 119, 216
Calcium carbonate, 1, 17, 54, 100, 121, 177, 203, 222, 240
Calcium fluoride, 88, 94, 95, 99, 101, 104, 115–118, 253
Calcium nitrate, 79, 120, 121, 124–126, 128, 131
Calcium silicate hydrates, 1–6, 12, 13, 16, 30, 31, 44, 48, 49, 51, 73, 79, 80, 82, 93, 99–101, 103, 109, 113, 115, 119, 120, 123–125, 131, 144–148, 178, 215, 217, 218, 220–222, 228, 239–242, 250
Carbonization, 15, 17, 121, 122, 177, 189, 203, 204

K. Baltakys and T. Dambrauskas, *Synthetic Gyrolite*,
https://doi.org/10.1007/978-3-031-49259-4

Cations, 2, 4, 46, 80, 92, 93, 99, 143, 145, 147, 148, 153, 189, 193
Celitement, 11
Cement chemistry, 2, 12, 13
Cement hydrates, 220
Cement stone, 94, 220, 222
Chegemite, 1, 8, 12–14, 31
Chemical precipitation, 92, 140, 196, 243
Chemisorbent, 187, 242
Chemisorption, 25, 29, 156, 158, 169, 211, 224
Chromium ions, 83
Circular economy, 85–87
Clay minerals, 143, 145
CO_2 emissions, 213–217
Cobalt ions, 180
Complexing agents, 196, 197
Compressive strength, 5, 93–95, 222, 238–240, 242, 243
Constant C_{BET}, 25, 29, 63, 128, 166
Crystal lattice, 2, 46–48, 56, 65, 67, 68, 71, 80, 82, 99, 123, 143, 149, 154, 161, 164, 173, 178, 184, 185, 191, 224–226, 237, 242, 249
Crystalline, 2, 48, 50, 51, 60, 81, 83, 94, 97, 99, 101, 109, 115, 117, 147, 220, 224
Crystallinity, 5, 51, 56, 63, 67, 82, 95, 100, 103, 104, 121–123, 146, 149, 159, 162, 164, 167, 169, 204
Crystallites, 23, 24, 27, 51, 144
Crystallite size, 16, 41, 51, 53, 54, 56, 65, 73, 159, 162, 164, 169
Crystallization, 1, 5, 11, 12, 14, 18, 43, 48–50, 73, 79, 81, 91, 95, 102, 107, 108, 119, 123, 131, 217, 219, 221
Crystallographic parameters, 42
Crystal stability, 51
Crystal structure, 4, 9, 13, 14, 41, 43, 45, 47, 54, 65, 73, 80, 102, 145, 147, 189, 194
C-S-H gel, 3, 4, 44, 48, 50, 81, 220
C-S-H(I), 2, 3, 5, 6, 79, 80, 82, 83, 99–102, 104, 105, 109, 111, 116, 117, 119–123, 155
Cumulative heat, 22, 220, 221, 223, 228, 229, 248, 249
Cuspidine, 7, 11, 12, 115, 116, 118, 119
Cylindrical-shape pores, 68, 70, 71, 129, 167–169, 174

D

Deceleration, 219, 220, 223, 249, 251
Degree of hydration, 236, 250
Dehydration, 96, 100, 102, 158, 240
Desorption, 22, 23, 28, 57, 59–62, 64, 67–69, 128, 129, 154, 161, 165–167, 177, 224, 245, 246
Desorption isotherms, 23, 61, 166
Differential distributions of the volume pore sizes, 26, 30, 71
Dimeric with polymeric chains model, 3
Dreierketten arrangement, 3
Dry fluorite method, 88

E

Early stage of hydration, 223, 242
Economically attractive, 87, 90, 140
Electrocoagulation, 92, 93
Electrodialysis, 92, 93
Endothermic effect, 18, 54, 55, 96, 100, 102, 122, 123, 154, 158, 177, 189, 225, 226, 249
Environmentally friendly, 2, 11, 22, 86, 211, 212, 215
Environmentally friendly cements, 214, 215
Ethylenediaminetetraacetic Acid (EDTA), 196–201, 204
Exothermic effect, 19, 54, 55, 100, 103, 104, 123, 158, 177, 189, 200, 202, 226
Exothermic reaction, 217, 218, 225, 226, 229, 242, 249, 250
Extraction, 84, 196–198

F

Fluoride, 42, 88, 91–95, 97, 99, 115, 143
Fluorine ions, 79, 88, 89, 91, 93–96, 119
Formation mechanism, 44, 107

G

Galuskinite, 8, 13, 31
γ-Ca_2SiO_4, 7, 10
γ-Ca_2SiO_4 group, 10
Gibbs energy, 107
Gypsum additive, 82
Gyrolite, 3, 4, 6, 7, 10, 31, 41–54, 56, 57, 60–68, 70–73, 79–83, 99–105, 107–109, 111–120, 123–131, 139, 147–173, 176–196, 199–205, 211, 223–256
Gyrolite group, 10, 51, 113, 119

H
Hardened cement, 221–223
Heat evolution rate, 223, 224, 226, 227, 242, 248, 251
Heavy metal ions, 139, 141, 143, 145, 146, 149, 150, 177, 178, 184, 186, 187, 189, 190, 194–196, 203, 211, 212, 223, 224, 243, 256
Heavy metals, 139–141, 143, 198, 204
Hexafluorosilicic acid, 88–91
High-temperature in-situ XRD, 17, 19, 22
Hillebrandite, 2, 3, 5, 6, 14, 15
Hydration, 2, 11–13, 22, 94, 95, 99, 204, 213, 217–239, 241, 242, 248, 250–253, 256
Hydration kinetic, 211, 238, 248, 250, 256
Hydration models, 233
Hydration products, 11, 12, 99, 217, 218, 220, 223, 226, 229, 232, 237, 239, 240, 242, 243, 251, 252
Hydraulic activity, 21, 22
Hydrothermal conditions, 2, 5, 10–12, 44, 48–50, 80, 81, 95, 101, 105
Hydrothermal synthesis, 1, 4, 5, 11, 13, 15, 19, 41, 44, 48, 51–57, 60, 61, 63, 65, 67, 69, 70, 73, 79–83, 99, 101, 103–106, 108, 109, 114–116, 119–127, 130, 145, 159–164, 184, 225, 253
Hydroxyledgrewite, 1, 5, 8, 12–31
Hysteresis loop
  type H1, 59
  type H2, 59
  type H3, 23, 30, 59, 60, 70
  type H4, 60

I
Immobilization, 96, 213, 253
Induction, 94, 219–221, 223, 225, 242, 248, 249
Initial, 1, 12, 44, 49, 51, 57, 79, 82, 83, 94, 96, 101, 107, 110, 120, 121, 123, 126, 146, 151–153, 155, 161, 171, 178, 181, 184, 185, 201–203, 219–221, 223–225, 228, 232, 233, 237, 248, 250
Intercalation, 130, 131, 147, 152, 155, 158, 185, 186, 190, 202, 203, 205, 224, 247
Interlayer, 3, 9, 45, 46, 50, 56, 63, 67, 119, 123, 125, 143, 162, 169
Interlayer sheets, 139, 147
Interlayer water, 41, 54, 56, 63, 102
Interplanar distance, 5, 52, 149
Ion exchange, 80, 81, 92, 93, 140, 143, 146–149, 196, 243
Iron, 171, 173, 178, 184, 185, 188, 237, 242
Irreversibility, 193, 202
Irreversible, 154, 156, 158, 160, 169, 173, 179, 186, 189, 224
Isotherm
  type I, 58
  type II, 27, 58, 60
  type III, 59
  type IV, 59, 63
  type V, 59
  type VI, 59
Isothermal calorimetry, 223

J
Jander's model, 235
Jennite, 3–6, 9, 10, 14, 30, 222
Jennite group, 5

K
Kelvin equation, 25, 26, 29, 67, 68, 129, 166
Kilchoanite, 2, 7, 10, 215
Kinetic equation, 150–152, 173, 233, 246
Kinetic models, 151–153, 156, 162, 176, 179–181, 183, 187, 189, 194
Kinetic parameters, 149–151, 153, 156, 157, 162, 173, 176, 180, 181, 183, 187, 188, 202, 233–236, 242, 250–252
Kinetics of cement hydration, 221, 232

L
Landfilling, 85, 86, 211, 256
Leaching, 95, 156, 158, 169, 173, 179, 186, 193, 198, 202, 204, 212, 224, 244, 253
Loss on ignition, 18, 54, 241, 242

M
Macropores, 23, 24, 27, 30
Macroporous, 58–60, 164
Manganese ions, 181, 182
Material exchange process, 43, 44
Mesopores, 23, 24, 26, 27, 29, 30, 60, 128
Mesoporous, 59–61, 66, 67, 70, 128, 164
Metajennite, 5, 10, 197

Metal chelates, 197
Metastable, 10, 19, 48, 50, 51, 56, 73, 100, 101, 113, 115, 123, 250
Microcalorimetric analysis, 219, 229, 248
Microcalorimetric curve, 220, 248
Micropores, 25–27, 29, 58, 63, 66, 67, 166, 168
Micropore volume, 25, 26, 58, 66, 67, 166
Morphology, 47, 51, 57, 86, 115–117, 221, 229

N
Nanofiltration, 92
Natural gyrolite, 41–44, 46, 79
Nesosilicates, 14
Nickel ions, 178, 179
Nitrogen adsorption-desorption, 22, 60, 127, 164, 225

O
Octahedral layer, 45–47
Optical parameters, 42
Optical properties, 14, 43, 45
Ordinary Portland cement, 99, 141, 204, 211, 212, 217, 223, 248, 256

P
Particle size, 47, 96, 97, 106
Period, 94, 201, 219–221, 223, 225, 229, 235, 242–244, 248, 249, 251
Phyllosilicates, 45, 46
Physical properties, 44, 96, 212
Point B, 25, 58–60, 63, 128, 164
Pore diameter, 23, 24, 59, 60, 71, 131
Pore size distribution, 57, 58, 67, 70, 169–172
Portlandite, 2, 3, 15, 16, 18, 20, 82, 219–222, 224–231, 239–241, 249, 250
Pozzolanic, 94, 95, 221
Pozzolanic reaction, 222
Principal technological scheme, 253–256
Pseudo-first-order adsorption kinetics model, 150
Pseudo-second-order adsorption kinetic model, 150, 157, 158, 180
Purification, 2, 91, 93, 139, 140, 143, 197–199, 256

Q
Quartzite, 44

R
Reaction rate, 47, 233–236, 251–253
Recrystallization, 9, 19, 48, 49, 52, 54, 56, 100, 102, 104, 109, 111, 113, 114, 119, 123, 125, 155, 200, 220, 222, 242, 250
Released heat, 21, 22, 218, 219, 225, 226, 228, 249
Removal, 10, 18, 92, 93, 95, 123, 140, 145, 177, 178, 189, 197, 243
Reverse osmosis, 92
Rietveld method, 239

S
Scherrer equation, 53
Secondary pollution, 93, 140, 141, 204, 211, 212, 243, 256
Selectivity, 139, 141, 145–147, 149, 182, 185
Selectivity sequence, 139, 146, 185
Semicrystalline calcium silicate hydrate, 49, 94, 99, 120, 200, 228, 250
Shape of pores, 23, 57, 69, 164
Silica gel, 13, 79, 88, 91, 93–96, 99, 101, 115, 142, 143, 253
Silicate chain, 3, 4, 12, 50
Silicate layer, 4, 45–47, 50
Siliceous limestone, 44
Silicic acid salt, 1
Simultaneous adsorption, 139, 149, 185, 187, 189, 190
Slit-shape pores, 26, 27, 29, 30, 59–61, 70, 71, 128–131, 164, 166, 168, 169, 174
Slow continued reaction, 219, 220, 223
Sodium ions, 46, 80, 81, 146
Solidification, 212, 213
Solid waste, 79, 83, 85–87, 93, 139, 211
Solubility, 10, 49, 50, 80, 91, 222, 225, 249, 253
Specific heat capacity, 20, 21
Specific surface area, 23, 24, 29, 41, 47, 61, 64, 65, 71, 128, 129, 131, 141, 164, 166, 168
Spent adsorbent, 211–213, 256
Spurrite, 8, 13
Stabilization, 99, 212
Structural model, 2
Substitution reaction, 154, 157, 159, 161, 162, 169, 173, 179, 182, 186, 194, 195, 244

Supplementary cementitious materials, 214, 217, 218, 228
Synthetic β-wollastonite, 147
Synthetic gyrolite, 41, 48, 53, 56, 60–68, 71, 72, 79–83, 104, 115, 117, 119, 123, 125, 127, 129–131, 139, 147–159, 161, 162, 164, 166, 167, 169–173, 175–196, 199–205, 211, 223–244, 246, 248–253, 255, 256
Synthetic tobermorite, 145, 146, 149
Synthetic xonotlite, 146–149

T

Thermal stability, 17, 20, 43, 115
Thermodynamically stable, 106, 111, 113, 221
Thermodynamic calculations, 48, 52, 73, 79, 107, 109, 113, 226, 242
Tilleyite, 7, 9, 11, 13
Tobermorite, 2–6, 9, 30, 44, 48–51, 79, 80, 82, 83, 108, 111, 112, 114, 145–147, 222
Tobermorite gel, 49
Tobermorite group, 5, 145
Total pore volume, 26, 41, 67, 70, 71, 129–131, 166, 169, 174
Total released heat, 220
T-plot model, 25, 29, 66
Trabzonite, 1, 5, 8, 12
Transition metal ions, 147, 196, 204
Treatment processes, 140, 243
Truscottite, 2, 6, 10, 46, 48, 52–54, 56, 73, 79, 108, 114, 116, 117, 119, 125

W

Waste, 21, 51, 79, 83–88, 93, 95, 97, 99, 119, 139, 145, 211, 212, 216, 217, 253
Waste management hierarchy, 85
Wastewater, 92, 93, 139–141, 147, 149, 178, 184, 185, 189, 190, 196, 204, 256
Water vapor adsorption, 96, 98
Wollastonite, 4–6, 9, 12, 13, 46, 54, 83, 100, 102, 104, 115, 117, 119, 123, 147, 148, 155, 158, 177, 200, 216
Wollastonite group, 5

X

X interlayer, 46, 47, 63, 65
Xonotlite, 2, 3, 5, 6, 9, 10, 44, 48, 49, 51, 80–83, 108, 114, 146, 147

Z

Zeolites, 41, 43, 86, 140, 143, 145
Zinc, 82, 149–162, 164, 166, 169, 176, 184, 190, 201
Zinc ions, 81, 149, 150, 152–156, 158–162, 164, 166, 167, 169, 176, 190, 195, 201
Z-phase, 6, 10, 46, 48–50, 79, 81, 83, 100–103, 105, 108, 109, 111, 112, 116, 121–126

The manufacturer's authorised representative in the EU is Springer Nature Customer Service Centre GmbH, Europaplatz 3, 69115 Heidelberg, Germany. If you have any concerns regarding our products, please contact ProductSafety@springernature.com

Printed and bound by CPI Group (UK) Ltd, Croydon, CR0 4YY

02/10/2026

02227132-0007